Advances in Environmental Science

D.C. Adriano and W. Salomons, Editors

Acidic Precipitation

Volume 1

Case Studies

Edited by D.C. Adriano and M. Havas

Springer-Verlag
New York Berlin Heidelberg
London Paris Tokyo

Volume Editors:

D.C. Adriano
Savannah River Ecology Laboratory
University of Georgia
Aiken, SC 29801
USA

M. Havas
Institute of Environmental Studies
University of Toronto
Toronto M5S 1A4
Canada

Library of Congress Cataloging-in-Publication Data
Acidic precipitation / volume editors, D.C. Adriano and M. Havas.
    p.   cm.—(Advances in environmental science)
  Bibliography: v. 1, p.
  Includes index.
  Contents: —v. 1. Case studies.
  1. Acidic deposition—Environmental aspects—Case studies.
I. Adriano, D.C. II. Havas, M. III. Series.
TD196.A25A26 1989
363.7'386—dc19   88-37418

Printed on acid-free paper

Typeset by McFarland Graphics.
Printed and bound by Quinn-Woodbine, Inc., Woodbine, New Jersey.
Printed in the United States of America.

9 8 7 6 5 4 3 2 1

ISBN 0-387-96929-2 Springer-Verlag New York Berlin Heidelberg
ISBN 3-540-96929-2 Springer-Verlag Berlin Heidelberg New York

Preface to the Series

In 1986, my colleague Prof. Dr. W. Salomons of the Institute for Soil Fertility of the Netherlands and I launched the new *Advances in Environmental Science* with Springer-Verlag New York, Inc. Immediately, we were faced with a task of what topics to cover. Our strategy was to adopt a thematic approach to address hotly debated contemporary environmental issues. After consulting with numerous colleagues from Western Europe and North America, we decided to address *Acidic Precipitation*, which we view as one of the most controversial issues today.

This is the subject of the first five volumes of the new series, which cover relationships among emissions, deposition, and biological and ecological effects of acidic constituents. International experts from Canada, the United States, Western Europe, as well as from several industrialized countries in other regions, have generously contributed to this subseries, which is grouped into the following five volumes:

Volume 1 *Case Studies*
    (D.C. Adriano and M. Havas, editors)

Volume 2 *Biological and Ecological Effects*
    (D.C. Adriano and A.H. Johnson, editors)

Volume 3 *Sources, Depositions, and Canopy Interactions*
    (S.E. Lindberg, A.L. Page, and S.A. Norton, editors)

Volume 4 *Soils, Aquatic Processes, and Lake Acidification*
    (S.A. Norton, S.E. Lindberg, and A.L. Page, editors)

Volume 5 *International Overview and Assessment*
    (T. Bresser and W. Salomons, editors)

From the vast amount of consequential information discussed in this series, it will become apparent that acidic deposition should be seriously addressed by many countries of the world, in as much as severe damages have already been inflicted on numerous ecosystems. Furthermore, acidic constituents have also been shown to affect the integrity of structures of great historical values in various places of the world. Thus, it is hoped that this up-to-date subseries would increase the "awareness" of the world's citizens and encourage governments to devote more attention and resources to address this issue.

The series editors thank the international panel of contributors for bringing this timely series into completion. We also wish to acknowledge the very insightful input of the following colleagues: Prof. A.L. Page of the University of California, Prof. T.C. Hutchinson of the University of Toronto, and Dr. Steve Lindberg of the Oak Ridge National Laboratory.

We also wish to thank the superb effort and cooperation of the volume editors in handling their respective volumes. The constructive criticisms of chapter reviewers also deserve much appreciation. Finally, we wish to convey our appreciation to my secretary, Ms. Brenda Rosier, and my technician, Ms. Claire Carlson, for their very able assistance in various aspects of this series.

Aiken, South Carolina   *Domy C. Adriano*
                                                                   Coordinating Editor

Preface to *Acidic Precipitation*, Volume 1 (*Advances in Environmental Science*)

As a result of pioneering research in the 1960s and because of the perceived and real environmental effects described during the ensuing years, the terms *acidic rain, acidic deposition,* or *acidic precipitation* have become commonplace in the scientific and popular literature. In the last decade, governments throughout the world have responded to public pressure and to the concerns of the scientific community by establishing research programs on national and international scales. These programs have been designed to enhance our understanding of the important links between atmospheric emissions and their potential environmental effects in both industrialized and developing nations. Acidic precipitation was studied initially because of its effects on aquatic systems. However, because reports from Western Europe in the early 1980s suggested a link with forest decline, acidic precipitation is now considered a potential environmental stress in terrestrial systems as well as in aquatic systems. Most recently, scientists viewed acidic precipitation as part of a larger "global change" issue along with other issues such as warming climate, increasing carbon dioxide in the atmosphere, and atmospheric ozone depletion.

As has been the case with many environmental issues of the twentieth century, acidic precipitation has its origin in emissions to the atmosphere of numerous compounds from both natural and man-made sources. This volume of the subseries *Acidic Precipitation* emphasizes some of the classical interactions between acidic deposition and ecological effects. It covers the cycling, behavior, and effects of acidic components in nature. Included are the effects of acidic deposition on soil chemistry, soil solution chemistry, aquatic chemistry, forest productivity, and fish populations. Several major ecological consequences, such as a decline in forest productivity, soil and water acidification, depletion of fish populations, and slower litter decomposition are highlighted. A whole chapter is devoted to the comparative biogeochemistry of aluminum, encompassing several ecosystems in North America and Europe.

With the National Acidic Precipitation Assessment Program of the United States nearing a 1990 completion date, and with programs in Canada and many European countries accelerating to reach a consensus on the role that atmospheric emissions and acidic precipitation play in the environment, publication of this series is timely.

The editors wish to thank the contributors to this volume for their excellent discussions of some of the most relevant ecosystem studies dealing with acidic deposition. The authors also wish to thank reviewers for the volume chapters: Drs. Lindsay Boring and John Dowd of the University of Georgia, Dr. Jack Waide of the USDA Forest Hydrology Laboratory, Dr. Christopher Cronan of the University of Maine, and Professor Egbert Matzner of the University of Göttingen. And, finally, we are grateful for the expert assistance of my secretary, Ms. Brenda Rosier, and my technician, Ms. Claire Carlson, that made our task bearable.

Aiken, South Carolina   *Domy C. Adriano*

Toronto, Canada   *Magda Havas*

Contents

ALBIOS: A Comparison of Aluminum Biogeochemistry in Forested Watersheds Exposed to Acidic Deposition 113
C.S. Cronan and R.A. Goldstein

Long-Term Acidic Precipitation Studies in Norway 137
G. Abrahamsen, H.M. Seip, and A. Semb

Chemistry of Rocky Mountain Lakes 181
J.T. Turk and N.E. Spahr

Influence of Airborne Ammonium Sulfate on Soils of an Oak Woodland Ecosystem in the Netherlands: Seasonal Dynamics of Solute Fluxes 209
N. van Breemen, P.M.A. Boderie, and H.W.G. Booltink

Contributors

G. Abrahamsen, Institute of Soil Sciences, Agricultural University of Norway, N-1432 Aas-NLH, Norway

J.J. Battles, Department of Natural Resources, Cornell University, Ithaca, NY 14850, USA

P.M.A. Boderie, Department of Soil Science and Geology, Agricultural University, P.O.B. 37, NL-6700 AA Wageningen, The Netherlands

H.W.G. Booltink, Department of Soil Science and Geology, Agricultural University, P.O.B. 37, NL-6700 AA Wageningen, The Netherlands

C.S. Cronan, Department of Botany and Plant Pathology, University of Maine, Orono, ME 04469, USA

R.A. Goldstein, Environmental Science Department, Electric Power Research Institute, 3412 Hillview Avenue, Palo Alto, CA 94303, USA

M. Hauhs, Institute for Soil Science and Forest Nutrition, Büsgenweg 2, D-3400 Göttingen, FRG

A.H. Johnson, Department of Geology, University of Pennsylvania, Philadelphia, PA 19104, USA

D.W. Johnson, Environmental Sciences Division, Oak Ridge National Laboratory, Oak Ridge, TN 37831-6038, USA

F.T. Last, Department of Soil Science, The University, Newcastle-upon-Tyne NE1 7RU, UK

S.E. Lindberg, Environmental Sciences Division, Oak Ridge National Laboratory, Oak Ridge, TN 37831-6038, USA

E. Matzner, Research Center Forest Ecosystems/Forest Decline, University of Göttingen, D-3400 Göttingen, FRG

H.M. Seip, Center for Industrial Research, P.O. Box 124, Blindern, N-0314 Oslo 3, Norway

A. Semb, Norwegian Institute for Air Research, P.O. Box 64, N-2001 Lillestrøm, Norway

T.G. Siccama, Yale School of Forestry and Environmental Studies, New Haven, CT 06511, USA

W.L. Silver, Yale School of Forestry and Environmental Studies, New Haven, CT 06511, USA

N.E. Spahr, U.S. Geological Survey, Bldg. 53, MS 415, Denver Federal Center, Lakewood, CO 80225, USA

J.T. Turk, U.S. Geological Survey, Bldg. 53, MS 415, Denver Federal Center, Lakewood, CO 80225, USA

N. van Breemen, Department of Soil Science and Geology, Agricultural University, P.O.B. 37, NL-6700 AA Wageningen, The Netherlands

Acidic Deposition on Walker Branch Watershed[*][†]

D.W. Johnson[‡] and S.E. Lindberg[‡]

Abstract

After nearly a decade of atmospheric deposition research on Walker Branch Watershed, an experimental watershed in eastern Tennessee, we are able to draw several conclusions as to the fate and effects of atmospheric inputs and begin to extrapolate some of the more important processes regulating element fluxes to other ecosystems. On Walker Branch Watershed we have found that dry deposition is a very important component of atmospheric deposition, accounting for as much as half of the input of N and S to the ecosystem. Both N and S accumulate within the watershed, but the mechanisms of accumulation differ considerably: N accumulation is due to biological uptake, whereas S accumulation is due to sulfate adsorption onto soils.

We estimate that the leaching of K^+, Ca^{2+}, and Mg^{2+} from the forest canopy has been increased by $\simeq 50\%$ due to acid deposition, but we are unable to determine the effects, if any, of this increase on the vitality of the forest at this time. We find no evidence of negative effects of artificial acid irrigation on decomposition or nutrient mineralization, but any chronic, long-term effects of atmospheric deposition on decomposition remain unknown. On the basis of the proportion of total cations balanced by sulfate in soil solutions, we estimate that atmospheric S deposition has increased cation leaching from the soils of Walker Branch Watershed by 50 to 100%. Sulfate-mediated leaching is not at its current

[*]Research sponsored primarily by the Office of Health and Environmental Research, U.S. Department of Energy, under Contract No. DE-AC05-84OR21400 with Martin Marietta Energy Systems, Inc. Support was also provided by the Electric Power Research Institute (RP-1813-1 and RP-1907-1), the National Science Foundation Ecosystem Studies Program (DEB-7824395), the National Acid Precipitation Program under U.S. Environmental Protection Agency (Agreement No. 79DX0533), and the Biomass Energy Technology Division, U.S. Department of Energy. Research conducted on the Oak Ridge National Environmental Research Park.

[†]Publication No. 3138, Environmental Sciences Division, Oak Ridge National Laboratory.

[‡]Environmental Sciences Division, Oak Ridge National Laboratory, Oak Ridge, TN 37831-6038, USA.

potential, however, because $\simeq 50\%$ of the incoming sulfate is either retained or reduced and volatilized within the watershed. We hypothesize that SO_4^{2-} is retained by adsorption onto subsurface horizons of the soils, which occupy most of the watershed. We have compared the sulfate adsorption of Walker Branch Watershed soils with that of soils from a variety of other watersheds (i.e., Hubbard Brook, NH; Coweeta, NC; Camp Branch, TN; Chesuncook, ME) and find a consistent regional pattern of S accumulation in watersheds and relatively high sulfate adsorption in Ultisols in the southeastern United States versus a lack of S retention in watersheds and relatively low sulfate adsorption in Spodosols in the northeastern United States.

Despite the accelerated rate of leaching, we calculate that most soils of Walker Branch Watershed have sufficient exchangeable base cation reserves to preclude significant acidification due to acid deposition in the near future. However, we have noted some depletion of exchangeable Ca^{2+} and Mg^{2+} in subsurface horizons of certain very poor, ridgetop soils. The Ca^{2+} depletion is most likely due to the very high rates of Ca accumulation in the oak-hickory vegetation at these sites, but the Mg^{2+} depletion cannot be accounted for in this way and may be due to leaching.

I. Introduction

The Walker Branch Watershed Project was initiated in the mid-1960s with the goal of understanding the basic biological, chemical, and physical processes, and their interactions, that govern the cycling of materials in forested landscapes (Johnson and Van Hook, 1988). The project was begun before public concerns about acidic precipitation had arisen. Because of the existence of the project and the nature of the research being conducted, we have been in a position to respond quickly to, and assist in, the resolution of this issue.

Since the early characterization of nutrient cycles on the watershed as part of the International Biological Program, research on the effects of atmospheric deposition has been the dominant activity on Walker Branch Watershed. In 1980 Walker Branch Watershed was selected as a site in the National Atmospheric Deposition Program and in 1981 as a site in the Multistate Atmospheric Power Production Pollution Study (MAP3S) precipitation chemistry network. Because of the extensive research experience of the investigators and the long-term data bases being developed on various characteristics of Walker Branch Watershed (water quality, soil chemistry, productivity, and mineral cycling), the Electric Power Research Institute initiated two research projects in 1981 on the effects of atmospheric pollutants (acidic precipitation) on canopy processes and soil chemistry. These projects provided the first opportunity to apply the Walker Branch Watershed data sets to the evaluation of anthropogenic impacts on forested ecosystems. Throughout the history of these studies, terrestrial nutrient cycling has been the predominant focus, because the watershed overlies dolomite bedrock and aquatic effects have never been a concern.

Unfortunately, acid deposition effects are especially difficult to study because there are no comparable control watersheds nearby that are not subject to the same inputs. Our primary approach to the problem has been to make both ecosystem- and process-level measurements of the inputs, canopy interactions, and internal cycling, to determine the ultimate fate of atmospherically deposited substances, and, from that information, to deduce the effects of such deposition on the ecosystem. This approach, using both watershed- and plot-level studies, tells us the fate of atmospherically deposited pollutants and also allows us to make a fairly reasonable assessment of the effects of acid deposition on nutrient leaching (i.e. by comparing SO_4^{2-} and NO_3^- fluxes with cation fluxes). It does not, however, allow us to investigate the secondary effects of acid deposition, for example, upon decomposition and nutrient mineralization. To investigate the latter, we have employed small-scale acidification experiments in the field.

II. Site Description

Walker Branch Watershed is located on the U.S. Department of Energy's Oak Ridge Reservation in Anderson County, Tennessee, near Oak Ridge National Laboratory (Figure 1-1). The climate is humid mesothermal with a mean annual temperature of 14.1°C and a mean annual precipitation of 143 cm, most of which falls as rain. The watershed, which occupies a total of 97.5 ha, consists of two subwatersheds—the west catchment being 38.4 ha and the east catchment 59.1 ha. The catchment basin is bounded on the north by Chestnut Ridge, which reaches an elevation of 350 m and slopes rapidly southward to an elevation of 265 m in the valley at the confluence of the two forks. The Walker Branch Watershed basin is underlain by the Knox Group, a 610-m-thick sequence of siliceous dolomite, divided into four formations on the basis of minor variations in lithology and chert characteristics. The bedrock geology is largely obscured by a mantle of weathering products from the underlying dolomite and is thickest along drainage divides. Chert is interspersed throughout the weathered mantle, but is concentrated at the ground surface and the bedrock-weathered material interface. The depth to bedrock is highly variable but often reaches 30 m on ridgetops.

The soils formed over the dolomitic substrate are primarily Ultisols. Small areas of Inceptisols are found in alluvial areas adjacent to the streams. The soils have a silt loam texture, are generally well drained, and have a high infiltration capacity. The predominant clay mineral found in these soils is kaolinite, with lesser amounts of vermiculite, hydrous micas, and quartz forming the complement. Soils of the Fullerton and Bodine Series occupy >90% of the watershed. Fullerton soils (clayey, kaolinitic thermic Typic Paleudults) occupy the ridgetops and upper-slope positions, and Bodine soils (loamy-skeletal, siliceous thermic Typic Paleudults) are found on intermediate and lower slopes. Both the Bodine and Fullerton soils are acidic (pH 4.2–4.6) and low in exchangeable bases, N, and P. Claiborne soils (fine-loamy, mesic Typic Paleudults) occupy minor areas in the major stream bottom of the east catchment on alluvial or colluvial deposits washed

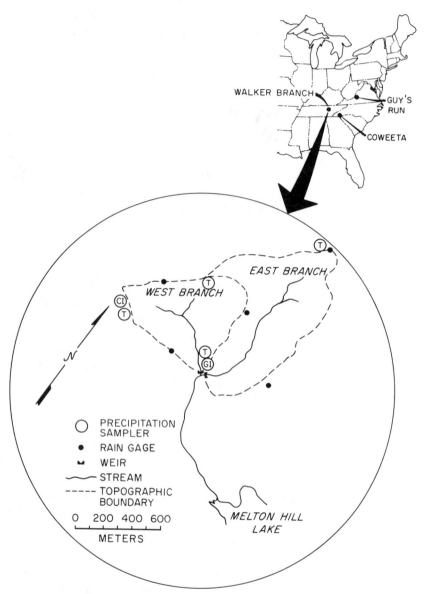

Figure 1-1. Location of Walker Branch Watershed.

from the uplands. Tarklin soils (fine-loamy, siliceous Typic Fragiudults) occupy sinkholes and depressions in ridgetop and upper-slope positions. Both Claiborne and Tarklin soils are less acidic and richer in N and P than the Fullerton or Bodine soils. Stonyland and Rockland occur on lower slopes near the weirs, where the dolomite substrate tends to outcrop or lies near the soil surface.

The overstory vegetation is predominantly an oak-hickory association, with lesser amounts of pine-oak-hickory and pure pine present. Small areas of mixed mesophytic vegetation are found in sheltered coves and stream valleys. Four major forest associations were originally defined in a vegetation analysis by Grigal and Goldstein (1971): pine (principally *Pinus virginiana* and *P. echinata*), yellow-poplar (principally *Liriodendron tulipifera*), oak-hickory (primarily *Carya tomentosa, C. glabra, Quercus prinus, Q. alba,* and *Acer rubrum*), and chestnut oak (principally *Q. prinus*). A small (2.5-ha) loblolly pine (*Pinus taeda*) plantation located in the northwest corner of the watershed was later designated a fifth forest type. Other overstory species include *Oxydendrum arboreum, Nyssa sylvatica,* and, in cove sites, *Fagus grandifolia.* Understory species include *Cornus florida, Sassafras albidum, A. rubrum,* and *N. sylvatica. Locinera japonica* and *Rhus radicans* occur as ground vegetation in pine stands and in open areas.

The watershed provides an ideal field laboratory in which to examine deposition and cycling of airborne contaminants derived from atmospheric emissions related to energy production. The catchment is located within $\simeq 20$ km of two major coal-fired power plants and one small plant (Figure 1-2). As indicated by the annual wind rose, these point sources are frequently upwind of the watershed. The total annual emissions during 1977 from coal combustion at the largest power plant [Kingston Power Plant operated by the Tennessee Valley Authority (TVA)] were as follows: suspended particulates $= 35 \times 10^3$ Mg (or tonnes), sulfur oxides $= 154 \times 10^3$ Mg, nitrogen oxides $= 30 \times 10^3$ Mg. These emissions have been reduced since 1977, however, by improvement in TVA emission controls throughout the region. For example, from 1977 to 1982, the TVA's systemwide SO_2 emissions were reduced by 50%.

These sources of airborne contaminants to the southwest and northeast of the watershed may be particularly important because of the influence of the local topography, which is dominated by parallel ridges. These ridges run west-southwest to east-northeast and strongly influence the channeling of winds through the parallel valleys, as shown by the annual wind rose in Figure 1-2. The potential influence of these point sources on local air quality is apparent.

The possibility of regional effects on local air quality must not be neglected, however. In addition to these local sources, the watershed is located $\simeq 20$ km west of greater Knoxville, Tennessee, within 260 km of three major regional urban centers (Chattanooga and Nashville, TN, and Atlanta, GA), and within 350 km of 22 coal-fired power-generating stations with a combined generating capacity of 2×10^4 MW (Figure 1-2).

III. Process-Level Studies

A. Atmospheric Deposition

In forests such as those on Walker Branch Watershed, both plant nutrients and acidic pollutants are deposited from the troposphere by a number of processes: by particles and vapors carried by winds to plant surfaces during dry periods (dry

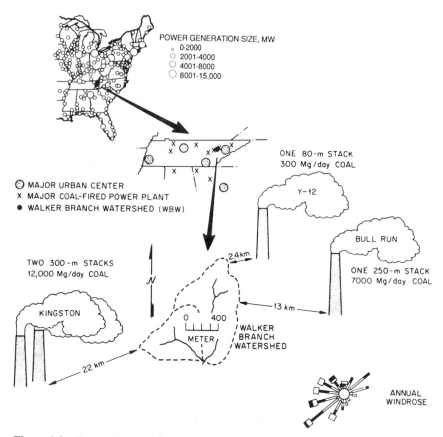

Figure 1-2. Geographical location of Walker Branch Watershed in relation to continental, regional, and local sources of atmospheric emissions due to fossil fuel utilization. Shown are the spatial distribution of centers of coal- and oil-fired power generation in the eastern United States, the location of specific regional coal-fired power plants and major urban centers, and the proximity of three local power plants to the watershed.

deposition) and by the scavenging of these materials from the air by precipitation (wet deposition). Intensive sampling of dry deposition, of precipitation before and after its interception by the forest canopy, and of the concentrations of airborne particles and vapors are all needed in order to make accurate estimates of atmospheric input to the canopy and forest floor and to understand the interactions of acidic deposition with the plant canopy (NAS, 1983).

Atmospheric deposition has often been determined from samples of bulk precipitation collected monthly or weekly in continuously open containers situated near ground level (e.g. Likens et al., 1977). This approach collects some unknown portion of the dry deposition, subjects the sample to contamination by soil dust, plant debris, and insects, and is now acknowledged to produce unreliable results

(NAS, 1983). This is particularly true for forests in industrialized regions, where acidic deposition is intensive, because canopies having large surface areas can interact with anthropogenic particles and vapors to produce significant dry deposition fluxes of S and N oxides.

Atmospheric chemistry, deposition, and canopy interactions of several plant nutrients and strong acids were intensively studied during 1981 to 1983 in a ridgetop oak (*Quercus prinus, Q. alba*) stand located near the western edge of Walker Branch Watershed. The sampling area was dominated by mature mixed oak and hickory stands with 20- to 25-m-high canopies in the middle of which we erected two meteorological towers extending into and above the canopy. To obtain several estimates of deposition, we developed and employed a number of techniques for measuring precipitation, particles, and vapors. Automatic samplers were installed in a forest clearing and beneath the trees to collect wet deposition and throughfall on an event basis (Lindberg, 1982; Lovett and Lindberg, 1984). Continuously open containers were used to collect bulk deposition in the clearing for comparison, and plastic collars were used for collecting stemflow (Richter et al., 1983). We sampled 99% of the recorded precipitation, which averaged 128 cm/year during this period. Rain and throughfall solutions and extracts of solid aerosol samples were analyzed by standard atomic absorption and ion chromatography procedures (Lovett and Lindberg, 1984; Richter et al., 1983).

Dry deposition was determined by a combination of techniques. Coarse (>2 μm) dry-deposited particles were collected during 26 dry periods of 4,000 h duration, using replicate petri dish deposition plates situated on towers extending above the canopy and attached to tree branches in the upper canopy. Further details and direct comparisons of the fluxes, chemistry, size distributions, and morphologies of particles deposited on adjacent oak leaves (*Q. prinus*) and deposition plates during eight of these field experiments have been published (Lindberg and Lovett, 1985). The dry deposition rates of coarse particles to the plates were scaled to the full canopy on a seasonal basis, using factors determined from statistical analysis of Ca^{2+} fluxes in throughfall as published by Lovett and Lindberg (1984). Linear regression was used to separate throughfall into its component parts of foliar exchange and dry deposition washoff, assuming leaching to be proportional to the amount of precipitation in a storm event and dry deposition to be proportional to the duration of the antecedent dry period (Lindberg and Lovett, 1985). Calcium was chosen because it exists primarily in the coarse size class of airborne particles and has no vapor component in the atmosphere. The scaling factors were 2.7 for the growing season and 1.0 for the dormant season, representing the ratio of the mean dry deposition rate of coarse particles to the whole canopy divided by the measured mean dry deposition rate to individual plates. Hence, the measured coarse particle dry deposition fluxes to the plates were multiplied by a scaling factor of 2.7 during the growing season to estimate the coarse particle flux to the canopy.

Vapor and fine particle (≤ 2 μm) deposition rates were calculated from atmospheric concentration data collected above the canopy and from appropriate long-term deposition velocities measured at this site (Huebert, 1983; Hicks, 1984) or taken from the literature (Garland, 1983; Hicks, 1984; Fowler, 1980). We obtained continuous measurements of SO_2, measured HNO_3 vapor concentrations for 15 dry periods of 2,000 h, and sampled airborne particles for 26 dry periods totaling 4,000 h. Atmospheric vapor concentrations were measured, using a filter pack (Lovett and Lindberg, 1984; Huebert, 1983) for HNO_3 vapor and flame photometry (in a nearby clearing) for SO_2 (TVA, 1982). The concentrations and size distributions of suspended particles were measured above the canopy, using filtration and impactor methods (Lindberg and Harriss, 1983). The deposition velocities were as follows: during the growing season, $SO_2 = 0.5$ cm $\cdot$ s^{-1}, HNO_3 vapor $= 2$ cm $\cdot$ s^{-1}, and fine particles $= 0.2$ cm $\cdot$ s^{-1}; during the dormant season, $SO_2 = 0.2$ cm $\cdot$ s^{-1}, HNO_3 vapor $= 0.5$ cm $\cdot$ s^{-1}, and fine particles $= 0.05$ cm $\cdot$ s^{-1}.

The data collected over this 2-year period suggest that dry deposition is a significant mechanism in the total annual flux of each of these ions to the forest (Table 1-1). Although the total atmospheric input of SO_4^{2-} and NH_4^+ occurs primarily by means of precipitation scavenging, dry deposition of vapors is the most important mechanism for NO_3^- and H^+, and dry deposition of coarse particles is the most important one for K^+ and Ca^{2+}. The contribution of dry deposition ranges from $\simeq 30\%$ of the total atmospheric deposition for NH_4^+, to $\simeq 50\%$ for SO_4^{2-} and H^+, to $\simeq 60\%$ for K^+ and NO_3^-, to $\simeq 70\%$ for Ca^{2+}. Dry deposition is primarily controlled by vapor uptake for SO_4^{2-}, NO_3^-, and H^+ (70, 76, and 97% of the total dry input, respectively), by fine particles for NH_4^+ (63%), and by coarse particles for K^+ and Ca^{2+} ($\simeq 95\%$ each). In comparing these deposition rates, recall that any such estimates are subject to considerable uncertainty (Hicks et al., in press). The standard errors given in Table 1-1 provide only a measure of uncertainty in the calculated sample means relative to the population means; hence, additional uncertainties in analytical results, hydrology, scaling factors, and deposition velocities must be included. We estimate the overall uncertainty for wet deposition fluxes to be on the order of 20 to 30%, and those for dry deposition to be $\simeq 50\%$ for SO_4^{2-}, Ca^{2+}, K^+, and NH_4^+ and $\simeq 75\%$ for NO_3^- and H^+.

The contribution of each ion to the total input by each deposition process reflects the chemistry of the contributing parent compounds (Figure 1-3). Precipitation at this site is dominated by SO_4^{2-} and H^+, with sufficient SO_4^{2-} to account for all of the free acidity as H_2SO_4 (Lindberg, 1982). However, if one corrects for sea salt SO_4^{2-}, on the basis of ion ratios in seawater, and considers the probable contributions of sulfate salts of Ca and ammonium, the resulting estimates suggest that nitric acid could contribute up to 30% of the free acidity in precipitation. The fine-particle portion of dry deposition is primarily a mixture of H_2SO_4, $(NH_4)_2SO_4$, and intermediates, with at most 30% of the SO_4^{2-} occurring as acid sulfate, based on ion ratios. However, the coarse particles in dry deposition are apparently dominated by Ca salts of NO_3^- and SO_4^{2-}.

Interelement correlations support the presence of these parent compounds. The concentrations of H^+, NH_4^+, SO_4^{2-}, and NO_3^- are all significantly intercorrelated in precipitation; those of H^+, SO_4^{2-}, and NH_4^+ are similarly correlated in fine aerosols; and those of Ca^{2+} and NO_3^- are correlated in extracts of coarse particles

Table 1-1. Total annual atmospheric deposition of major ions to an oak forest at Walker Branch Watershed, Anderson County, Tennessee, measured during 1981 to 1983.

	Atmospheric deposition [kmol(+ or −) · ha^{-1} · year^{-1}][a]					
	SO_4^{2-}	NO_3^-	H^+	NH_4^+	Ca^{2+}	K^+
Precipitation	0.70 (0.05)	0.20 (0.02)	0.69 (0.05)	0.12 (0.01)	0.12 (0.02)	0.009 (0.001)
Dry deposition	0.88 (0.08)	0.34 (0.04)	0.88 (0.08)	0.057 (0.0015)	0.31 (0.03)	0.013 (0.002)
Relative contribution of each type of dry deposition (%)						
Fine particles	0.08	0.003	0.02	0.63	0.03	0.08
Coarse particles	0.22	0.24	0.006	0.14	0.97	0.92
Vapors[b]	0.70	0.76	0.97	0.23	0	0
Total deposition	1.6 (0.09)	0.52 (0.04)	1.6 (0.09)	0.18 (0.02)	0.43 (0.04)	0.022 (0.003)

[a] Values given are means and associated standard errors based on the means of all sampled events for 2 years of data.

[b] Includes SO_2, HNO_3, and NH_3. We assumed complete conversion of deposited SO_2 to H_2SO_4 and of NH_3 to NH_4^+ in determining the vapor input of H^+. The deposition of NH_3 was not measured, but was estimated from the literature (Tjepkema et al., 1981).

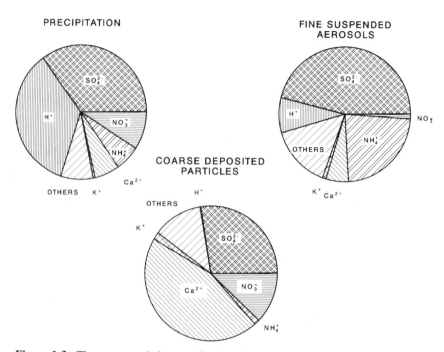

Figure 1-3. The average relative contribution of several major ions to the total analyzed chemistry of wet and dry deposition on a charge-equivalent basis. *Others* include Cl^-, PO_4^{3-}, Na^+, and Mg^{2+}.

deposited on inert surfaces ($r \geq 0.6$, $P \leq 0.01$ in each case). Reactions among NH_3 and gaseous and particulate oxides of S and N in the atmosphere can account for many of these relationships in precipitation and in the fine aerosol fraction (NAS, 1983). The presence of coarse-particle nitrate has been attributed to reactions between HNO_3 vapor and airborne crustal constituents such as Ca^{2+}, which predominate in the coarse size fraction (Wolff, 1984).

The comparability between wet and dry deposition of SO_4^{2-} (Table 1-1) has been suggested by atmospheric models (Shannon, 1981), by large-scale budget approaches (Galloway and Whelpdale, 1980), and by theory (Fowler, 1980) but, to our knowledge, has not been confirmed by extensive field measurements over several seasons. Dry deposition of S oxides has been attributed in most studies primarily to deposition of SO_2 plus fine-particle deposition to foliage (NAS, 1983). At this forested site, the mean annual air concentration of S oxides is 10.4 $\mu g \ S \cdot M^{-3}$, 75% of which is attributed to SO_2 and an additional 22% of which is attributed to fine-particle sulfate. The importance of SO_2 uptake is confirmed by our results (Table 1-1). However, the measurements suggest that the remaining fraction of the airborne sulfate associated with coarse particles is also important in dry deposition, contributing a disproportionate share of the S input to this forest.

Coarse-particle sulfate contributes $\simeq 20\%$ of the total dry-deposited S while comprising only 3% of the airborne S oxide mass (Lindberg et al., 1986). Similarly, 73% of the estimated particle sulfate flux results from particles >2 μm in diameter, although an average of only 12% of the airborne sulfate mass is in this size range.

The importance of coarse-particle deposition of SO_4^{2-} has only recently been realized (Davidson et al., 1985). This fraction is important because of its efficient removal from the atmosphere by sedimentation and impaction. Our measurements suggest a mean annual dry deposition velocity for coarse-particle SO_4^{2-} to this forest canopy of 0.4 cm $\cdot$ s^{-1}, as calculated from the measured rates of dry deposition to inert plates during each season (Lindberg and Lovett, 1985), extrapolated to the canopy (Lovett and Lindberg, 1984), and normalized to the airborne sulfate concentrations. This value is similar to deposition velocities for fine-particle SO_4^{2-} of 0.2 to 0.7 cm $\cdot$ s^{-1} measured for forests in short-term field studies but much larger than the value of <0.1 cm $\cdot$ s^{-1} calculated on the basis of wind tunnel studies [see the review by Hicks (1984)]. Recent measurements of coarse-particle SO_4^{2-} flux in other areas of the United States (Davidson et al., 1985) and Europe (Lindberg, 1987) using our methods have yielded mean dry deposition rates to inert surfaces similar to those measured at this site.

The importance of dry deposition is even more pronounced for NO_3^- than for SO_4^{2-} (Table 1-1). Dry deposition of fine-particle NO_3^- is negligible compared to the flux of coarse-particle NO_3^-, which accounts for 99% of the particle flux to the canopy and 26% of the total NO_3^- dry deposition. However, most of the airborne NO_3^- at this site is not in particle form but exists as HNO_3 vapor, a species not considered in deposition studies until quite recently (Huebert, 1983). The mean annual concentration of HNO_3 vapor above this forest is 0.73 μg N $\cdot$ m^{-3} (SE = 0.06) and that of particle NO_3^- is 0.17 μg N $\cdot$ m^{-3} (SE = 0.05). The ratio of HNO_3 vapor to particle NO_3^- exhibits a strong seasonal dependence; the mean ratio is 3.7 in the winter but 12 in the summer, reflecting the role of temperature in controlling the equilibrium between the two forms as well as the role of sunlight in influencing the conversion of NO_x (NO_2 + NO) to HNO_3 (NAS, 1983). The predominance of HNO_3 vapor over particle NO_3^- has been reported at a number of sites (e.g. Kelly et al., 1984; Lindberg and Johnson, in press).

Nitric acid vapor is a major sink for atmospheric NO_x and is a highly soluble gas that is efficiently scavenged by precipitation. However, its solubility apparently also results in efficient deposition to plant canopies during dry periods. Measurements at this site suggest that dry deposition of HNO_3 vapor contributes 46% of the total annual atmospheric flux of NO_3^- to the canopy, making it the single most important nitrogen oxide species studied here (Table 1-1).

Because of the apparent importance of dry deposition in the input of NO_3^- to this forest and because of the acknowledged uncertainty in determining dry deposition in general (Hicks et al., 1987), we used three other approaches to estimate it: (1) a statistical analysis of NO_3^- fluxes in throughfall; (2) application of published deposition velocities to our air concentration data; and (3) collection of dry-deposited HNO_3 plus particle NO_3^- on reactive surfaces situated throughout the canopy. These three methods yielded values of total NO_3^- dry deposition of 0.13, 0.24, and 0.56 kmol $(-) \cdot ha^{-1} \cdot year^{-1}$, respectively (Lovett and Lindberg, 1986). These values bracket our best estimate of 0.32 kmol$(-) \cdot ha^{-1} \cdot year^{-1}$ (Table 1-1) and provide an indication of the degree of uncertainty involved in dry deposition estimates such as these.

Dry deposition of NO_x also contributes to the total dry flux of NO_3^-, assuming oxidation of NO_x in the plant canopy. Dry deposition of NO_x at Walker Branch Watershed was estimated, in a concurrent study (Kelly and Meagher, 1985), to be 0.14 kmol $(-) \cdot ha^{-1} \cdot year^{-1}$. This study utilized a chemiluminescent NO_x detector above the canopy to determine the concentrations and published deposition velocities to estimate the flux; however, to what extent these measurements of NO_x include HNO_3 vapor is not clear.

Despite the uncertainty in these estimates and knowing that our lowest value in the above range of NO_3^- dry deposition rates is an underestimate because of irreversible NO_3^- uptake in the plant canopy (Lovett and Lindberg, 1986), we can conclude that dry deposition is probably the dominant mechanism in the airborne flux of NO_3^- to this forest, contributing up to 80% of the total.

For the major cations, K^+ and Ca^{2+}, the dry-deposition contribution to the total flux is on the order of 60 to 70% (Table 1-1). This would be predicted from their aerosol size distributions at this site: 34% of the airborne K^+ and 55% of the Ca^{2+} are associated with particles >2 μm in diameter (Lindberg et al., 1986). The smaller fraction of coarse-particle K^+ in the atmosphere may explain the somewhat lower relative contribution of dry deposition to its total input compared with that of Ca^{2+}.

Dry-deposited Ca^{2+} and K^+ may not represent entirely new inputs to the forest but, rather, resuspended surface material, possibly of local origin. However, our measurements indicate that both aerosol concentrations and deposition rates of Ca^{2+} to inert surfaces in and above the canopy increase with increasing height above the forest floor (Lindberg et al., 1984), indicating an airborne source and a canopy sink. Although the source of Ca^{2+} is apparently particles in the air over this forest, the ultimate source of these particles is primarily resuspended soil and road dust (originating outside of the forest sampling plots) plus a minor contribution from local coal combustion. Similar measurements of K^+ concentration and deposition in and above the canopy indicate that the highest values occur within the canopy itself, decreasing both above and below, suggesting an in-canopy source of K^+. This material consists of weathered leaf cuticle, deciduous leaf hairs, and pollen (Coe and Lindberg, 1987). Because most of the dry deposition of K^+ occurred during the growing season, as discussed below, as much as 50% of the annual K^+ deposition may represent an internal forest cycle (Lovett and Lindberg, 1984). No attempt has been made in this study to determine the influence of insects and other phyllosphere organisms on either dry deposition washoff or total throughfall fluxes. These effects could be important for K^+, but probably do not significantly impact the other ions studied here.

Wet deposition clearly dominates the input only of NH_4^+, as might be suggested by the atmospheric chemistry of the NH_3-NH_4^+ system. The vapor is highly soluble in acidic rain droplets and cloudwater, and can account for a significant portion of dissolved NH_4^+ in such solutions. In part because of its solubility, NH_3 has been reported to represent a minor fraction ($\simeq 5$ to 20%) of the airborne burden of NH_3 + NH_4^+ in other studies in the United States (Tjepkema et al., 1981; Daum et al., 1984). The low vapor levels and the fact that 85% of the particulate NH_4^+ exists in the fine size range at this site (Lindberg et al., 1986) may explain why removal by precipitation is favored over dry deposition.

The atmospheric deposition of free hydrogen ions by wet and dry processes is difficult to quantify because of the reactivity of H^+. The chemistry of precipitation and of particle extracts necessarily represents the end result of any acid-base reactions occurring prior to analysis. In addition, acid and base precursor vapors may experience varying degrees of oxidation once deposited, thus representing varying amounts of potential free acidity. We calculated potential H^+, assuming complete oxidation of deposited SO_2 and protonation of deposited NH_3 (Table 1-1). Our analyses suggest that the total input of H^+ to this forest canopy is strongly influenced by the dry deposition of the acidifying vapors, which represent ~60% of the total deposition. Nitric acid vapor and SO_2 represent 97% of the total dry deposition of potential free acidity, the remainder being attributable to fine-particle deposition. This is reflected in the air concentrations at this site; the mean total acidic vapor concentration (as potential H^+) is 0.54 $\mu mol(+) \cdot m^{-3}$, whereas that of particle-associated H^+ is 0.046 $\mu mol(+) \cdot m^{-3}$, 80% of which occurs in the fine size range (Figure 1-3).

With some simple assumptions regarding SO_2 oxidation (up to 20% could be adsorbed onto leaf surfaces and degassed prior to oxidation) (Taylor et al., 1983), Ca^{2+} dry deposition (all or none could be $CaCO_3$), and NH_4^+ dry deposition (all or none of the NH_4^+ in aerosol and deposited particle extracts could have originated from adsorbed NH_3), we can bracket our best estimate (Table 1-1) of atmospheric deposition of free acidity to this ecosystem. The results of these assumptions yield an estimated range of total atmospheric input of 0.9 to 2.0 kmol(+) $H^+ \cdot ha^{-2} \cdot year^{-1}$, leaving little doubt that the net deposition to this forest is acidic. This flux of H^+, and many other ions as well, was seriously underestimated by the standard bulk deposition collectors used at this site in a concurrent study (Richter and Lindberg, 1988). Comparison of the results in Table 1-1 with deposition estimates from bulk precipitation collected in a nearby clearing at ground level suggests that bulk deposition most significantly underestimates total deposition for those ions with a major vapor or fine-particle component in the atmosphere. Total deposition of H^+, NO_3^-, SO_4^{2-}, and NH_4^+ was underestimated by 50 to 70% (Lindberg et al., 1986). These data suggest that biogeochemical cycling studies in which system inputs are based solely on measurements of bulk deposition must be interpreted cautiously.

Although our results are generally supported by published calculations, there are few if any directly comparable field studies. Dry deposition has been identified as being potentially important in the atmospheric flux of S and N to forests in southern Sweden from published air concentrations and deposition velocities (Grennfelt et al., 1980). Dry deposition was estimated to contribute 40% (NO_3^-) to

60% (SO_4^{2-}) of the total input (0.3 and 1.4 kmol($-$) $\cdot$ ha^{-2} $\cdot$ year^{-1}, respectively) and was predicted to be dominated by HNO_3 and SO_2 (70 and 80% of dry deposition, respectively). In the Federal Republic of Germany, total SO_4^{2-} deposition was estimated to be 0.31 and 0.46 kmol($-$) $\cdot$ ha^{-1} $\cdot$ year^{-1} to beech and spruce forests, with dry deposition contributing 50 and 70% of the flux, respectively (Mayer and Ulrich, 1982). However, wet and dry deposition were not measured separately but were estimated, assuming that soluble ions in bulk deposition represented the wetfall contribution and that the wet:dry deposition ratio estimated from winter throughfall beneath a leafless beech canopy was representative of summer conditions in both canopies. These assumptions are tenuous; both dry-deposited SO_2 and particle S yield SO_4^{2-} ions to solution, and our data suggest that wet:dry deposition ratios are not constant between seasons (Figure 1-4). In a coastal spruce forest in Scotland, total deposition was estimated from bulk collectors with and without overhanging mesh nets (Miller and Miller, 1980). Total SO_4^{2-} deposition was estimated at 0.21 kmol ($-$) $\cdot$ ha^{-1} $\cdot$ year^{-1}, 30 to 40% of which was attributed to the aerosol filtering action of the mesh. The filtered material was attributed primarily to relatively large sea-salt aerosols and did not include SO_2 or a significant fraction of submicron particles. In addition,

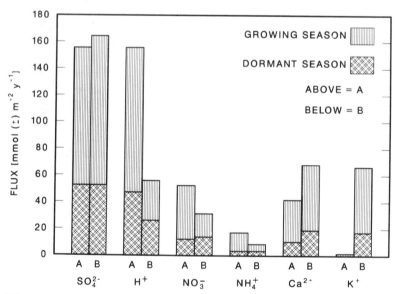

Figure 1-4. The seasonal ion fluxes for total wet plus dry deposition above the canopy and combined throughfall plus stemflow fluxes below the canopy. The growing season at this site is defined as the months of April through October. Uncertainties of the deposition estimates are given in the text; for below-canopy fluxes, they are $\simeq$15% for SO_4^{2-}, NO_3^-, and H^+; $\simeq$25% for Ca^{2+} and K^+; and $\simeq$35% for NH_4^+.

some of this filtered material may have been derived from windblown mists that are not efficiently sampled by the bulk collectors.

In the United States, micrometeorological methods were used in a 1-month study to measure the velocities of HNO_3 vapor deposition to a grassy field (Heubert, 1983). Measured values of 1 to 4 cm $\cdot$ $^{-1}$ resulted in a dry deposition estimate of 0.024 kmol $\cdot$ ha^{-1} $\cdot$ month, comparable to the wet deposition measured nearby. Particle deposition was not determined. In the northeastern United States, total SO_4^{2-} deposition to a deciduous forest was estimated using automatic wet-dry samplers for large particles and precipitation (Johannes et al., 1981), and SO_2 was estimated from literature values. The total deposition in the Johannes study was similar to that measured in this study: 60% was wetfall and 30% was attributed to SO_2. The results of these studies, using widely varying approaches, suggest that the general contribution of dry removal processes to total atmospheric deposition of SO_4^{2-} and NO_3^- is in the range of 30 to 70%.

B. Interactions between Deposition and the Canopy

The fate of material deposited on the forest canopy by wet and dry processes is reflected in the ion flux in throughfall and stemflow solutions collected below the canopy during periods of rain. During the winter when the canopy is leafless, the above-canopy (wet plus dry deposition) and below-canopy (throughfall plus stemflow) fluxes are generally comparable for SO_4^{2-}, NO_3^-, and NH_4^+, indicating that the exposed bark surface has only a moderate influence on internal leaching, assuming that deposited dry material is efficiently washed off and deposited on the forest floor by precipitation (Figure 1-4). This is not the case for K^+ and Ca^{2+}, which are exchanged with H^+ at nearly a 1:1 ratio during the interception of wet and dry deposition by the canopy. Approximately 0.2 kmol(+) $\cdot$ ha^{-1} of H^+ is removed and $\simeq 0.2$ kmol(+) $\cdot$ ha^{-1} of K^+ plus Ca^{2+} is released in excess of the amounts deposited to the canopy, and the other anions and other cations pass through the canopy with little change.

During the growing season, the canopy has a significant influence on deposition reaching the forest floor for all ions except SO_4^{2-} (Figure 1-4). Absorption by the canopy decreases the fluxes of atmospheric H^+, NO_3^-, and NH_4^+ (by $\simeq 50$ to 70%), and loss from the canopy increases those of Ca^{2+} (by 55%) and K^+ (by a factor of 24). The flux of SO_4^{2-} is increased by <5%. Thus, during the summer months, the chemistry of total atmospheric deposition to the canopy is acidic, consisting primarily of H_2SO_4 and HNO_3, with lesser amounts of NH_4^+ and Ca^{2+} salts of these strong acid anions. However, below the canopy, the flux of ions to the forest floor in throughfall plus stemflow is dominated by Ca^{2+} and K^+ salts of SO_4^{2-}, with only a minor contribution by the mineral acids.

Studies of bulk and wet-only precipitation have indicated an in-canopy removal of 30 to 40% of the free H^+ in precipitation (Cronan and Reiners, 1983; Hoffman et al., 1980), whereas our data suggest that the Walker Branch Watershed canopy apparently removes >70% of the total deposition of free H^+ during the growing season. The removal is thought to involve both ion exchange and weak-base-buffering reactions (Lovett et al., 1986). The absorption of H^+ from wet or dry deposition could result in a reduction in canopy nutrient pools through exchange of H^+ for other cations. This is reflected in the growing season data for Ca^{2+} and K^+ (Figure 1-4). The excess flux of these cations below the canopy compared with that above it [≈ 0.2 kmol(+) $\cdot$ ha^{-1} for Ca^{2+} and ≈ 0.5 kmol(+) $\cdot$ ha^{-1} for K^+] is interpreted to represent leaching of internal plant nutrients from the foliage. Based on an analysis of the complete ion charge balance of the deposition-canopy exchange, including measurement of mobile organic anions in throughfall, we have reported that 40 to 60% of the leaching of K^+, Ca^{2+}, and Mg^{2+} from the forest canopy at Walker Branch Watershed can be attributed to exchange for deposited airborne acids (Lovett et al., 1986). The effects of the loss of nutrient cations from the canopy through exchange for H^+ depend on the ability of the tree to replenish these pools and, depending on soil nutrient status, could be either positive or negative (Lovett et al., 1986).

The uptake of N by the forest canopy is expected in an N-deficient ecosystem and apparently involves dry-deposited N compounds because of their longer residence time on the foliage relative to precipitation. For example, the difference between the atmospheric flux of NO_3^- above and below the canopy during the growing season [a canopy uptake of 0.22 kmol($-$) $\cdot$ ha^{-1}] is nearly equal to the dry deposition of HNO_3 to the canopy during this period. However, foliar absorption of NO_3^- and NH_4^+ in precipitation may also occur to some extent (Cole and Rapp, 1981). The fate of absorbed NO_3^- in the canopy is not well understood. If nitrate reductase is present, the ion may be assimilated by plant cells. Fertilizer studies suggest this to be the case for some tree species (Eberhardt and Pritchett, 1971).

There is no apparent retention of deposited S by foliage in this S-rich ecosystem (Figure 1-4). Instead, the data suggest a small amount of leaching of internal plant SO_4^{2-} [0.06 kmol($-$) $\cdot$ ha^{-1} $\cdot$ year^{-1}] and that incident precipitation plus washed-off dry deposition accounts for $\approx 95\%$ of the below-canopy flux. Although 40% of the total annual S deposition occurs by SO_2 deposition to the canopy, much of this may be washed off and deposited on the forest floor during subsequent rain events. Evidence suggests that SO_4^{2-} in throughfall at this site consists of two fractions exhibiting different mechanisms of removal from the canopy—one being surface washoff and the other being diffusion through leaf membranes (Lovett and Lindberg, 1984). We interpret the source of SO_4^{2-} ions in surface washoff to be soluble SO_4^{2-} in dry-deposited particles on the foliage (Lindberg and Lovett, 1985). The SO_4^{2-} lost by the diffusion process could have originated from atmospheric SO_2 originally absorbed by the foliage and then lost by diffusion to rain on the leaf surface. However, the exact source of any particular SO_4^{2-} ion removed from the canopy can be positively identified only by isotope tracer studies. In recent $^{35}SO_4^{2-}$ whole-tree labeling experiments at Walker Branch Watershed, Garten et al. (1988) estimated that leaching of internal SO_4^{2-} comprised 1 to 20% (mean 6%) of the total below-canopy flux of SO_4^{2-} in maple and tulip poplar stands.

Evidence from independent studies of tree fertilization and biogeochemical cycling of S at this site supports the idea that rainfall leaching of internally cycled plant S is a minor contributor to the total flux of SO_4^{2-} to the forest floor. One study involved urea-nitrogen fertilization of forest plots, which theoretically should bind internal plant SO_4^{2-} into proteins, thereby reducing their leachability by precipitation (Richter et al., 1983). Annual SO_4^{2-} fluxes in throughfall plus stemflow beneath adjacent treated and untreated plots (three each) were not significantly different. Assuming atmospheric deposition of SO_4^{2-} to be comparable among plots, the results suggest little or no internal plant leaching of SO_4^{2-} in the untreated plots.

The second study employed a mass balance approach using measurements of foliar S content and S fluxes in litterfall, throughfall, and stemflow to estimate foliar leaching of internal S (Meiwes and Khanna, 1981). This method assumes that the difference between the flux of S to the forest floor in litterfall and the S content of the living foliage prior to litterfall must be accounted for by leaching of S from foliage during precipitation, primarily as SO_4^{2-}. (Internal translocation is assumed to be zero.) This approach yields a value for foliar leaching of 0.09 $kmol(-) \, SO_4^{2-} \cdot ha^{-1} \cdot year^{-1}$ for this same period, similar to the value that was determined at Walker Branch Watershed.

The measurements of seasonal flux illustrated in Figure 1-4 indicate that the relative contributions of various deposition processes are not constant throughout the year, as some have assumed (Mayer and Ulrich, 1982). Dry deposition tends to dominate the input of all ions during the summer growing season when the canopy is fully developed, providing a significant surface area for interaction with suspended particles and vapors. Wet deposition is generally the dominant process during the winter dormant season when the canopy is barren and when the atmospheric concentrations of most particle constituents are at a minimum. Although vapor concentrations peak in the winter at this site, the mean winter values of SO_2 and HNO_3 exceeded those measured during the summer by only 15 to 20% (Lindberg et al., 1984). As a result of these combined factors, the summer:winter ratios of total atmospheric deposition range from $\simeq 2$ to 3 for all ions. Hence, atmospheric input to this ecosystem peaks during the growing season when foliage exposure and vegetation sensitivity to airborne pollutants are at a maximum.

C. Decomposition and Nutrient Mineralization

Concern has been expressed that acid deposition might affect the rates of decomposition and nitrogen mineralization in forest soils (Baath et al., 1980). Abrahamsen (1980) reported that irrigation with sulfuric acid in Norwegian coniferous forests stimulated N mineralization, nitrification, and nitrate leaching. There is also concern that Al mobilization by acid deposition may reduce P availability in soils. As N is the major limiting nutrient and P supplies are very low on Walker Branch Watershed, we initiated studies to determine whether irrigating with sulfuric and nitric acid would affect the status of the soil's available N and P.

In one study (Johnson and Todd, 1984), we found that after 1 year of irrigation with H_2SO_4 and HNO_3 at 2 and 10 times the current inputs on a Tarklin soil, no effect of any treatment was noted on the level of soil mineral N, P, or Al^{3+}, on N mineralization, or on CO_2 evolution. Large seasonal variations in all of the above parameters were noted, however, and the existence of these variations must be considered when sampling for long-term changes in soils (e.g. Ulrich et al., 1980).

In another study, two and ten times the annual ambient inputs of SO_4^{2-} and H^+ were applied as K_2SO_4 and $HKSO_4$ to the forest floor of a Fullerton soil on Walker Branch Watershed to study the effects of acid and S inputs on soil microarthropods and on the availability of P (Craft and Webb, 1984). Previous studies have indicated that soil microarthropods played a key role in P cycling on Walker Branch Watershed (McBrayer, 1977). As was the case in the previous study on Tarklin soil, neither two nor ten times the annual ambient inputs of SO_4^{2-}, applied as neutral salt (K_2SO_4) or acidic salt ($KHSO_4$), had any effect on exchangeable Al^{3+}, pH, or extractable P. However, the K_2SO_4 treatment that was ten times the annual ambient input caused chronic reduction in soil microarthropods and CO_2 evolution. It was hypothesized that the reduction of microarthropods was due to the osmotic effects produced by the high level of K^+ in the treatment.

Although neither of these studies indicated that acute inputs would significantly affect decomposition or soil biota, chronic effects of lower-level inputs have not been studied and therefore cannot be evaluated.

D. Soil Leaching

The potential for increased soil leaching due to acid deposition on Walker Branch Watershed is mitigated considerably by the immobilization and/or reduction and gaseous losses of S and N within the ecosystem: inputs exceed outputs by >50% for S (Shriner and Henderson, 1978) and by >75% for N (Henderson and Harris, 1975). Thus, the potentials for sulfate- and nitrate-mediated leaching are reduced accordingly.

The mechanisms for S and N retention within the terrestrial ecosystem of Walker Branch Watershed are quite different. For S, we have considerable evidence that adsorption to Fe and Al oxides is the dominant mechanism (Johnson and Henderson, 1979; Johnson et al., 1981, 1985), whereas biological uptake is clearly dominant for N (Henderson and Harris, 1975). We have as yet no information on the possible reduction of SO_4^{2-} or NO_3^- and gaseous losses. Further details on these S and N retention mechanisms are provided in the following sections on S and N cycling.

In order to accurately assess the role of acid deposition in soil leaching processes, we must quantify natural leaching by carbonic and organic acids (Johnson et al., 1977). Intensive studies of soil leaching processes were initiated in 1980 at two sites on Walker Branch Watershed: a chestnut-oak forest on a Fullerton soil and a yellow poplar forest on a Tarklin soil. We found that carbonic acid was the major natural leaching agent at both sites, but that SO_4^{2-} was the dominant anion in soil solution, even though soil SO_4^{2-} adsorption onto Fe and Al oxides was occurring at the chestnut-oak site (Richter et al., 1983). As shown in Figure 1-5, SO_4^{2-} concentrations are reduced by >50% between the A1 and B21

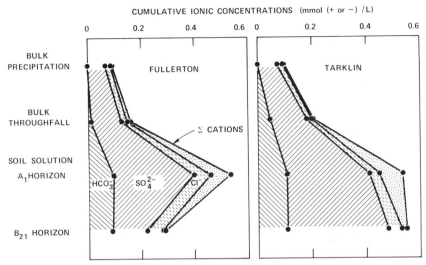

Figure 1-5. Mean volume-weighted concentrations of major anions and total cations in precipitation, throughfall, and soil solutions at the chestnut-oak (Fullterton soil) and yellow poplar (Tarklin soil) sites on Walker Branch Watershed. (Source: Richter, D. D., D. W. Johnson, and D. E. Todd. 1983. J Environ Qual 12:263–270.)

horizons of the chestnut-oak site, whereas SO_4^{2-} concentrations actually increase slightly between the A1 and B21 horizons of the yellow poplar site. These differences in SO_4^{2-} mobility are due to differences in SO_4^{2-} adsorption capacity (Figure 1-6), which are in turn due to differences in Fe-oxide content between the two soils (Table 1-2). The contrast in leaching rates between these two sites that receive basically the same atmospheric S inputs clearly illustrates the potential of SO_4^{2-} adsorption by soil for mitigating the effects of acid deposition on soil leaching. Further aspects of leaching and ion budgets in these sites are described in Section IV.B.

IV. Ecosystem-Level Studies

A. Nitrogen and Sulfur Cycling

For both N and S, atmospheric inputs exceed stream water outputs. We feel that both S and N are accumulating within the terrestrial ecosystem, but the mechanisms of accumulation differ strikingly. According to our most recent estimates, total inorganic N inputs total $9.8 \ kg \cdot ha^{-1} \cdot year^{-1}$ (Table 1-1), and streamwater outputs total $3.1 \ kg \cdot ha^{-1} \cdot year^{-1}$, indicating that approximately two-thirds of incoming N is retained in the ecosystem. This accumulation is most likely due to biological uptake: the vegetation increment alone ($15 \ kg \cdot ha^{-1} \cdot year^{-1}$; Figure

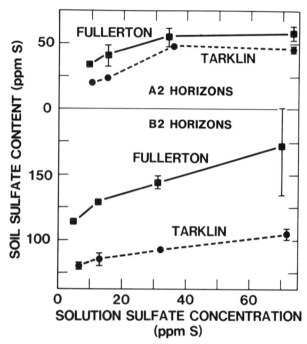

Figure 1-6. Sulfate adsorption isotherms for the Fullerton and Tarklin soils. (Source: Johnson, D. W., D. D. Richter, G. M. Lovett, and S. E. Lindberg. 1985. Can J For Res 15:773–782.)

Table 1-2. Extractable (citrate-dithionite) Fe and Al (± standard errors) in Fullerton and Tarklin soils from the chestnut-oak and yellow poplar acid deposition sites.

Horizon	Depth (cm)	Fe (%)	Al (%)
		Fullerton soil	
A1	0–7	0.94 ± 0.08	0.39 ± 0.03
A2	7–38	0.49 ± 0.03	0.17 ± 0.01
B1	38–50	1.10 ± 0.09	0.20 ± 0.02
B2	50–80	2.49 ± 0.14	0.33 ± 0.001
		Tarklin soil	
A1	0–7	0.51 ± 0.01	0.15 ± 0.004
Ap	7–18	0.47 ± 0.01	0.18 ± 0.0001
B1	18–32	0.63 ± 0.03	0.18 ± 0.006
B2	32–50	0.75 ± 0.05	0.17 ± 0.01

Source: Johnson, D. W., G. S. Henderson, and D. E. Todd. 1981. Soil Sci 132:422–426.

1-7) could account for all of the observed N accumulation in the watershed, and N uptake by vegetation (124 kg · ha^{-1} · year^{-1}) equals nearly 15 times the inorganic N input. These observations are not surprising, as forests in this region are generally N-deficient (Farmer et al., 1970). In addition to uptake by vegetation, immobilization of atmospherically deposited N by heterotrophic organisms in the soil is probably substantial, based on the results of fertilization studies on the watershed (Kelly and Henderson, 1978a, 1978b). Finally, there is the possibility of denitrification in wetter (e.g. riparian) portions of the watershed, a possibility that has not as yet been explored.

In contrast to the N cycle, the vegetation increment of S (2.3 kg · ha^{-1} · year^{-1}) equals <10% of the S input, and the uptake (23 kg · ha^{-1} · year^{-1}) approximately equals atmospheric S input (Figure 1-7). As noted previously, we have evidence that SO_4^{2-} accumulated in soils by adsorption onto Fe and Al hydrous oxides in subsurface B horizons (Johnson and Henderson, 1979; Johnson et al., 1981, 1982). The deep (up to 30 m), highly weathered soils on Walker Branch Watershed provide a very large sink for SO_4^{2-} adsorption, whereas the surface soils (A2 horizon) contain little adsorbed SO_4^{2-} but large amounts of soluble and

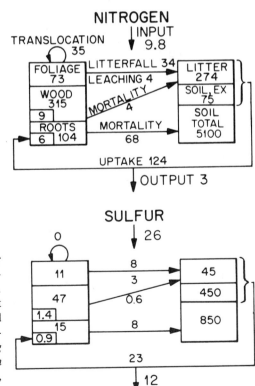

Figure 1-7. Nitrogen and sulfur cycling on Walker Branch Watershed. (Source: Johnson, D. W., and G. S. Henderson. Nutrient cycling. *In* D. W. Johnson and R. I. Van Hook, eds. 1989. *Analysis of biogeochemical cycling processes in Walker Branch Watershed.* Springer-Verlag, New York.)

readily mineralizable SO_4^{2-} (Figure 1-8). The distribution of adsorbed versus soluble SO_4^{2-} in the profile corresponds to the distribution of hydrous iron oxides (Johnson and Henderson, 1979).

Although a large proportion of soil S is in the form of SO_4^{2-}, most of it (66%) is in the form of organic S. Swank et al. (1984) recently suggested that the incorporation of S into soil organic matter may account for S retention in ecosystems at the Coweeta watershed in North Carolina. Incorporation of S into the organic matter in the soil may well account for some S retention in Walker Branch Watershed as well, but the patterns of accumulation and export suggest that adsorption is the primary S-retention mechanism. Comparison of the S cycles in the previously mentioned chestnut-oak (Fullerton soil) and yellow poplar (Tarklin soil) stands supports the contention that S accumulates by adsorption to B

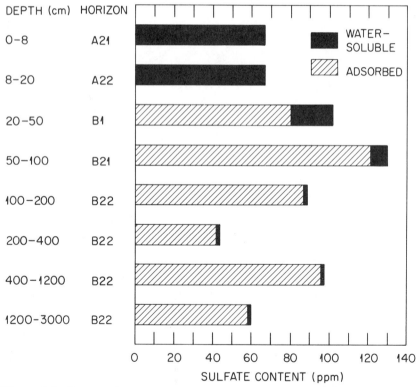

Figure 1-8. Content of water-soluble and adsorbed (i.e. extractable with 0.016 M NaH_2PO_4) SO_4^{2-} in soils of Walker Branch Watershed. (Source: Adapted from Johnson, D. W., and G. S. Henderson. 1979. Soil Sci 128:34–40.)

horizons in Walker Branch Watershed. Differences in S uptake by vegetation between the two sites are small relative to differences in S accumulation by the ecosystems, and in neither case is the vegetation increment significant relative to the ecosystem's inputs and outputs of S (Figure 1-9). The differences in the SO_4^{2-} content of the soil and in S retention by the ecosystem are clearly due to B horizon SO_4^{2-} adsorption. This is evidenced by the pattern of solution SO_4^{2-} concentration (decreasing concentrations in B horizon solutions from the Fullerton but not the Tarklin soil; Figure 1-5) and by soil SO_4^{2-} adsorption isotherms (Figure 1-6). As is the case for N, there is the possibility that SO_4^{2-} is reduced and volatized in wetter (e.g. riparian) areas of the watershed. This possibility has not yet been explored.

In an attempt to test the hypothesis that SO_4^{2-} adsorption onto Fe and Al hydrous oxides is a major factor in controlling the S flux in an ecosystem, we conducted a laboratory study of the SO_4^{2-} adsorption capacities of a number of soils from watersheds with known S budgets. We hypothesized that S retention in an ecosystem is related to the SO_4^{2-} adsorption capacity of the soil, which, in turn, is negatively related to the soil's organic matter content (organic matter is thought to block SO_4^{2-} adsorption sites) and positively related to its Fe and Al hydrous oxide

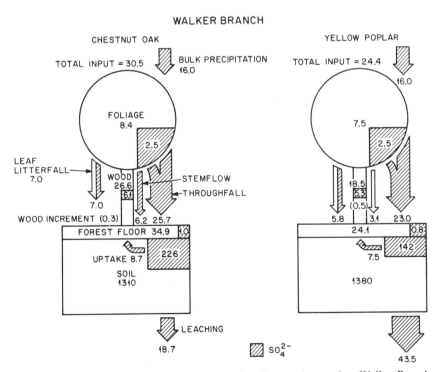

Figure 1-9. Sulfur cycling in a chestnut-oak and yellow poplar stand on Walker Branch Watershed. (Source: Johnson, D. W., D. D. Richter, H. Van Miegroet, D. W. Cole, and J. M. Kelly. 1986. Water Air Soil Pollut 30:965–979.)

content. If this is true, it should be possible to relate the SO_4^{2-} adsorption capacity of the soil (and the ecosystem's S-retention capacity) to the U.S. Department of Agriculture (USDA) soil classification system and use existing soil maps to delineate geographical areas hypothesized to have soils that retain SO_4^{2-} and those that do not. Such a regional extrapolation would be useful not only in explaining S retention per se, but also as one of several indices of sensitivity to leaching by acid deposition.

The overall results of this study (Johnson et al., 1980; Johnson and Todd, 1983) are depicted in Figure 1-10. As a general rule, Spodosols had lower SO_4^{2-} adsorption capacities than did Ultisols, which appeared to be related to the higher organic matter content and lower quantity of crystalline Fe hydrous oxides (Fe_c) in subsurface horizons. These findings are generally consistent with ecosystem S budgets: in most cases, ecosystems with Spodosols show little or no net S retention, whereas ecosystems with Ultisols usually show marked S retention (Johnson et al., 1980; Rochelle et al., 1987).

In summary, N cycling and accumulation on Walker Branch Watershed are controlled almost entirely by biological processes, whereas S cycling and especially accumulation are greatly influenced by geochemical processes. The

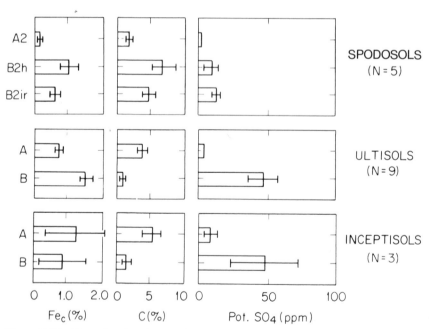

Figure 1-10. The ratio of oxalate- to dithionite-extractable Fe (Fe_o/Fe_d), percent crystalline iron hydrous oxides (Fe_c), percent carbon (C), and potential sulfate (SO_4) adsorption in Podzols (Spodosols and heavily podzolized soils), Ultisols, and Inceptisols. (Source: Adapted from Johnson, D. W., and D. E. Todd. 1983. Soil Sci Soc Amer J 47:792–800.)

contrasting mechanisms of accumulation for S and N in the watershed are apparent from the proportions of organic versus inorganic components of these nutrients in the soil: approximately one-third of total soil S is in inorganic form (SO_4), whereas <2% of total soil N is in inorganic form ($NH_4 + NO_3$) (Figure 1-7; Henderson and Harris, 1975).

B. Cation Cycling

The cycles of Ca, K, and Mg on Walker Branch Watershed as a whole have been described in detail elsewhere (Henderson et al., 1978; Johnson and Henderson, 1988), and only a few brief comments will be made here. Because the watershed is underlain by dolomite, bedrock dissolution is the dominant mechanism by which Ca and Mg are exported (on a watershed scale), and exports of these elements greatly exceed atmospheric inputs. Exports exceed inputs for K also, but to a lesser degree than for Ca and Mg, indicating that bedrock dissolution is much less of a factor in the export of K (Figure 1-11). The relatively large net annual accumulation of Ca in vegetation is also noteworthy, especially in terms of its effects on changes in the supply of exchangeable Ca in the soil, as described later (Section V.B.).

As noted previously, bedrock is up to 30 m deep on ridgetop sites in Walker Branch Watershed, and the overlying soil is highly weathered and depleted of Ca and Mg. For these reasons, we felt it necessary to examine the cycling and export of cations on a plot scale and construct budgets for that part of the soil (60 to 100 cm) that constitutes the bulk of the active rooting zone. Thus, intensive ecosystem-level studies of acid deposition were initiated in 1980 on the two forested sites (chestnut oak and yellow poplar). Process-level studies on deposition and soil leaching were combined with more traditional analyses of nutrient cycling at these sites in order to assess the effects of acid deposition on Ca, K, and Mg cycling there.

Coincidentally, each ecosystem received a total estimated H^+ input of $\simeq 2.1$ $kmol(+) \cdot ha^{-1} \cdot year^{-1}$, 70% of which was from atmospheric deposition and 30% from internal formation of carbonic acid (Figure 1-12; Johnson et al., 1985). The vegetation increment, or the net annual accumulation of base cations in vegetation, effectively added another 0.9 and 1.1 $kmol(+) \cdot ha^{-1} \cdot year^{-1}$ of internal acidification potential (defined as depletion of exchangeable base cations) in the yellow poplar and chestnut-oak sites, respectively.

As noted previously, the yellow poplar (Tarklin) site had higher concentrations of SO_4^{2-} and base cations in soil solution than did the chestnut-oak site (Figure 1-5). Consequently, the total cation leaching rates were greater in the former site $[2.7 \, kmol(+) \cdot ha^{-1} \cdot year^{-1}]$ than in the latter site $[1.7 \, kmol(+) \cdot ha^{-1} \cdot year^{-1}]$ (Figure 1-12). Although both sites show net annual losses of base cations, the chestnut-oak site shows a net annual gain of Ca^{2+} from atmospheric deposition (Figure 1-13). This was due to both the previously mentioned differences in SO_4^{2-} retention and the lower exchangeable Ca^{2+} supplies in the chestnut-oak soil (Fullerton) (Figure 1-13). The differences in the supplies of exchangeable Ca^{2+}

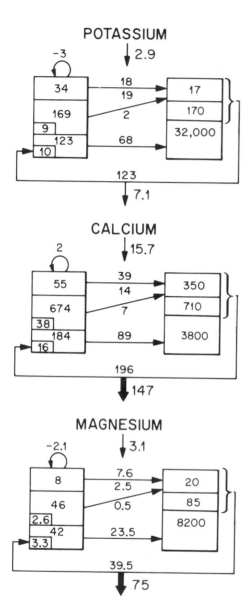

Figure 1-11. Cycles K, Ca, and Mg on Walker Branch Watershed. (Source: Johnson, D. W., and G. S. Henderson. Nutrient cycling. *In* D. W. Johnson and R. I. Van Hook. 1989. eds. *Analysis of biogeochemical cycling processes in Walker Branch Watershed.* Springer-Verlag, New York.)

may well be due to differences in the accumulation of Ca in aboveground organic matter. The chestnut-oak site has over twice as much Ca in vegetation and forest floor components [≈84 and 30 kmol(+) · ha⁻¹, respectively] as in soil exchange sites [50 kmol(+) · ha⁻¹], whereas the vegetation and forest floor contain about half as much Ca [≈37 and 9 kmol(+) · ha⁻¹, respectively) as the soil exchange complex [80 kmol(+) · ha⁻¹] in the yellow poplar site (Figure 1-12).

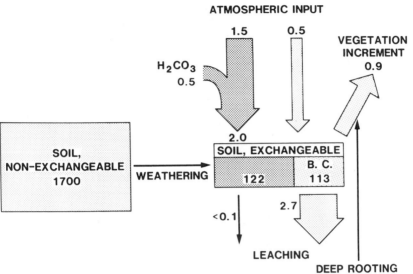

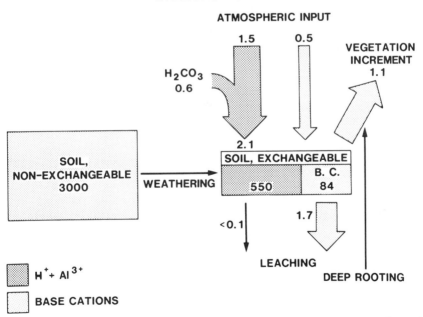

Figure 1-12. Total cation contents and fluxes in soils from the yellow poplar and chestnut-oak sites [kmol(+) · ha^{-1} or kmol(+) · ha^{-1} · year^{-1}]. (Source: Johnson, D. W., D. D. Richter, G. M. Lovett, and S. E. Lindberg. 1985. Can J For Res 15:773–782.)

CALCIUM [kmol (+)/ha or kmol (+) · ha⁻¹·year⁻¹]

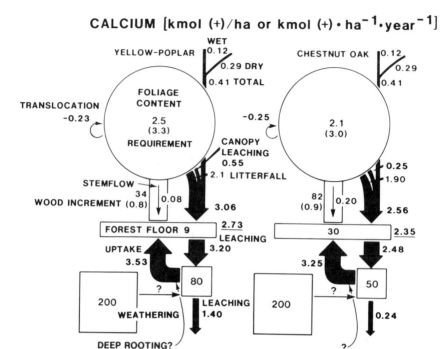

Figure 1-13. Calcium cycles in the yellow poplar and chestnut-oak sites. (Source: Johnson, D. W., D. D. Richter, G. M. Lovett, and S. E. Lindberg. 1985. Can J For Res 15:773–782.)

Differences in Ca accumulation in vegetation appeared to have a major effect on Ca, K, and Mg export through leaching versus whole-tree harvesting in chestnut-oak and loblolly pine sites near Walker Branch Watershed. In these sites, which were on the same ridge (Chestnut Ridge) as Walker Branch Watershed, the total base cation leaching rates were virtually identical, but Ca^{2+} leaching was much greater in the loblolly pine stand than in the chestnut-oak stand (Figure 1-14). The differences in leaching rates were apparently due to differences in the supplies of exchangeable Ca^{2+} in the soils of the two sites (lower in the chestnut-oak soil than in the loblolly pine soil), which were, in turn, thought to be due to differences in Ca accumulation in vegetation (greater in the chestnut-oak stand than in the loblolly pine stand; Figure 1-14). Interestingly, Mg^{2+} leaching was greater in the chestnut-oak site than in the loblolly pine site, even though the exchangeable Mg^{2+} supplies were similar (Figure 1-14). This situation is a logical consequence of the ratio of Ca^{2+} to Mg^{2+} on exchange sites. A generalized formulation of the Gaines-Thomas equation (Gaines and Thomas, 1953) describing soil-soil solution exchange is

$$\frac{(M^{a+})^b [M^{b+}]^a}{(M^{b+})^a [M^{a+}]^b} = Kgt \qquad (1)$$

where parentheses ( ) denote exchange phase; brackets [ ] denote solution phase; M^{a+} and M^{b+} are ions of change a and b, respectively; and Kgt is the Gaines-Thomas selectivity coefficient, a constant.

For $Ca^{2+} - Mg^{2+}$ exchange, we have

$$\frac{(Mg^{2+}) [Ca^{2+}]}{(Ca^{2+}) [Mg^{2+}]} = Kgt \qquad (2)$$

Rearranging yields:

$$\frac{[Ca^{2+}]}{[Mg^{2+}]} = Kgt \frac{(Ca^{2+})}{(Mg^{2+})} \qquad (3)$$

We see from Equation 3 that the ratio of Ca^{2+} to Mg^{2+} in solution (specifically $[Ca^{2+}]/[Mg^{2+}]$) is defined by the ratio of these ions on the exchange sites $[(Ca)/(Mg^{2+})]$ rather than the ratio of the absolute amounts. The same principles apply to all other major cations. If the total amount of exchangeable base cations is

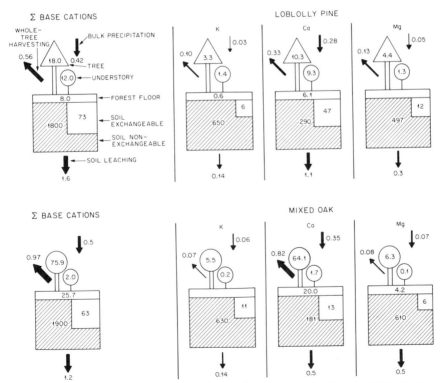

Figure 1-14. Total base cation, K, Ca, and Mg budgets for a mixed oak and loblolly pine stand on Chestnut Ridge, approximately 8 km from Walker Branch Watershed. (Source: Johnson, D. W., and D. E. Todd. 1987. Plant Soil 102:99–109.)

very low, an ecosystem may accumulate all base cations from atmospheric deposition and release Al^{3+} into soil solution (Reuss, 1983; Ulrich et al., 1980).

V. Long-Term Trends

A. Deposition and Its Interactions with the Canopy

All of our data support the hypothesis that dry deposition plays a major role in atmosphere-canopy interactions in a deciduous forest in the eastern United States. Chronic exposure of the canopy to dry-deposited particles and vapors increases the opportunity for deposition-foliage interactions because of the longer in-canopy residence time of dry deposition relative to precipitation (Lindberg et al., 1986). These interactions can result in both uptake and loss of ions. We estimate that atmospheric deposition supplies nutrients at a rate comparable to 40% of the annual needs of the Walker Branch Watershed forest for Ca and N, and >100% of its annual need for S, based on measurements of the nutrient content of the annual woody increment (nutrient content of new wood tissue produced each year) (Lindberg et al., 1986; Johnson, 1984). Even the forest's total requirement for S (total requirement = annual woody increment plus annual foliage production) is exceeded by deposition, whereas inorganic N deposition is only 5 to 10% of the total requirement (Cole and Rapp, 1981). The deposition and canopy interactions of organic N in these canopies has not been studied. If these proportions are raised because of increased industrial and automotive emissions, the forest may satisfy increasing portions of its nutrient requirements by assimilation of airborne material while simultaneously being exposed to increasing levels of airborne trace contaminants (Lindberg et al., 1982). The effects of excess S, N, and trace metal deposition by atmosphere-dominated element cycles are possibly already being manifested in high-elevation forests in the eastern United States and Europe (Johnson and Siccama, 1983).

B. Soils

The long-term effects of acid deposition are likely to include a tendency toward enriched N and S status and, in soils that are not naturally extremely acid, a decrease in base cation status (to a greater degree than is natural). The degree to which an ecosystem can be enriched in N must ultimately depend on its organic matter content, because N does not significantly accumulate in soils in inorganic forms. Organic matter content can be enriched in N concentration (up to a point), and the ecosystem's organic matter content itself can be increased by increased primary production if N had been previously limiting to growth.

The ability of a forest ecosystem to accumulate S is not solely dependent on organic matter accumulations because SO_4^{2-} adsorbs and, under certain conditions, precipitates in soils. Thus, forests growing on soils with high SO_4^{2-} adsorption capacities (i.e. those enriched in Fe and Al hydrous oxides) can accumulate atmospheric S well beyond their biological need for, or their ability to benefit from, such inputs. Furthermore, soil acidification enhances SO_4^{2-} adsorption (Harward and Reisenauer, 1966), providing a further buffer against depletion of base cations from the soil by H_2SO_4 inputs as well as enhancing the ecosystem's capacity for S retention. The SO_4^{2-} content of soil can constitute one-third to one-half of the total ecosystem S content in forests having SO_4^{2-}-adsorbing soils that have been subjected to prolonged inputs of atmospheric S at elevated levels (Johnson, 1984). As noted in the previous section on S cycling, Walker Branch Watershed falls into the category of a high-SO_4^{2-} ecosystem with about one-third of the soil's S content in SO_4^{2-} form. The question then arises: What proportion (if any) of the current sulfate content of the soil has accumulated in modern times (i.e. within the last $\simeq 100$ years).

To address this question, we sampled soils from beneath and adjacent to an old house erected ca. 1890. The house was selected from among several others for the following reasons: (1) the foundation was laid directly on the ground without excavation; (2) the roof was intact, and leakage was minimal; (3) adjacent to the house were relatively undisturbed areas (i.e. cleared, but not farmed or pastured) with slope and aspect comparable to those beneath the house; and (4) the site was located on Fullerton soil very similar in characteristics to that on Walker Branch Watershed.

Differences in total S and soluble sulfate content of the soils beneath and adjacent to the old house were not statistically significant, but B horizon samples from adjacent areas contained significantly greater amounts of adsorbed sulfate than similar samples from beneath the house, implying that sulfate had accumulated since 1890 (Table 1-3). The absence of vegetation and rainfall on the soil beneath the house no doubt reduced its S losses through uptake and leaching, thereby implying that the differences between the sulfate content of the soil beneath the house and that of adjacent soils is a minimal estimate of the actual sulfate accumulation in soil on exposed sites over the last 90 years.

Table 1-3. Soluble, adsorbed, and total S in B horizon samples beneath and adjacent to a house erected ca. 1890 (means ± standard errors).

Depth from top of B horizon[a] (cm)	Soluble SO$_4$-S	Adsorbed insoluble SO$_4$-S (ppm)	Total S
	Beneath house		
0–15	20 ± 4	67 ± 23	1520 ± 85
15–30	12 ± 1	107 ± 25	1507 ± 138
	Adjacent to house		
0–15	17 ± 3	195 ± 41[b]	1642 ± 63
15–30	15 ± 1	152 ± 50	1758 ± 153

[a] B horizon begins $\simeq 30$ cm from soil surface.

[b] Significantly greater than in soils from beneath the old house (t test, $\alpha = 0.05$).

By far the foremost concern over acid deposition effects in terrestrial ecosystems is the potential for increases in soil acidity. This concern has stimulated considerable interest in documenting long-term changes in soil acidity and nutrient status (Ulrich et al., 1980; Linzon and Temple, 1980; Malmer, 1976; Tamm and Hallbacken, 1987). In assessing the causes of changes in soil acidity, it is important to consider both leaching and vegetation uptake. Preferably, studies of long-term changes in the soil should be accompanied by detailed element-cycling studies. This is seldom the case, but two notable exceptions are the study by Ulrich and associates (1980) in a beech forest in the Federal Republic of Germany, where acid deposition is thought to have produced marked changes in both soil and soil solution chemistry, and the study by Van Miegroet and Cole (1984), where nitrification coupled with nitrate leaching in an N-fixing red alder stand caused soil acidification.

Walker Branch Watershed has been the subject of several intensive element-cycling studies, including analyses of the effects of acid deposition on soil leaching in the two forest ecosystems (chestnut-oak and yellow poplar stands) described previously. These studies showed that although acid deposition very likely had accelerated the rate of soil leaching by at least a factor of 2 [to the current rates of 1.7 and 2.7 $kmol(+) \cdot ha^{-1} \cdot year^{-1}$ in the chestnut-oak and yellow poplar sites, respectively], the reserves of exchangeable cations in the soil were sufficiently large [84 and 113 $kmol(+) \cdot ha^{-1}$] that no perceptible changes in the base cation status or acidity of the soil were expected for many decades or even centuries (Richter et al., 1983; Johnson et al., 1985). Nevertheless, both these and previous nutrient-cycling studies on Walker Branch Watershed found that the exchangeable Ca^{2+} supplies in the soil were very low [20–30 $kmol(+) \cdot ha^{-1}$] relative to the rather large rate of Ca uptake and accumulation by vegetation [1.6–2.0 $kmol(+) \cdot ha^{-1} \cdot year^{-1}$], especially in forests dominated by oaks (*Quercus* spp.) and hickories (*Carya* spp.) (Henderson et al., 1978; Cole and Rapp, 1981, Johnson et al., 1985; Figure 1-11). The budgets suggested that the sequestering of Ca in the woody tissues of trees could, in itself, cause a complete depletion of the soil's supply of exchangeable Ca^{2+} in oak-hickory and chestnut-oak forests in 10 to 20 years unless Ca uptake by vegetation was supplemented by weathering and/or deep rooting (beyond 60 cm). Clearly, weathering and deep rooting must occur in these forests, or the trees would not survive. However, the combination of low pools of Ca^{2+} in the soil and high rates of uptake by vegetation led us to hypothesize that decreases in the soils' supply of exchangeable Ca^{2+} would occur.

Given the historical data base (and, in many cases, the original samples) from early nutrient-cycling studies on Walker Branch Watershed (Henderson and Harris, 1975; Henderson et al., 1978), we were able to test the validity of some of the projections described above (as well as lay the groundwork for future sampling) by a systematic resampling of the original study plots (Johnson et al., 1988). There are no soil-leaching data for these plots, but one plot was adjacent (20 m) to the intensively studied chestnut-oak stand described above, and the potential effects of the vegetation increment on soil changes could be assessed from the 1973 and 1982 biomass and nutrient content data on all the plots.

The results showed that seasonal variations in exchangeable Ca^{2+}, Al^{3+}, K^+, and extractable P were more significant than any long-term trends in the surface (0 to 15 cm) soils. Seasonal variations in all but extractable P were minor and statistically insignificant in subsoils (45 to 60 cm), however, and the marked changes in the nutrient contents of these horizons (especially the decreases in Ca^{2+} and Mg^{2+}) deserve further scrutiny. The most consistent patterns in the subsoils were the decreases in exchangeable Ca^{2+} and Mg^{2+}, which were most notable (60 to 90% reductions) in subsoils from those cherty, ridgetop plots (91, 107, 179) with the lowest supply of these nutrients (Figure 1-15). There were also marked decreases (70 to 75%) in exchangeable Mg^{2+} in subsoils from plots 281 and 237, a marked increase (300%) in exchangeable K^+ in the subsoil of plot 107, and a marked decrease in extractable phosphorus in the subsoil of plot 91 (Figure 1-15).

There are several possible explanations for the apparent changes in the nutrient content of the subsoil from 1972 to 1982. Vegetation increment could have caused the decreases in the exchangeable Ca^{2+} content of the subsoil in the three cherty, upland chestnut-oak and oak-hickory forests (plots 91, 107, and 179). Vegetation

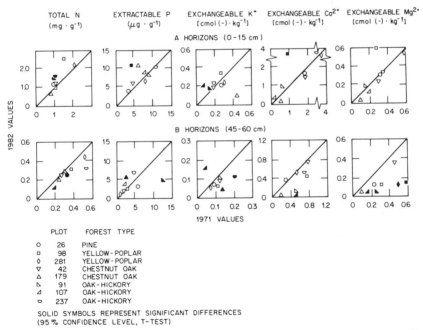

Figure 1-15. 1982 versus 1971 average values for total N, extractable P, exchangeable Ca^{2+}, K^+, and Mg^{2+} in surface (0 to 15 cm) and subsurface (45 to 60 cm) soils from selected plots on Walker Branch Watershed. Solid symbols represent significant differences (95% confidence level, t-test) between 1971 and 1982 values. (Source: Johnson, D. W., G. S. Henderson, and D. E. Todd. Changes in nutrient distribution in forests and soils of Walker Branch Watershed, Tennessee, over an 11-year period. Biogeochemistry, in press.)

increment per se could not have accounted for the decreases in the exchangeable Mg^{2+} content of the subsoil in these three plots or in the other two plots (281 and 237) in which decreases occurred. However, the total uptake by vegetation (which exceeds the net vegetation increment by five- to tenfold) from subsoils, and the subsequent return via litterfall to the soil surface, may have resulted in a redistribution of nutrients from subsoils to surface soils, as suggested by Thomas (1967) in his study of the accumulation and cycling of Ca by dogwood trees on Walker Branch Watershed. Without a full accounting of changes in all soil horizons, this hypothesis can be neither confirmed nor denied.

It is safe to assume that soil leaching, which, as noted previously, is thought to have been increased approximately twofold due to acid deposition on Walker Branch Watershed, has played some role in the observed decline in the supply of exchangeable base cations in the subsoil. This would apply especially to the reductions in exchangeable Mg^{2+}, which cannot be accounted for by net vegetation increment.

The changes observed in plot 42 are of particular interest in light of the forecasts made from element budgets in the nearby intensive acid deposition study plot (the chestnut-oak site described previously). These studies indicated that the exchangeable acidity and base cation reserves in the soil were sufficiently large in relation to the flux rates that little change in exchangeable acidity or base cations would be expected in any time frame less than several decades. These forecasts are supported in part by the data from plot 42 in that little change in exchangeable base cations was noted (Figure 15). Despite the constancy in the supply of exchangeable base cations, however, there were decreases in pH and marked increases in exchangeable Al^{3+} (Johnson et al., 1988). This apparent anomaly may be due to the dissolution of interlayered polyhydroxy Al, as suggested by Ulrich and associates (1980) for the Solling site in the Federal Republic of Germany.

It should be repeated that Walker Branch Watershed is not an undisturbed ecosystem, but a patchwork of former farms and woodland pastures that has been allowed to revert to forest since 1942. Thus, there is no reason to expect that either the vegetation or the soils should be in a steady-state condition, and there is every reason to expect that changes will occur as the new forest develops and ages. It is our intention to maintain these long-term monitoring plots for future resampling and research to ascertain the continuing changes in this forest ecosystem.

VI. Conclusions

After nearly a decade of research on Walker Branch Watershed, we can draw several conclusions as to the effects of acid deposition on nutrient cycling at that particular site:

1. Dry deposition is an important component of the atmospheric inputs of N, S, Ca, K, and Mg to the watershed. In terms of the latter three elements, however, this deposition is probably of local rather than long-distance origin.

2. The watershed accumulates atmospherically deposited N by biological uptake and atmospherically deposited S by sulfate adsorption onto the soil.
3. The forest canopy takes up or absorbs H^+, NO_3^-, and NH_4^+ but releases K, Ca, and Mg as part of the natural nutrient-cycling process. Our best estimate is that the leaching of the three latter cations has been increased by $\simeq 50\%$ due to acid deposition. Sulfate is relatively unaffected by interactions with the forest canopy.
4. Decomposition and nutrient mineralization are relatively unaffected by artificial acid irrigation at relatively high rates for short durations (1 to 2 years). These results suggest that the current inputs of acid deposition have little effect on decomposition and nutrient mineralization, but a caveat must be added in that the effects of long-term chronic inputs have not been investigated.
5. On the basis of the ratio of total cations to sulfate in soil solution, we estimate the soil leaching has been increased by $\simeq 50$ to 100% due to acid deposition. However, sulfate adsorption to Fe and Al oxides in the soil reduces the potential leaching due to acid deposition by $\simeq 50\%$.
6. Intensive studies at two sites on Walker Branch Watershed indicated that, despite accelerated leaching due to acid deposition, a significant change in soil acidity in less than many decades to centuries was very unlikely. There was a potential for reductions in the supply of exchangeable Ca in the soil due to high rates of accumulation at oak-hickory vegetation at some sites, however, and such changes have been noted in subsoils from poor ridgetop oak-hickory forest sites over the period 1971 to 1982. Reductions in exchangeable Mg in the same sites cannot be accounted for by uptake by vegetation, however, and may be due to leaching.

VII. Future Research

These results cannot be extrapolated to other sites in a quantitative sense, but the process-level information can be extrapolated to other sites in a qualitative sense (i.e. as a model or hypothesis to be tested elsewhere). The results of the studies on Walker Branch Watershed are currently being used at a number of other forested sites as part of the Electric Power Research Institute's Integrated Forest Study (Johnson et al., 1986). This project employs methods developed at WBW to determine the effects of atmospheric deposition on forest nutrient cycling at 13 sites in North America and Europe. We have extrapolated the hypothesis that sulfate adsorption controls the retention of S in forest ecosystems with some success, finding a general, regional relationship between the amount of Fe and Al oxides in the soil, the soil classification system, and the retention of S in forested watersheds (Johnson et al., 1980; Johnson and Todd, 1983). We have also determined that conclusions regarding deposition rates, the importance of dry deposition, and canopy interactions for N, S, and base cations from data at WBW apply to other southeastern forests (Lindberg and Turner, 1988; Lindberg and Johnson, 1989) and that deposition rates are much higher in mountain forests than in forests at lower elevations (Lindberg et al., 1988). Thus, in the final analysis, we recognize that the most important product we can produce from a site-specific study such as this one is information on the processes controlling the retention and movement of ions in the ecosystem and the effects of acid deposition on these processes.

References

Abrahamsen, G. 1980. *In* D. S. Shriner, C. R. Richmond, and S. E. Lindberg, eds. *Atmospheric sulfur: Environmental impact and health effects,* 397–416. Ann Arbor Science, Ann Arbor, Michigan.

Baath, E. B., M. Berg, B. Lohm, B. Lundgren, H. Lundkvist, T. Rosswall, and A. Wiren. 1980. Pedobiologia 20:85–100.

Coe, J. M., and S. E. Lindberg. 1987. J Air Pollut Control Assoc 37:237–243.

Cole, D. W., and M. Rapp. 1981. *In* E. E. Reichle, ed. *Dynamic properties of forest ecosystems,* 341–409. Cambridge Press, London.

Craft, C. B., and J. W. Webb. 1984. J Environ Qual 13:436–440.

Cronon, C. S., and W. A. Reiners. 1983. Oecologia 59:216–223.

Davidson, C. I., S. E. Lindberg, J. A. Schmidt, L. G. Cartwright, and C. R. Landis. 1985. J Geophys Res 90:2123–2130.

Daum, P. H., S. E. Schwartz, and L. Newman. 1984. J. Geophys. Res. 89:1447–1458.

Eberhardt, P. J., and W. L. Pritchett. 1971. Plant Soil 34:731–740.

Farmer, R. E., G. W. Bengston, and J. W. Curlin. 1970. Forest Sci 16:130–136.

Fowler, D. 1980. *In* D. Drablos and A. Tollan, eds. *Proceedings, International Conference on Ecological Impacts of Acid Precipitation,* 22–32. SNSF Project, Sandefjord, Norway.

Gaines, G. L., and H. C. Thomas. 1953. J Chem Phys 21:714–718.

Galloway, J. N., and D. M. Whelpdale. 1980. Atmos Environ 14:409–417.

Garten, C. T., E. Bondetti, and R. D. Lomax. 1988. Atmos. Envir. 22:1425–1432.

Garland, J. A. 1983. *In* J. Lobel and W. R. Thiel, eds. *Acid precipitation: Origin and effects,* 83. VDI Berichte 500, Verein Deutscher Ingenieure, Dusseldorf, Federal Republic of Germany.

Grennfelt, P., C. Bengtson, and L. Skarby. 1980. *In* T. C. Hutchinson and M. Havas, eds. *Acid precipitation effects on terrestrial ecosystems,* 29–40. Plenum Press, New York.

Grigal, D. E., and R. A. Goldstein. 1971. J Ecol 59:481–492.

Harward, M. E., and H. M. Reisenauer. 1966. Soil Sci 101:326–335.

Henderson, G. S., and W. F. Harris. 1975. *In* B. Bernier and C. H. Winget. eds. *Forest soils and land management,* 179–193. Les Presses de l'Universite Laval, Quebec.

Henderson, G. S., W. T. Swank, J. B. Waide, and C. C. Grier. 1978. Forest Sci 24:385–397.

Hicks, B. B. 1984. *In* A. P. Altshuller, ed. *The acidic deposition phenomenon and its effects.* EPA-600/8-83-016A. U.S. Environmental Protection Agency, Washington, D.C.

Hicks, B. B., M. L. Wesely, S. E. Lindberg, and S. M. Bromberg. *Proceedings, NAPAP Workshop on Dry Deposition, Harpers Ferry, West Virginia, 25–27 March, 1986.* ATDD Report No. 86-25.

Hoffman, S. E., S. E. Lindberg, and R. R. Turner. 1980. J Environ Qual 9:95–100.

Huebert, B. J. 1983. *In* H. R. Pruppacher, R. G. Semonin, and W. G. N. Slinn, eds. *Precipitation scavenging, dry-deposition, and resuspension.* Elsevier, New York.

Johannes, A. H., E. R. Altwicker, and N. L. Clesceri. 1981. *Characterization of acidic precipitation in the Adirondack region.* EPRI EA-1826. Electric Power Research Institute, Palo Alto, California.

Johnson, A. H., and T. G. Siccama. 1983. Environ Sci Technol 17:294A–305A.

Johnson, D. W. 1984. Biogeochemistry 1:29–43.

Johnson, D. W., D. W. Cole, S. P. Gessel, M. J. Singer, and R. V. Minden. 1977. Arct Alp Res 9:329–343.

Johnson, D. W., and G. S. Henderson. 1979. Soil Sci 128:34–40.

Johnson, D. W., and G. S. Henderson. 1989. Nutrient cycling. *In* D. W. Johnson and R. I. Van Hook, eds. *Analysis of biogeochemical cycling processes in Walker Branch Watershed.* Springer-Verlag, New York.

Johnson, D. W., G. S. Henderson, D. D. Huff, S. E. Lindberg, D. D. Richter, D. S. Shriner, D. E. Todd, and J. Turner. 1982. Oecologia 54:141–148.

Johnson, D. W., G. S. Henderson, and D. E. Todd. 1981. Soil Sci 132:422–426.

Johnson, D. W., G. S. Henderson, and D. E. Todd. 1988. Changes in nutrient distribution in forests and soils of Walker Branch Watershed, Tennessee, over an eleven-year period. Biogeochemistry 5:275–294.

Johnson, D. W., J. W. Hornbeck, J. M. Kelly, W. T. Swank, and D. E. Todd. 1980. *In* D. S. Shriner, C. R. Richmond, and S. E. Lindberg, eds. *Atmospheric sulfur deposition: Environmental impact and health effects.* Ann Arbor Press, Ann Arbor, Michigan.

Johnson, D. W., S. E. Lindberg, E. A. Bondietti, and L. F. Pitelka. 1986. *The integrated forest study on effects of atmospheric deposition,* 3–14. *In* TAPPI Proceedings, 1985 Annual Meeting, TAPPI Press, Atlanta, Georgia.

Johnson, D. W., D. D. Richter, G. M. Lovett, and S. E. Lindberg. 1985. Can J For Res 15:773–782.

Johnson, D. W., D. D. Richter, H. Van Miegroet, D. W. Cole, and J. M. Kelly. 1986. Water Air Soil Pollut 30:965–979.

Johnson, D. W., and D. E. Todd. 1983. Soil Sci Soc Amer J 47:792–800.

Johnson, D. W., and D. E. Todd. 1984. Soil Sci Soc Amer J 48:664–666.

Johnson, D. W., and D. E. Todd. 1987. Plant Soil 102:99–109.

Johnson, D. W., and R. I. Van Hook, eds. 1989. *Analysis of biogeochemical cycling processes in Walker Branch Watershed.* Springer-Verlag, New York.

Kelly, J. M., and G. S. Henderson. 1978a. Soil Sci Soc Amer J 42:963–966.

Kelly, J. M., and G. S. Henderson. 1978b. Soil Sci Soc Amer J 42:972–976.

Kelly, J. M., and J. F. Meagher. 1986. *Nitrogen input/output relationships for three sites in eastern Tennessee.* RP-1727. Electric Power Research Institute, Palo Alto, California.

Kelly, T. J., R. L. Tanner, and L. Newman. Trace gas and aerosol measurements of a remote site in the northeast. Atmos Environ (in press).

Likens, G. E., F. H. Bormann, R. S. Pierce, J. S. Eaton, and N. M Johnson. 1977. *Biogeochemistry of a forested ecosystem.* Springer-Verlag, New York.

Lindberg, S. E. 1987. Report of research in Göttingen, F.R.G. ORNL-FTR 2716, Oak Ridge National Laboratory, Oak Ridge, TN.

Lindberg, S. E. 1982. Atmos Environ 16:1701–1709.

Lindberg, S. E., and R. C. Harriss. 1983. J Geophys Res 88:5091–5100.

Lindberg, S. E., R. C. Harriss, and R. R. Turner. 1982. Science 215:1609–1611.

Lindberg, S. E., and O. W. Johnson. 1989. Annual group leader reports of the Integrated Forest Study. ORNL-TM #11052, Oak Ridge National Laboratory, Oak Ridge, TN (in press).

Lindberg, S. E., and G. M. Lovett. 1985. Environ Sci Technol 19:238–244.

Lindberg, S. E., G. M. Lovett, and J. M. Coe. 1984. *Acid deposition/forest canopy interactions.* Final Report for Project RP-1907-1 to the Electric Power Research Institute. EPRI, Palo Alto, California.

Lindberg, S. E., G. M. Lovett, D. D. Richter, and D. W. Johnson. 1986. Science 231:141–145.

Lindberg, S. E., and R. R. Turner. 1988. Water Air Soil Pollut 39:123–156.

Lindberg, S. E., D. Silsbee, D. A. Schaefer, J. G. Owen, and W. Petty. 1988 A comparison of atmospheric exposure conditions at high- and low-elevation forests in the southern Appalachian Mountains. *In* M. Unsworth, ed. *Processes of Acidic Deposition in Mountainous Terrain,* NATO Advanced Workshop, Edinburgh, UK, Kluwer Publ., London.

Linzon, S. N., and P. J. Temple. 1980. *In* D. Drablos and A. Tollan, eds. *Ecological impact of acid precipitation.* Johs. Grefslie Trykkeri, Mysen, Norway.

Lovett, G. M., and S. E. Lindberg. 1984. J Appl Ecol 21:1013–1027.

Lovett, G. M., and S. E. Lindberg. 1986. Biogeochemistry 2:137–148.

Lovett, G. M., S. E. Lindberg, D. D. Richter, and D. W. Johnson. 1986. Can J For Res 15:1055–1060.

Malmer, N. 1976. Ambio 5:231–233.

Mayer, R., and B. Ulrich. *In* H. W. Georgii and J. Pankrath, eds. *Deposition of Atmospheric Pollutants.* 195–200. Reidel, New York.

McBrayer, J. F. 1977. *In* W. J. Mattson, ed. *The role of arthropods in forest ecosystems,* 70–77. Springer-Verlag, New York.

Meiwes, K. J., and P. K. Khanna. 1981. Plant Soil 60:369–375.

Miller, H. G., and J. D. Miller. 1980. *In* D. Drablos and A. Tollan, eds. *Ecological impact of acid precipitation.* Johs. Grefslie Trykkeri, Mysen, Norway.

National Academy of Sciences (NAS). 1983. *Acid deposition-atmospheric processes in eastern North America.* National Academy Press, Washington, D.C.

Reuss, J. O. 1983. J Environ Qual 12:591–595.

Richter, D. D., D. W. Johnson, and D. E. Todd. 1983. J Environ Qual 12:263–270.

Richter, D. D., and S. E. Lindberg. 1988. J Environ Qual 17:619–622.

Rochelle, B. P., M. R. Church, and M. B. David. 1987. Water Air Soil Pollut 33:78–83.

Shanley, J. B. Dry deposition to spruce foliage and petri dishes in the Black Forest, West Germany. Atmos Environ (in review).

Shannon, J. D. 1981. Atmos Environ 15:1155–1163.

Shriner, D. S., and G. S. Henderson. 1978. J Environ Qual 7:392–397.

Swank, W. T., J. W. Fitzgerald, and J. T. Ash. 1984. Science 223:182–184.

Tamm, C. O., and Hallbacken. 1987. Water Air Soil Pollut 30:337–342.

Taylor, G. E., S. B. McLaughlin, D. S. Shriner, and W. J. Selvidge. 1983. Atmos Environ 17:789–796.

Tennessee Valley Authority (TVA). 1982. *Tennessee Valley Authority Monitoring Section, Air Resources Program, Ambient Air Quality Monitoring System Data Summary.* TVA/ONR/ARP-82/18. TVA, Muscle Shoals, Alabama.

Thomas, W. A. 1967. Ecol Mon 39:101–120.

Tjepkema, J. D., R. J. Cartica, and H. F. Hemond. 1981. Nature 294:445–446.

Ulrich, B., R. Mayer, and P. K. Khanna. 1980. Soil Sci 130:193–199.

Van Miegroet, H., and D. W. Cole. 1984. J Environ Qual 13:586–590.

Wolff, G. T. 1984. Atmos Environ 18:977–981.

Acidic Precipitation: Case Study Solling

E. Matzner*

Abstract

By using the flux-balance approach, the rates of deposition of major elements, including H^+, S, and N, and the effects on ion cycling in the canopy and mineral soil were obtained in two mature forest stands of the north German Solling region. Both stands are heavily impacted by acid precipitation, the spruce (*Picea abies* Karst.) stand receiving about twice the rate of acidity as the beech (*Fagus silvatica* L.) stand. The deposition of protons results in increased cation leaching from the leaves and needles with subsequent acidification of the rhizosphere. In both sites the mineralization of aboveground litter was significantly inhibited, and the amount of organic matter in the top layer almost doubled during the period of investigation. The organic top layer appears to be the major sink for deposited N. The Ca/Al and Mg/Al ratios of the soil solution have decreased in both stands to levels posing high risk of Al toxicity to tree roots. According to input-output budgets of the mineral soil sources of ions are obvious in the case of most major nutrients. In spruce the behavior of sulfate in the mineral soil changed from accumulation to release over the study period. Proton budgets reveal that the most important part of the acid load of the mineral soil stems from the deposition of strong acids. The prevailing buffer mechanism is the release of Al ions from hydroxides, sulfates, and exchangeable sites into the soil solution. The transfer of Al ions by seepage water to deeper soil layers results in a high acid load of those layers. Soil analysis showed deep-reaching acidification and emphasize the risk of groundwater and surface water acidification. The long-term effects of acid deposition on soil chemistry are superimposed by seasonal acidification phases resulting from natural excess nitrification under favorable climatic conditions. In acid soils these acidification phases cause pH reductions and additional Al releases.

I. Introduction

Being part of the International Biological Program, ecosystem research in the German Solling has been ongoing since 1966 and thus represents worldwide one of the longest continuous studies in this field of science.

---

*Research Center Forest Ecosystems/Forest Decline, University of Göttingen, D-3400, Göttingen, FRG.

Measurements of the elemental cycling revealed high rates of deposition of various air pollutants, especially sulfuric acid, in this remote area (Ulrich et al., 1973).

Based on the available data for the period of 1969 to 1975 Ulrich (Ulrich et al., 1979) was one of the first to point out the detrimental effects of acid deposition on soils and forest ecosystems. The major conclusions of Ulrich and associates (1979) were the evident acidification of soils, increasing risk of root damage due to Al toxicity, and the prognosis of forest decline caused by acid deposition.

Because of the recent forest decline in large parts of central Europe, research on the effects of air pollutants on soils and forest ecosystems has been expanded significantly. Many of the recognized symptoms are related to the nutritional status of the tree and to root disturbance. Processes of the elemental cycling like leaching of cations from leaves and Al mobilization in soils are now seen in a new light.

This paper summarizes the results of measurements of acid deposition effects on elemental cycling in forest canopies and soils of the Solling experimental forests from 1969 to 1985. Special emphasis is given to the elements H^+, Na, K, Ca, Mg, Mn, Al, Fe, S, P, Cl, and N. Heavy metal budgets of these sites are given by Mayer (1981). The data base used for this paper is available in Matzner and co-workers (1982) and Matzner (1988).

II. Research Area

The ecosystem research of the Solling project focused on two adjacent forest ecosystems represented by a 140-year-old stand of European beech (*Fagus silvatica* L.) "B 1" and by a 105-year-old stand of Norway Spruce (*Picea abies* Karst). "F 1." Both stands are located on the plateau of the Solling mountains in northwest Germany at about 500 m elevation (Figure 2-1). The beech stand developed from natural regeneration; the spruce stand was planted.

A detailed description of the sites, their history, and structure is given by Seibt (1981). Ellenberg and associates (1986) summarized the aims and organization of the Solling project. The mean annual temperature is 6.4°C and the long-term average precipitation is 1088 mm per year.

Soils have developed from weathered sandstone covered by loess sediments. The soils are strongly acidified, pH($CaCl_2$), reaching from 2.9 to 4.2, and the base saturation of the CEC is less than 5% in the whole soil profile; detailed chemical characteristics are given by Mayer and group (1986). The average annual SO_2 concentration of ambient air was 40 $\mu g/m^3$ in 1985-1986.

III. Methods

The measuring approach is demonstrated in Figure 2-2. The element fluxes marked with dotted circles were measured directly; the others were calculated.

Bulk precipitation and throughfall were sampled each by 15 standard rain

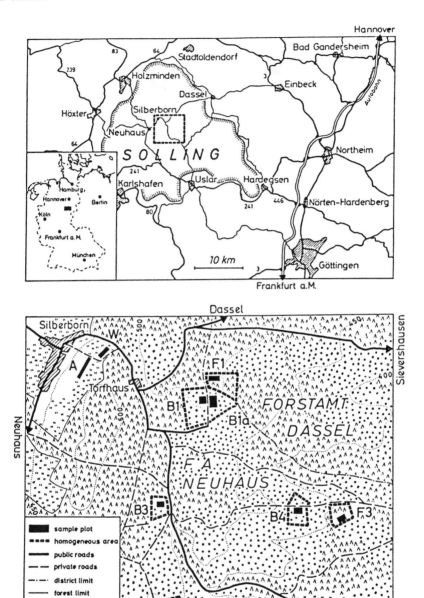

Figure 2-1. Situation of the research area and of the experimental plots on the Solling plateau, 55 km northwest of Göttingen. **B:** beech (*Fagus silvatica*) forests, deciduous; *B1* and *B1a*, old; *B3*, younger, *B4*, youngest stand. **F:** spruce (German "Fichte," *Picea abies*) forest, evergreen; *F2*, old, *F1*, younger, *F3*, youngest stand. **W:** meadow (German "Wiese," *Trisetetum flavescentis*, *Festuca rubra* facies), a mown grassland. **A:** arable field (with *Zea mays* and *Lolium multiflorum*, resp.).

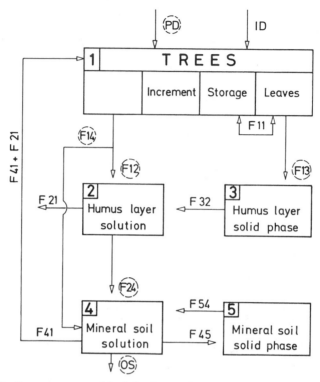

Figure 2-2. Compartment model of the fluxes of elements within a forest ecosystem. PD = precipitation-deposition; ID = interception-deposition (for definition of *PD* and *ID*, see the next section); and OS = output with seepage water. The ecosystem internal fluxes are F 11 = translocation of elements inside the plant; F 12 = canopy drip; F 13 = stemflow; F 13 = litterfall; F 24 = input to the mineral soil; F 21 = uptake by the stand out of the humus layer; F 32 = net mineralization; F 54 = release from the soil by desorption, dissolution, and weathering; F 45 = accumulation in the soil by adsorption, exchange and precipitation; and F 41 = uptake by the stand out of the mineral soil.

collectors made of Plexiglas (50 cm^2) that were placed in a systematic grid of 3 × 5 m. Samples were taken once or twice a week and were mixed and analyzed on a monthly basis. Bulk precipitation was collected on a meadow about 800 m away from the stands at the same altitude. During periods of snowfall, PVC buckets of 570 cm^2 were used.

Stemflow of three beech trees was collected during 1969 to 1976 in large vessels. Stemflow has no significance in the spruce stand. The fluxes of elements with stemflow were calculated by using the amounts of stemflow that were measured intensively on a large number of trees by Benecke (1984). From 1977 to 1984 no measurements of stemflow chemistry took place. During this period stemflow concentrations were calculated by a regression to monthly throughfall concentrations based on the available data from 1969 to 1976. Since 1984 the amount and chemistry of stemflow were again measured at four trees. Methods and results of increment studies are described by Seibt (1981) and Heller (1986). Nutrient content of the various biomass components were published recently by Mayer et al. (1986).

Seepage water underneath the rooting zone was collected by three (beech) and seven (spruce) ceramic suction lysimeters (plates 30 cm diameter) at a depth of about 90 cm. Monthly seepage water fluxes were taken from the hydrological models of Benecke (1984) and Salihi (1984).

Since 1981 soil solution was additionally collected by ceramic cups at depths of 10, 20, 40, and 80 cm. Six replicates were used per depth and three mixed samples were composed and analyzed. Inventories of the element storage of the mineral soil and organic top layer were done in 1966 (spruce in 1968), 1973, 1979, and 1983. Three to four mixed samples of three to four corings distributed over the area were analyzed. Under beech only two mixed samples each mixed from eight corings were analyzed in 1966.

The analytical methods used are listed below (for details, see Meiwes et al., 1986).

H^+ in solution: glass electrode
pH of the soil: 25 g dry soil + 25 ml 0.1 N $CaCl_2$, glass electrode
NH_4, NO_3, P: colorimetrically by continuous flow system
S: potentiometric titration, colorimetrically by continuous flow system
Cl: potentiometric titration, colorimetrically by continuous flow system
Na, K, Ca, Mg, Mn, Fe, Al: atomic adsorption spectroscopy
Exchangeable cations of the mineral soil: extraction by 1N NH_4Cl

IV. Results

A. Input of Major Elements by Atmospheric Deposition

There is some difficulty in determining total deposition rates from the atmosphere to terrestrial ecosystems. Precipitation input cannot be regarded as total deposition in most cases, because interception deposition due to surface properties of the vegetation is not taken into account (for definitions, see Figure 2-3). The flux of chemical elements in throughfall is a result of both deposition from the atmosphere (external input) and canopy processes (internal cycling). The canopy may act as a sink or as a source for special chemical constituents. Some constituents may, however, pass the canopy inertly.

The approach used in the Solling area to quantify the rates of deposition and element turnover in the canopy was developed by Ulrich and associates (1979; 1983b) and has recently been described by Bredemeier and co-workers (1988) and by Matzner (1988) in detail. Only a short description is given here.

The method is designed to assess long-term (e.g. yearly) total deposition rates in forests and is based upon measurements of major element fluxes in and out of forest canopies (Figure 2-3). Precipitation deposition (PD) can be sampled in bulk

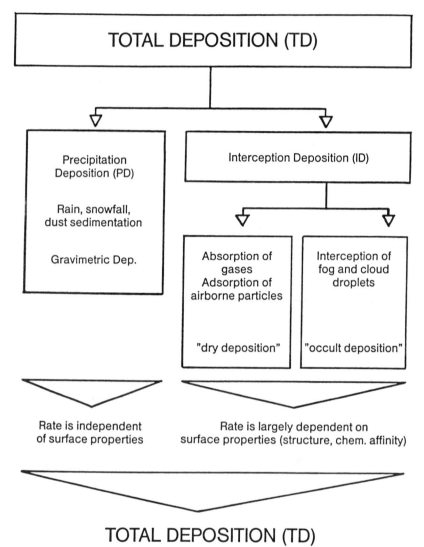

Figure 2-3. Partition of total deposition.

collectors. Interception deposition (ID) is the amount of elements additionally deposited upon the forest surface. The sum of both yields total deposition (TD) for each chemical constituent:

$$TD_x = PD_x + ID_x \tag{1}$$

where x = chemical element. The output of elements from the canopy is assumed to occur only by throughfall.

If throughfall (TF) is the only output flux from the canopy, the following equation is valid:

$$TF_x = PD_x + ID_x + S_x \tag{2}$$

where S indicates the canopy sink/source function (i.e. assimilation or leaching) with respect to the chemical element x.

Calculating interception (ID) and total deposition (TD) rates for all major elements starts with those elements for which the canopy can be assumed to act as an inert sampler, that is, for which S_x is 0. This is considered in the case of Na, Cl, and S:

$$ID_x = TF_x - ND_x \tag{3}$$

where x = Na, Cl, S. The assumption of inert flow of deposited Na, Cl, and S through the canopy is critically reviewed by Bredemeier and associates (1988) and by Matzner (1988).

Sodium occurs in the atmosphere only in aerosols, predominantly from sea spray. From the ID of Na^+, the particulate ID of other elements is calculated according to equation 4:

$$\frac{ID}{PD_{Na}} = \frac{ID}{PD_{x \, part}} \tag{4}$$

where x = H, K, Ca, Mg, Al, Fe, Mn, S. In addition to particulate interception deposition, some elements are also dry deposited as gases:

$$ID_x = ID_{x \, part} + ID_{x \, gas} \tag{5}$$

where x = S, (N). For Cl, a gaseous deposition fraction is neglected, because it should be important only in close vicinity to HCl sources. However, gaseous deposition sulfur (SO_2) is an important fraction of total deposition. Its rate can be calculated from the difference of total interception deposition (Eq. 3) and $ID_{x \, part}$ (Eq. 4).

Within the pH range found in precipitation samples and with ambient concentrations of O_2, sulfur dioxide will be oxidized, according to sulfuric acid.

Thus, the calculated S_{gas}-deposition rate means an equivalent deposition of protons. Total deposition of H^+ is therefore:

$$TD_H^+ = PD_H^+ + ID_{H \, part} + ID_{H \, gas} \tag{6}$$

Protons deposited in the canopy can partly be buffered by cation (especially K, Mn, Ca, Mg) exchange from tissues. In this case, deposited protons do not appear as measurable acidity in throughfall.

$$H^+\text{-buffering} = TD_H^+ - TF_H^+ \tag{7}$$

Cation leaching rates are calculated according to Equation 8:

$$CL_x = TF_x - TD_x \tag{8}$$

where x = Ca^{2+}, Mg^{2+}, K^+, Mn^{2+} and CL = cation leaching in the canopy.

1. Annual Rates of Deposition

The rates of precipitation deposition (PD = bulk precipitation) and of total deposition of the alkaline earth elements and of H^+, Fe, Mn, Al, S, and Cl are presented in Table 2-1.

The average element fluxes in throughfall and bulk precipitation are given together with those of seepage water output in the appendix.

The rates of interception deposition are significantly higher for the spruce stand than for the beech. This is caused by the larger filtering area of the spruce canopy, especially during winter.

The relation of TD/PD in both stands indicates that the rates of interception deposition of H^+ and S are high compared to the other elements. This effect can be attributed to the deposition of SO_2 with subsequent formation of sulfuric acid in the water films on the leaves, bark, and needles. Furthermore, the adsorption of unbuffered acid droplets may give the same result.

The range of the annual rates of deposition given in Table 2-1, which was observed during the 17 years of investigation, emphasizes the importance of long-lasting measurements for determining the deposition to a forest stand. The variation in total deposition from year to year may exceed 100%, whereas the variation in precipitation deposition is less. However, an overall temporal trend was not evident. Thus, the stands have been impacted by acid deposition for decades.

2. Seasonal Pattern of Deposition

Further information about the mechanisms of deposition are available from the seasonal pattern of the rates of deposition. Figure 2-4 therefore shows the mean monthly values of the rates of precipitation deposition of H^+ and S and the mean monthly values of the difference between throughfall and PD. This difference is called *canopy difference* (CD) and corresponds to the rate of interception deposition in the case of S, as no foliar leaching of S is assumed.

In the case of H^+ the CD does not correspond to the rate of interception deposition because protons are partly buffered on the plant surfaces. This process is discussed in detail later. The buffering of protons within the canopy will presumably influence the seasonal pattern of the CD only slightly for spruce, but the low values observed for beech during the vegetation period are partly caused by proton buffering.

The rates of PD of H^+ are about the same throughout the season, but a slight maximum of S deposition is found during the summer.

The CD of S and of H^+ shows a pronounced seasonal pattern indicating high rates of interception deposition during the winter. One mechanism causing an increase in the rates of interception deposition during winter may be the adsorption of acid droplets as indicated by the occurrence of fog. The mean number of days with fog is given in Figure 2-4. The maximum is also found during winter. Especially fog arising from long-range transport of clouds in the higher elevated sites may be enriched with pollutants, mainly H^+ and S (Schrimpff, 1983;

Table 2-1. Rates of deposition in the Solling area (kg · ha⁻¹ · yr⁻¹, mean and range values for 1969-1985).

	H	Na	K	Ca	Mg	Fe	Mn	Al	SO₄-S	Cl
PD	0.82	7.8	3.7	9.8	1.7	0.7	0.4	1.2	23.2	16.7
	0.61-1.27	4.9-12.1	2.4-5.6	6.5-21.8	1.3-3.9	0.2-1.1	0.1-0.9	0.6-2.1	19.6-27.2	10.5-25.5
TD Beech	2.00	14.1	6.6	17.1	2.9	1.3	0.7	2.1	50.0	32.5
	1.13-2.75	9.5-18.2	4.9-9.9	11.4-32.0	2.1-3.8	0.3-2.4	0.2-1.4	0.9-3.2	39.1-66.0	24.2-40.9
TD Spruce	3.79	17.0	8.0	21.1	3.9	1.6	0.9	2.5	83.1	38.6
	2.68-5.26	9.0-25.4	5.7-11.9	13.2-37.9	1.8-6.3	0.4-3.1	0.2-1.8	1.0-4.5	54.0-108	27.5-54.3

PD = Precipitation deposition.

TD = Total deposition.

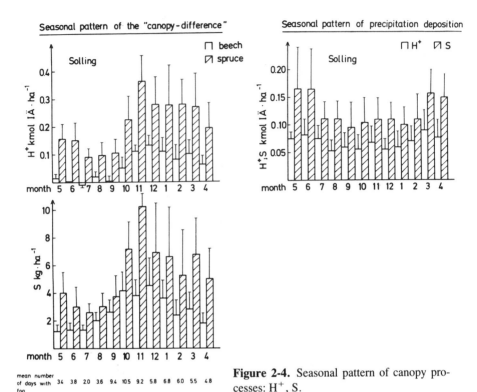

Figure 2-4. Seasonal pattern of canopy processes: H⁺, S.

Wisniewski, 1982). That is why the interception of fog may play a significant role in determining the rates of deposition in higher elevations. This was indicated earlier by Ulrich and associates (1979).

However, recent (1985 and 1986) measurements of atmospheric SO_2 levels (unpublished data) revealed high average concentrations during the winter, reaching 67 $\mu g/m^3$ from October to March, yet only 16 $\mu g/m^3$ was recorded from April to September. Monthly mean values during winter were as high as 156 $\mu g/m^3$, which increases the rate of interception deposition of S significantly.

3. Turnover of Elements in the Canopy

As mentioned above, the H^+ deposition is partly buffered during canopy passage by cation exchange. If the rate of interception deposition is low, one can find a net reduction of the H^+ load in throughfall as compared to the precipitation deposition (Ulrich et al., 1973; Cronan and Reiners, 1983). To keep the electroneutrality in the plant and the solution, the plant has to take up an amount of anions from precipitation that is equivalent to the protons buffered, or the buffering must be connected with the leaching of cations from the canopy. Table 2-2 gives an overview of the leaching and buffering processes within the canopy of the Solling stands.

Table 2-2. Turnover of elements in the forest canopy: H^+ buffering and leaching of cations (keq $\cdot$ ha^{-1} $\cdot$ yr^{-1}, mean and range values for 1969-1985).

	Beech	Spruce
H^+ buffering	0.66 (0.01–1.22)	0.64 (0.00–1.59)
Leaching of K^+	0.54 (0.21–0.80)	0.51 (0.35–0.88)
Leaching of Ca^{2+}	0.35 (0.0–0.80)	0.51 (0.0–0.84)
Leaching of Mg^{2+}	0.09 (0.0–0.17)	0.07 (0.0–0.13)
Leaching of Mn^{2+}	0.12 (0.09–0.14)	0.16 (0.12–0.24)
Σ leached cations	1.10	1.25
Leaching of dissolved organic acids* ($\bar{x}$ 1971-85)	0.28 (0.03–0.67)	0.49 (0.0–1.65)
$(1 - 2 + 3)$	-0.17	-0.12

*Calculated from the cation/anion balance of PD and throughfall.

The leaching of cations from the canopy may be accompanied by the leaching of dissolved organic anions. The flux of organic anions is calculated from the cation-anion balance of the elements measured in PD and throughfall. The deficit of anions (= amount of organic anions) is, according to Table 2-2, higher in throughfall than in the PD. That is why a positive value is given for the leaching of organic anions from the canopy of these stands.

The mean annual rates of H^+ buffering amount to about 33% of the total H^+ deposition for the beech stand (0.66 keq $\cdot$ ha^{-1}) and to about 17% of the total deposition (0.64 keq $\cdot$ ha^{-1}) for the spruce stand.

The cation leached most (when expressed in keq) is K for beech, followed by Ca, Mg, and Mn. In the spruce stand, K and Ca leaching is about the same. The range of values given for both sites indicates a strong annual variability, which was also discussed by Ulrich (1983) in detail.

Total amount of cations leached has been calculated as 1.10 for beech and 1.25 keq $\cdot$ ha^{-1} $\cdot$ yr^{-1} for spruce. The leaching cannot be totally balanced by the buffering of protons because leaching of cations is twice the buffering rate. Taking the leaching of organic anions into account, the processes of leaching of cations, buffering of H^+, and leaching of anions are nearly balanced. The remaining difference in the charge balance can be attributed to different processes; the calculation of the rates of interception deposition assumes no leaching of mineral anions. The leaching of about 2 to 3 kg S $\cdot$ ha^{-1} $\cdot$ yr^{-1} (which is very little compared to the fluxes with throughfall) would already be sufficient for balancing the leaching budget. Ulrich and associates (1979) calculated leaching of S from senescent leaves to be about 3.3 kg S $\cdot$ ha^{-1} $\cdot$ yr^{-1}.

Furthermore, the charge balance may be adjusted by the uptake of other cations beside H^+ into the leaf. The cation taken up may be NH_4^+. Again the uptake of 2 to 3 kg NH_4^+-N per ha per year would be sufficient to balance the charge deficit. The rates of leaching of K and Mg calculated for the beech stand correspond to the difference between the amount of elements stored in the green leaves and those

found in the litterfall (Matzner et al., 1982). The change in the element storage of
the leaves throughout the season therefore can be explained by leaching. As the
element storage of litterfall is higher than in the green leaves for Ca and about equal
for Mn, the leaching of Ca and Mn cannot be attributed to the changing element
contents of the leaves throughout the growing season.

The annual leaching of K in the spruce stand amounts to about half of the total
storage within the needle biomass. The ratios for Ca and Mg are 1:3 and for
Mn 1:4.

4. N Input

Determination of the rates of interception deposition and total deposition of N
proved to be difficult because several additional processes may occur in the
canopy, such as the assimilation of N by the plant or by canopy epiphytes.
Furthermore, one cannot account for hidden turnover of the various N
forms—NH_4^+, NO_3^-, N_{org}, NO_x, HNO_3 N_2, and NH_3 in the canopy or the sampler.

Despite these difficulties an estimate of the rates of N deposition to the stands
can be made, based on the flux measurements (see below).

The temporal pattern of the N fluxes with throughfall, which is given in Figure
2-5, is characterized by an increase from 1969 to 1976. From 1976 onwards the N
fluxes are more or less constant under beech and slightly decreased under spruce.

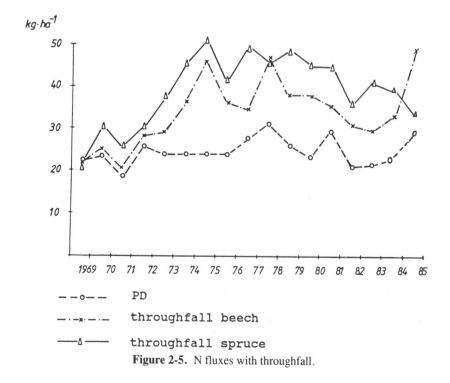

Figure 2-5. N fluxes with throughfall.

Nitrogen may be deposited in the form of particles or, similar to S, by adsorption of gases (NH_3, NO_x, HNO_3 vapor). Assuming that all N deposited will be washed off by precipitation and is measured in the throughfall, the ratio of total deposition:precipitation deposition of NH_4 and NO_3 would be only 1.3:1.2 for beech and 1.4:2.2 for spruce. These ratios are low when compared to other elements (e.g. for Na, which is deposited only in the form of particles, the ratio is 1.8 for beech and 2.4 for spruce). For S, ratios of 2.2 (beech) and 3.6 (spruce) were found. From these data one can conclude that parts of the N deposition do not reach the soil with throughfall but are presumably assimilated in the canopy. This conclusion is also confirmed by the seasonal pattern of the CD of NH_4 and NO_3 given by Matzner (1986). Although the annual NH_4 flux with throughfall of both stands is higher than with bulk precipitation, the CD becomes negative during the summer, indicating NH_4 assimilation in the canopy. No negative values of the CD are found for NO_3, but the CD is rather low during summer, and the process of nitrification of NH_4 with subsequent leaching of NO_3 is assumed to be negligible. High rates of interception may yield positive values despite NO_3 assimilation.

The rates of interception deposition (ID) of N should be estimated roughly by applying the ratio of ID:PD for Na. This estimate assumes that Na-containing aerosols are deposited in the same amount as N-containing aerosols, but it will underestimate the deposition of N because gaseous compounds are neglected.

Estimating the total deposition as described allows the calculation of the minimum rate of N assimilation in the canopy by subtracting the N flux with throughfall. This results in a rate of N assimilation in the canopy of about 8 kg $\cdot ha^{-1} \cdot yr^{-1}$ in both Solling stands, according to 20 to 25% of the N flux with throughfall. Because of a lot of uncertainties about the turnover of N forms in the canopy, no data can be given of the form of N uptake.

The fluxes of N with throughfall therefore underestimate the actual N input. Accepting the magnitude for N uptake in the canopy calculated above, the average total annual N input is about 40 (beech) to 50 (spruce) kg N $\cdot$ ha^{-1}; it is equal to the total N demand of the spruce stand and provides 70% of the N demand of the beech stand.

5. Discussion

Compared to other north German forests the Solling area is highly impacted by the deposition of air pollutants. This seems to be characteristic for elevated sites as indicated by others (Wiedey and Gerriets, 1986; Godt, 1986; Matzner et al., 1984) and is presumably caused by high rates of fog and cloud interception and by high SO_2 levels during winter.

The deposition of acidity and its effects on the element cycling in forest canopies have been the subject of several studies. The process of H^+ buffering in the canopy by ion exchange was discussed by Cole and Johnson (1977), Cronan and Reiners (1983), Johnson and associates (1985), Lovett and Lindberg (1984), van Breeman and co-workers (1986), and by Ulrich (1973 with associates; 1983a, 1983b). Quantification was only approached in the papers of Ulrich (1983a, 1983b), van Breeman and co-workers (1984), and Lovett and Lindberg (1984).

Cronan and Reiners (1983) distinguished between two reactions involved in this process, ion exchange and the protonation of Brønstedt bases. This concept is similar to the one put forward by Ulrich (1983a), which has been described in this paper. Cronan and Reiners (1983) and Ulrich (1983b) emphasize that the H^+ buffering in the canopy does not reduce the acid load of the soil. From the ecological point of view the acidity buffered in the canopy affects a rather critical space of the system: the apoplast of the root cortex and the rhizosphere. The possible impact of this process via soil acidification and acid toxicity on the stands may be evaluated at first by pH measurements close to roots. The acidification of the rhizosphere of agricultural crops has been demonstrated by Marschner and Römfeld (1983) and by Schaller and Fischer (1985). However, the importance of various processes for acidification may differ when comparing agricultural crops and forests. Unfortunately, no measurements on the pH close to roots are available in forest stands under field conditions so far.

The rates of proton buffering in forest canopies given by van Breeman and associates (1986) for different forests of the Netherlands are in the range of 0.32 to 1.06 keq $\cdot$ ha^{-1} $\cdot$ yr^{-1} and correspond well with those found in the Solling, despite different ways of calculation. Furthermore, they found the rates of buffering higher on calcareous sites compared to acid soils. This may be explained by the different nutrient status of the leaves and was also postulated by Ulrich and Matzner (1983).

The effects of proton buffering in the canopy accompanied by the increased leaching of the cations were the subject of experiments carried out by Horntvedt and co-workers (1980). They reported increasing leaching of Ca, Mg, and K by raising the load but did not find changes in the needle content of these elements, indicating a quick replacement by ion uptake. These findings correspond to experiments carried out by Kreutzer and Bittersohl (1986). By irrigating young trees of Norway spruce, they found rates of H^+ buffering of about 1.6 keq $\cdot$ ha^{-1} $\cdot$ yr^{-1}, which was in the same range as the sum of cation leaching. Cation content of the needles was not affected.

Johnson and associates (1985) concluded that the leaching of cations from forest canopies is raised by a factor of 2 to 3 by acid deposition and H^+ buffering and confirm the results from the Solling area.

Discussing the N input by deposition, it was mentioned that the N flux with throughfall underestimates the actual input because indications for N uptake in the canopy were found. The uptake of nutrients by leaves has been known for a long time and is practically used during leaf fertilization (Boynton, 1954). However, no data about the rates of uptake following leaf fertilization of forest stands are available. The rates of N uptake calculated for the stands under investigation of about 8 kg N $\cdot$ ha^{-1} $\cdot$ yr^{-1} therefore cannot be compared with the literature data. The process of N uptake in the canopy was also studied by Miller and Miller (1980), but they discussed only changes of N concentrations in precipitation and throughfall over a short period of time without quantifying the uptake rate.

The paper of Lang and co-workers (1976) focused on the influence of epiphytic lichens on the throughfall chemistry in a stand of Douglas fir. They report temperature-dependent uptake of N in the canopy (maximum during summer) and the turnover of mineral to organic N. The organic N is subsequently found in the throughfall. The results from the Solling sites are in good agreement with the experiments done by Lang and associates (1976), considering the seasonal pattern of the CD of NH_4^+ and the proportion of organic N in throughfall. The biomass of lichens in the Solling stand certainly is much lower than in the stand described by the Lang group (1976), but N turnover may also be possible by bacteria and algae on the leaves.

The long-term effects of the present N input in forest ecosystems are only poorly understood. Meyer (1984) attributed the root decline observed in damaged forest stands of Germany to a disturbance of the mycorrhiza by high levels of N deposition. Evaluation of this hypothesis must be the aim of future research as well as the effects of N uptake by leaves on the ionic status of the leaf and the rates of ion uptake by roots. A review of the present state of knowledge about the effects of N deposition in terrestrial ecosystems is given by Beese and Matzner (1986), who concluded that long-term changes of the composition of vegetation, nitrate leaching, and denitrification can be expected.

B. Effects of Acidic Precipitation on the Soil

1. Accumulation of the Organic Top Layer

The mineralization of root and leaf litter is the most important process of the internal cycling of elements in forest ecosystems, besides the process of ion uptake. The turnover of organically bound elements (especially N) into ionic and plant-available forms is a major supposition of high productivity under conditions of low atmospheric nutrient input.

Inhibition of litter decomposition, which leads to the accumulation of the organic top layer, thus may shorten the N supply of the stand considerably and subsequently result in an increased H^+ load to the mineral soil (see below).

Investigations on the rate of mineralization and its dependence on environmental factors have long been a subject of ecosystem research and have focused on the behavior of N (e.g. Zöttl, 1958, 1960; Runge, 1970).

In order to follow the long-term behavior of the organic top layer of the Solling stands, periodic inventories of the organic dry mass of the humus layer and its mineral element storage were conducted. The results are given in Table 2-3.

Because only two mixed samples were analyzed in 1966 in beech, no standard deviation (SD) is given in Table 2-3. However, the number of replicates (16) in 1966 was comparable to the sampling in the following years.

Comparison of the element storage of the organic top layer over the whole period from 1966 and 1983 in the beech stand clearly shows increasing N and C storage. The values recorded in 1983 are about 160% of those found in 1966. The same is true in the case of phosphorus, reaching in 1983 140% of the storage of 1966.

Constant storage or slight decrease is found in the case of K, Ca, and Mg; subsequently, the base cation content (in mg/g) of the organic matter decreases,

Table 2-3. Development of the element storage of the organic top layer.

		Org. matter	N	P	Na	K	Ca	Mg	Mn	Fe	Al	C/N	C/P
							Beech (kg · ha^{-1})						
1966 n = 2	x̄	29600	809	52	14	83	96	34	22	435	340	18	285
1973 n = 4	x̄	41800	953	60	7	67	117	24	23	385	259	22	348
	SD	6240	220	6	1	4	22	1	5	30	17		
1979 n = 3	x̄	44600	1010	60	5	71	116	29	25	472	287	22	372
	SD	2640	60	6	1	15	9	4	7	80	30		
1983 n = 4	x̄	48100	1269	71	8	79	113	29	28	540	281	19	339
	SD	1200	130	6	1	9	22	2	6	58	38		
							Spruce (kg · ha^{-1})						
1968 n = 4	x̄	49000	960	53	19	41	83	20	14	305	196	25	462
	SD	4320	83	5	3	10	8	4	1	44	48		
1973 n = 3	x̄	56400	1130	60	7	51	88	20	15	378	210	25	470
	SD	5680	300	6	0	2	35	1	4	21	10		
1979 n = 3	x̄	68600	1250	79	7	89	114	34	21	650	269	26	409
	SD	5360	16	2	0.2	4	12	2	3	44	133		
1983 n = 4	x̄	96300	2034	101	9	85	153	35	25	750	370	24	477
	SD	9800	220	9	1	1	8	4	4	90	56		

SD = standard deviation.

while the element content of the litter was constant. From 1966 to 1983, only two-thirds of the annual litterfall decomposed.

The separation of organic top layer and mineral soil is difficult, and the mineral content of the OH layer is considerable. Thus, the determination of changes of the storage of Al, Fe, and Mn is impossible, and no further interpretation is done here for these elements.

Under spruce the increase of the element storage of the organic top layer is even more dramatic than under beech. The storage of 1983 is about 180 to 210% of that in 1968. In contrast to the findings under beech, the accumulation occurs for all elements in more or less the same rate, and thus the element content of the organic matter (in mg/g) remains constant.

The mean annual rate of accumulation of organic matter from 1968 to 1983 is about 2900 kg $\cdot$ ha^{-1} $\cdot$ yr^{-1}, while the above ground litter production was about 4000 kg $\cdot$ ha^{-1} $\cdot$ yr^{-1}.

These relationships emphasize the degree of decomposition inhibition found in these stands, which is atypical for mature forests.

The process of accumulation of elements in the organic layer was also followed by the flux balance approach. Net mineralization rates calculated from flux measurement in throughfall and funnel lysimeters (see the Methods section) revealed much lower rates of mineralization than element fluxes by litterfall (Matzner, 1987).

The reason for the reduced rates of decomposition may be high heavy metal concentrations, as discussed in detail by Mayer (1981), who concluded that heavy metals (especially Cu and Pb) from atmospheric deposition accumulate in the organic top layer and reach concentrations that were shown to reduce decomposer activity in the literature. Similar conclusions were drawn by Zielinski (1984), who found reduced decomposition in a Polish pine stand. The observed retardation of N mineralization should have significant effects on the N status of the stand, because the rates of N mineralization in the mineral soil of these sites are very low (Runge, 1974a; Beese, 1986). However, no N deficiency symptoms are visible, and the N content of leaves and needles is sufficient (Matzner, 1985).

The incorporation of high amounts of deposited nitrogen into the N cycle directly via foliar uptake may explain this contradiction at least partly. Via N uptake and litterfall, the organic top layer seems to be at present the most important sink of major parts of the N input.

2. Results from Input-Output Studies

a. Development of Seepage Water Chemistry

Before discussing the input-output budget of the mineral soil and the calculated element turnover, some qualitative aspects arising from the chemistry of the seepage water should be stressed.

Figure 2-6 shows the development of the concentrations of H$^+$ (pH), K, Ca, Mg, Mn, Al, SO$_4$, and NO$_3$. Note that the concentration of NO$_3$ is given as NO$_3$-N.

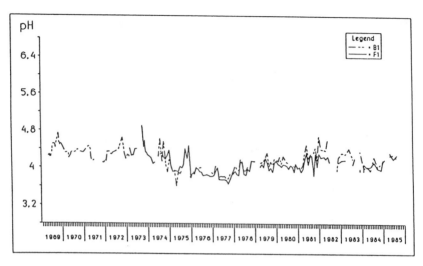

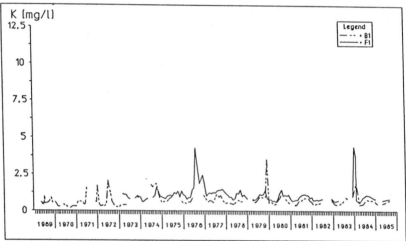

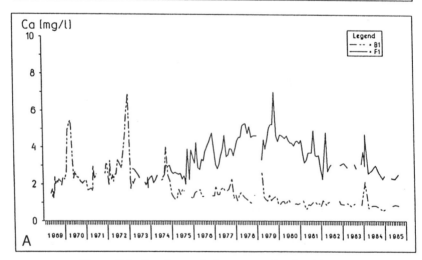

Figure 2-6. Development of seepage water chemistry.

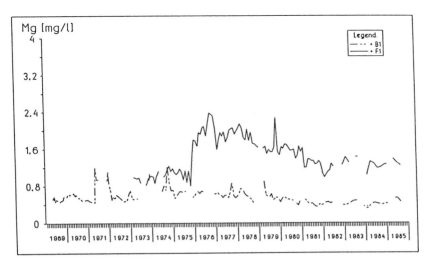

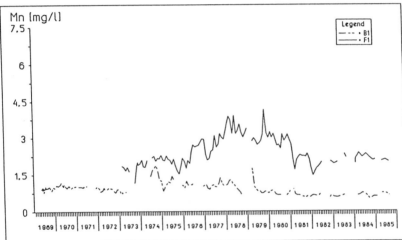

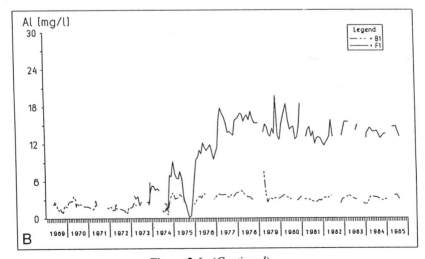

Figure 2-6. (*Continued*)

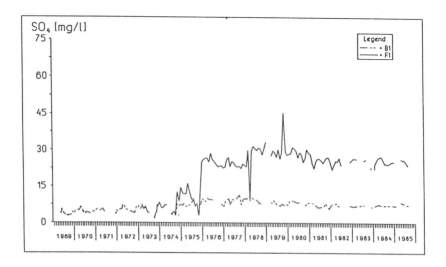

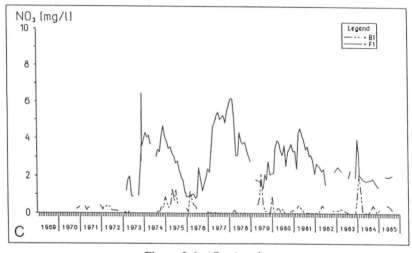

Figure 2-6. (*Continued*)

Almost no differences in the pH of the seepage water is visible when comparing beech and spruce. The pH was around 4.5 until 1975, decreased to 4.0 during 1975 to 1978, and finally stabilized at 4.5 since 1979. Distinct changes of seepage water chemistry occur in the case of Al and SO_4. Under beech the Al concentrations increase in 1975 according to the decrease in pH and reach mean values of about 4 mg/L; only about 2 mg/L have been recorded before. The Al concentrations slightly decreased since 1978 without reaching the low levels at the beginning of the measurement period. The percentage of Al ions of the sum of all cations in solution was 43% in 1969, rising to 57% in the last years, an indication of seepage water acidification.

Even more dramatic is the development under spruce. The Al concentrations increased from about 3 mg/L in 1973 to more than 15 mg/L in 1977 and have since stabilized in the range of 12 to 18 mg/L.

The increasing acidification of the seepage water under spruce is again reflected by the percentage of Al ions of the sum of cations that changed from 46% in 1973 to 70% since 1977. The development of the Al concentration of the seepage water corresponds to SO_4 being the dominant anion. The relationship of Al and SO_4 is discussed in detail in the section on Input-Output Budgets.

The Ca concentrations under beech decrease continuously from 1969 to 1985, with current values of about 1 mg Ca/L. The development under spruce is more complicated because the Ca concentrations increased in 1976, presumably caused by the increase of the Al concentrations and subsequent Ca exchange. Since 1982 the Ca levels are comparable to those before 1976.

The same behavior is found under spruce in the case of Mg. The concentrations increased in 1976 to about 100% of the earlier values and then slowly decreased again to the level of 1975. This may also be explained by Al exchange.

Several differences between beech and spruce are obvious when looking at the NO_3-N concentrations, which are almost close to the detection limit under beech, and about 4 mg/L was recorded on the average under spruce, with maximum values of 6 mg/L.

The Ca/Al ratio of the soil solution was shown to be a useful tool to describe the risk of Al toxicity to tree roots (Rost-Siebert, 1985). Rost-Siebert found root damage to Norway spruce seedlings as the Ca/Al ratio decreased to values less than 1. Figure 2-7 shows the development of the molar Ca/Al ratio and Mg/Al ratio in the seepage water of both stands. In laboratory experiments, Mg/Al ratios of less than 0.2 caused Mg-deficiency symptoms to Norway spruce (Jorns and Hecht-Buchholz, 1985).

The Ca/Al ratio as well as the Mg/Al ratio decreased with time under both sites. Furthermore, the variation became much less with time. At the beginning of the measurements the Ca/Al ratio was in the range of 0.5 to 1.5 and then shifted to values of about 0.2, leading to increased risk of Al toxicity to roots. The same is true in the case of the Mg/Al ratios. Values of less than 0.2, which are now found under spruce, should reduce the uptake of Mg significantly due to antagonistic effects (Jorns and Hecht-Buchholz, 1985; Rost-Siebert, 1985). Symptoms of Mg deficiency have been visible since about 1981 on older needles of the spruce stand.

b. Input-Output Budgets

Input-output budgets of the mineral soil were calculated according to Equation 9:

$$C5 = TD - IC - K3 - OS \qquad (9)$$

where C5 = change of the element storage of the mineral soil; TD = total deposition; IC = increment of the stand (above ground); K3 = change of the element storage of the organic top layer; and OS = output by seepage water.

Annual values of TD and OS are given by Matzner and associates (1982) and Matzner (1988); long-term average values of K3 (see Table 3) and IC (see Matzner

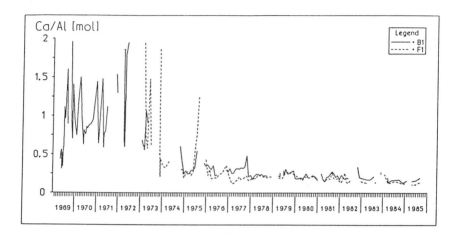

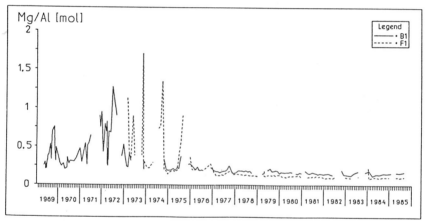

Figure 2-7. Ca/Al and Mg/Al ratio of seepage water.

and co-workers, 1982) were used for calculation. The results are given in Tables 2-4 and 2-5.

Because some fluxes can only be estimated (e.g. increment) and others can only be calculated (e.g. total deposition) by using assumptions, the significance of the budgets can also be only roughly estimated. Estimates of mean errors of the whole budget were developed by Matzner (1988) and are given in Tables 2-4 and 2-5.

Inert behavior of the mineral soil is to be expected in the case of Na and Cl. Despite small sink (beech) and source (spruce) functions of the mineral soil, this is confirmed by the results given in Tables 2-4 and 2-5. The changes of the Na and Cl storages of the soil are too small to be significant, as compared to the turnover of these elements and to possible systematic errors of the budget (Matzner, 1988).

Table 2-4. Input-output budgets of the mineral soil: beech (kg · ha^{-1}).

	Na	K	Ca	Mg	Fe	Mn	Al	S	P	Cl	N$_{total}$
1969	0.0	-5.0	-5.3	-2.8	1.5	-9.2	-12.3	24.9	-3.3	-9.4	-27.6
1970	-4.1	0.0	-15.1	-6.1	1.8	-11.3	-25.2	0.6	-1.6	-15.0	-22.0
1971	7.4	0.4	11.9	-1.4	0.9	-6.5	-3.6	22.1	-2.6	12.4	-24.1
1972	-1.1	-1.3	9.3	-2.5	1.2	-7.4	-6.4	14.6	-2.2	2.5	-15.4
1973	6.4	-4.7	-9.9	-3.8	0.7	-7.7	-9.5	9.8	-3.2	13.7	-18.5
1974	-4.3	-7.0	-9.0	-5.3	1.4	-11.4	-21.8	8.5	-2.4	-24.6	-10.2
1975	2.5	-2.7	1.7	-2.9	1.6	-7.6	-11.9	16.1	-2.7	4.2	3.1
1976	7.5	-2.6	-0.1	-2.4	1.1	-6.3	-9.1	17.3	-2.3	21.3	-7.0
1977	4.1	-0.1	4.1	-3.2	0.2	-6.4	-13.6	12.3	-2.4	8.3	4.0
1978	3.3	-2.0	1.2	-3.4	0.8	-7.8	-17.5	2.0	-2.3	11.5	4.8
1979	2.9	-4.7	2.0	-2.7	1.6	-6.9	-14.4	2.0	-2.5	5.6	-6.4
1980	2.8	-3.7	0.2	-3.1	1.4	-7.9	-19.5	-2.5	-2.4	5.8	-4.3
1981	-1.2	-4.1	-1.8	-4.2	0.2	-9.8	-32.5	-27.3	-2.7	-6.5	-8.9
1982	5.1	-3.2	6.9	-2.1	0.3	-4.9	-12.4	5.0	-2.5	12.1	-10.4
1983	4.7	-2.6	-1.1	-3.3	-0.2	-6.5	-18.9	-7.5	-2.1	6.4	-11.7
1984	1.4	-2.8	1.4	-3.4	0.0	-7.2	-18.1	-5.4	-2.4	0.0	-10.6
1985	1.6	-3.1	2.9	-3.9	0.0	-6.4	-16.8	5.0	-2.3	8.9	7.1
Sum	+39.0	-49.2	-0.7	-56.5	+14.5	-132	-264	+97.5	-41.9	+57.3	-158
ME	±48	±33	±66	±14	±4.0	±19	±44	±165	±11	±111	±226

ME = Mean estimated error of the budget.

Table 2-5. Input-output budget of the mineral soil: spruce (kg · ha^{-1}).

	Na	K	Ca	Mg	Fe	Mn	Al	S	P	Cl	N$_{total}$
1973	16.7	-5.0	-4.3	0.8	1.1	-6.2	-8.7	55.2	-4.3	27.7	-65.0
1974	-0.4	-9.4	-8.9	-2.7	1.2	-12.9	-32.4	39.5	-4.4	-0.1	-73.7
1975	5.9	-4.9	-1.4	-2.4	1.9	-7.8	-16.0	52.2	-4.5	17.8	-54.9
1976	2.0	-5.5	-6.2	-4.1	1.1	-7.0	-24.6	-2.8	-4.1	13.5	-54.8
1977	-3.3	-2.6	-0.8	-5.1	0.1	-9.4	-53.6	6.9	-4.1	-13.1	-62.2
1978	-13.9	-7.1	-17.7	-8.0	0.5	-16.3	-66.7	-41.7	-4.4	-25.7	-67.5
1979	-6.3	-4.8	-6.9	-4.7	2.1	-12.8	-50.3	-37.4	-4.3	-5.6	-55.0
1980	-7.1	-5.4	-15.6	-6.6	1.8	-16.8	-80.7	-63.0	-4.5	-13.8	-67.2
1981	-15.7	-5.8	-21.5	-9.0	0.1	-18.9	-118	-148	-4.5	-30.1	-80.8
1982	5.8	-5.1	3.2	-2.0	0.3	-4.9	-28.8	3.8	-4.5	14.9	-61.4
1983	6.4	-2.7	-8.2	-4.2	-0.2	-9.5	-58.8	-41.5	-4.4	11.0	-61.6
1984	-5.9	-5.3	-10.4	-5.8	-0.1	-12.3	-62.9	-52.4	-4.7	-23.7	-63.1
1985	-9.0	-6.0	-5.3	-5.5	-0.2	-9.1	-50.3	-46.5	-4.7	-13.7	-68.5
Sum	+24.8	-69.6	-104	-59.3	+9.7	-144	-652	-276	-57.4	-40.9	-836
ME	±51	±24	±58	±14	±4.4	±20	±100	±250	±7.3	±110	±250

ME = Mean estimated error of the budget.

Under spruce a source function of the mineral soil is evident for K, Ca, Mg, Mn, Al, S, P, and N.

Losses of K are found each year and are mainly caused by the accumulation of K in the increment and organic top layer (9.5 kg ha^{-1} yr^{-1}). The average K output by seepage water (3.7 kg ha^{-1} yr^{-1}) is much less than the annual K deposition and thus is of minor importance for the observed K losses. The stands' increment and organic top layer accumulation accounts for a loss of 11 kg ha^{-1} yr^{-1} in the case of Ca. However, the output by seepage water is still higher and reaches 14 kg ha^{-1} yr^{-1} in the average. Compared to these large output fluxes, the Ca budget of the mineral soil is only slightly negative caused by high rates of Ca deposition.

Comparable to Ca, the source function of the mineral soil for Mg is mainly caused by seepage water outputs that reach an average of 5.8 kg ha^{-1} yr^{-1}. Only 2.6 kg ha^{-1} yr^{-1} are stored in the annual increment and by accumulation of the humus layer.

The picture is somewhat different in the case of Mn, as the uptake by the stand (2.3 kg ha^{-1} yr^{-1}) is much less than the losses by seepage water (9.9 kg ha^{-1} yr^{-1}). Because the Mn input by deposition is very small, the source function of the mineral soil is clearly obvious.

The Al budget is dominated by the Al flux with seepage water of 53 kg ha^{-1} yr^{-1}. The Al uptake by the stand is almost negligible and changes of the Al storage of the organic top layer are not considered because of analytical problems. As no significant output of phosphorus occurs by seepage water, the stands' increment and humus accumulation are the only driving processes yielding to the negative P budget of the mineral soil. Under spruce the P cumulation in the accumulating humus layer (4.0 kg ha^{-1} yr^{-1}) is most important. The P flux by throughfall is taken as P input to the soil when calculating the P budget.

The N budget of the mineral soil of the spruce stand is negative despite high rates of N deposition. About 80 kg ha^{-1} yr^{-1} of N are accumulated in the organic top layer. An additional 10 kg are taken up for increment, and 12 kg of N are lost in the average per ha and year by NO$_3$ leaching with seepage water.

Under beech a source function of the mineral soil was found in the case of K, Mg, Mn, Al, and P, while—in contradiction to spruce—the Ca budget is balanced by the high rates of Ca deposition given in the Solling area.

With respect to possible systematic errors, the changes of the soil storage for S and N are considered insignificant. The processes explaining the source functions for K, Mn, Al, and P under beech are equal to those related to the findings under spruce.

Sulfur Budget of the Mineral Soil of the Spruce Stand. As indicated by the annual S budgets of the mineral soil, the S behavior has changed dramatically throughout the period of investigation. Following a period of S accumulation from 1973 to 1975, reaching 60% of the S input, a period of nearly balanced S budget was observed from 1976 to 1977. Finally, the S budget of the mineral soil became clearly negative since 1978. The annual variation in S release is mainly caused by changing seepage water rates.

Because of its magnitude, the S dynamic can only be explained by the changing behavior of its inorganic species. The accumulation of sulfate in acid soils is possible by adsorption or by precipitation as Al sulfates.

Sulfate adsorption isotherms of the Solling soils were given by Meiwes (1979) from soil samples taken in 1977. These isotherms did not explain the observed S accumulation and cannot explain the sudden increase of the S concentration in the seepage water while sulfur inputs remained constant. The S losses quantified by the budget for a soil depth of 90 cm during the period 1978 to 1985 reached 427 kg ha^{-1}, corresponding to 45% of the sulfate storage in 0 to 50 cm depth found by Meiwes (1979) in 1977. Thus, the order of magnitude of the S release is considered reasonable.

By extracting the soil several times with water, Prenzel (1982) and Meiwes (1979) found that Al sulfates with the approximate composition of $AlOHSO_4$ play a major role in the S dynamics of these soils. However, Prenzel concluded that sulfate is bound by ion exchange on Al-hydroxo-aquo complexes and Al-hydroxide surfaces, resulting in the formation of amorphous sulfates containing Al hydroxide. The solubility of these sulfates is controlled by the actual composition of the surfaces. The surface may be approximately $AlOHSO_4$.

Förster (1986) was able to simulate the S dynamic of this site in a more or less qualitative way. He postulated a two-step reaction:

$$Al(OH)_3 + H_2SO_4 \rightarrow AlOHSO_4 + 2\,H_2O \qquad (10)$$

The second reaction takes place as long as the amount of quick-reacting $Al(OH)_3$ in the soil is high and leads to the accumulation of sulfate.

$$2\,AlOHSO_4 + H_2SO_4 \rightarrow 2\,Al^{3+} + 2\,SO_4 + 2\,H_2O \qquad (11)$$

As the storage of quick-reacting $Al(OH)_3$ is exhausted, $AlOHSO_4$ is used for buffer reactions and the sulfate previously stored is released. This kind of development starts—under the present deposition regime—at the top of the soil profile continuing downward and thus may result in a sudden increase of the Al and SO_4 concentrations according to a breakthrough curve.

The triggering mechanism for the release of sulfate is the kinetically controlled release of quick-reacting $Al(OH)_3$ from primary minerals. Indications for exhausted Al pools in the topsoil are given by the gradients of the soil solution concentration of Al, showing increasing values with depth despite lower pH values in the top soil (see Figure 2-8).

c. Total H^+ Load of the Mineral Soil

The effects of acid precipitation on terrestrial ecosystems and especially on soils must be evaluated with respect to natural production of strong acids. Several papers have been published recently focusing on this problem and have given a controversial picture (Sollins et al., 1980; Krug and Frink, 1983; Becker, 1984; Isermann, 1983; Driscoll and Likens, 1982; van Breemen et al., 1983, Nilsson, 1983, 1985; Nilsson et al., 1982; Matzner, 1984; Ulrich, 1981a; Reuss and

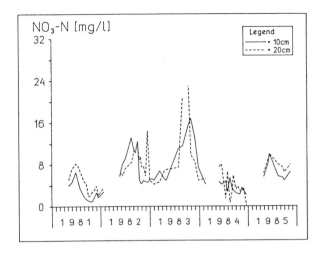

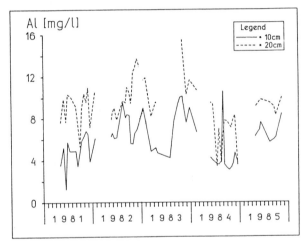

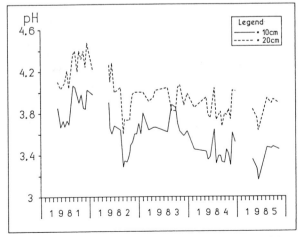

Figure 2-8. pH, nitrate, and Al concentration of the soil solution of the spruce stand.

Johnson, 1986; Ulrich, 1986). Contradicting conclusions on the implication of acid deposition on soils are often caused by neglecting the difference between intensity and capacity factors and the difference between the amount of acid and its strength and by omitting proton-consuming processes (Krug and Frink, 1983; Isermann, 1983). As van Breemen and associates (1983) have summarized all quantitatively important proton transfer processes in terrestrial ecosystems, no further general comments will be given here.

The method, first used by Ulrich (1980), to calculate the total H^+ load of the mineral soil for the Solling sites is based on the assumption that the electroneutrality of the solid phase is controlled by the ions Na^+, K^+, Ca^{2+}, Mg^{2+}, Mn^{2+}, Al^{3+}, Fe^{2+}, SO_4^{2-}, PO_4^{2-}, and Cl^-. The total input and turnover of protons must therefore be related to the change of storage of these elements and thus can be calculated from the budgets given in Tables 2-4 and 2-5 according to Equation 12.

$$H_{MS}^+ = C^- - A^+ \qquad (12)$$

where H_{MS}^+ = total amount of H^+ buffered in the mineral soil; A^- = sum of all sources and sinks of the conservative anions (keq); and C^- = sum of all sources and sinks of the conservative cations (keq).

Equation 4 postulates that source functions of the mineral soil of cations must be related to source functions of anions or to equivalent proton sinks. By using this approach one can account for the following proton sources:

Excess production and leaching of strong acid anions (NO_3, SO_4, $R\text{-}COO^-$); *excess production* means that the formation and dissociation rate is higher than the consuming processes of ion uptake, denitrification, S reduction, and mineralization of organic acids. Because the pH values in the soil profile are well below pH 4.5, carbonic acid production is neglected.

Accumulation of excess cations by increment and accumulating organic top layer; *Excess cations* equals alkalinity of the biomass.

H^+ turnover caused by the deposition of N. As shown above, significant amounts of ionic nitrogen is deposited in both ecosystems. The turnover of nitrogen within the ecosystem causes H^+ transfers (for an overview, see Reuss and Johnson, 1986), which are calculated according to Equation 13.

$$H_N^+ = NH_{4in} + NO_{3out} - NH_{4out} - NO_{3in} \qquad (13)$$

where NH_{4in} = NH_4 flux by throughfall; NO_{3in} = NO_3 flux by throughfall; NH_{4out} = NH_4 flux with seepage water; NO_{3out} = NO_3 flux with seepage water.

The calculation of the total H^+ load remaining in the mineral soil according to Equation 12 reveals rates of 2.5 keq ha^{-1} yr^{-1} for the beech and 5.2 for the spruce site. The contribution of the various processes adding to the total H^+ load is given in Table 2-6.

The accumulation of the organic top layer caused no proton turnover under beech because of equal cation and anion accumulation, and a small proton consumption is given for the spruce stand (anion accumulation > cation accumulation). The proton load caused by excess cation accumulation by the increment of

Table 2-6. Total H^+ load of the mineral soil $(keq \cdot ha^{-1} \cdot yr^{-1})$.

	Beech ($\bar{x}$ for 1969-1985)	Spruce ($\bar{x}$ for 1973-1985)
Total deposition	2.0	3.9
Accumulation of organic top layer	0.0	−0.1
Increment of the stand	0.6	0.4
N deposition	0.2	0.9
Net leaching of organic anions	0.2	0.5
H^+ load remaining in the mineral soil	2.5	5.2
Total H^+ load	3.0	5.6

the stands is about the same for beech and spruce, but differences are obvious with respect to the effects of N deposition. The average NH_4/NO_3 ratio of the N input is about 1.0 for spruce and 1.2 for beech. The complete uptake of the N deposited would have no effect on the proton budget in case of spruce and a small H^+ load would be given for beech. The NO_3 losses by seepage water under beech are small (1.5 kg ha^{-1} yr^{-1}), and thus the remaining proton load from N deposition approaches only 0.2 keq ha^{-1} yr^{-1}. The incoming nitrate is almost completely lost by seepage water under spruce. The proton consumption related to nitrate uptake does not take place, and thus the remaining proton load from N deposition comes up to 0.9 keq ha^{-1} yr^{-1}.

The dissociation and leaching of organic anions are of minor importance under beech (0.2 keq ha^{-1} yr^{-1}), but this process adds significantly (0.5 keq) to the proton load of the soil under spruce.

The proportion of the amount of H^+ remaining in the soil that is caused by the deposition of H^+ and N reaches 70% for beech and 85% for the spruce stand.

These relations clearly emphasize the importance of the deposition of air pollutants for these ecosystems.

d. Buffering of Protons in the Soil

Cation Release by Weathering of Primary Minerals. The release of alkali and earth alkali cations (Na, K, Ca, Mg) from weathering of silicates (silicate weathering) is of major importance for the elemental cycling in forest ecosystems. Plant-available cations, removed from the soil by harvesting practices or leaching with seepage water, may be replaced by silicate weathering. Furthermore, silicate weathering neutralizes at least partly the H^+ load entering the soil from ecosystem external and internal sources.

Under the present conditions the rate of silicate weathering therefore largely determines the development of the chemical soil state in forest ecosystems. Unfortunately, very little is known about the actual rates of this process and its dependence on environmental factors and soil conditions.

Most commonly the flux balance approach is used to derive actual rates of silicate weathering in watersheds by comparing input and output of cations. This method has several shortcomings: Calculating budgets requires appropriate data on total deposition, uptake by vegetation, and output by runoff or seepage water of the cations concerned. Furthermore, the determination of dry deposition of cations to forests is very difficult (Lovett and Lindberg, 1984; Bredemeier et al., 1987). As the rates of dry deposition are significant, over- or underestimation of this process leads to large errors with respect to the rates of silicate weathering. Neglect of dry deposition causes overestimation.

The uptake of cations by vegetation, which cannot be measured directly, is of specific importance in ecosystems. Neglect or underestimation of this process leads to underestimation of the weathering rate. The measurement of runoff losses seems to be a minor problem compared to the quantification of total input and vegetation uptake.

Once the budget is established, further problems may arise. Because the rates of silicate weathering in a watershed should be related to the amount of soil involved in order to compare different results, information on the depth of the soil volume and the stone content is required. In most cases both parameters can only be roughly estimated.

Furthermore, source functions of the soil with respect to Na, K, Ca, and Mg may be caused by changes in the exchangeable pools. Without considering losses of exchangeable cations, rates of silicate weathering calculated by input-output budgets may be significantly overestimated.

The rates of silicate weathering given in the literature (Driscoll and Likens, 1982; Johnson et al., 1981; Paces, 1986; Reid et al., 1981; Hauhs, 1985) range from 0.2 to 2 keq $\cdot$ ha^{-1} $\cdot$ yr^{-1}. However, most data have a tendency to overestimate because either changes of exchangeable pools or vegetation uptake or dry deposition is not taken into account.

The values of 0.2 to 2 keq $\cdot$ ha^{-1} $\cdot$ yr^{-1} derived from the flux balance approach are well within the range of silicate weathering rates calculated by different methods (Fölster, 1985).

For the Solling sites, the rates of silicate weathering are calculated by subtracting the change in exchangeable cation storage of the soil that was measured by periodic soil inventories (Matzner, 1988) from the change in cation storage (Tables 2-4 and 2-5) derived from Equation 1.

Comparison of the peroidic soil inventories in 1973 and 1983 (Matzner, 1988) for the depth of 0 to 50 cm under spruce revealed losses of exchangeable Ca and Na exceeding the values found by the flux balance approach. Thus the rate of silicate weathering becomes zero for these cations. Losses of K mostly stem from exchangeable pools, but no change in exchangeable Mg content was found.

Under beech, losses of Na and K were balanced by reduction of the exchangeable pool, and Ca and Mg storage were constant. In Table 2-7 the rates of silicate weathering for both sites are given. The corresponding soil depth is 90 cm. The rates of silicate weathering under both stands correspond very well. Only in the case of Mg did the calculated rate of silicate weathering significantly exceed the estimated error. The rates in the loess-derived acid soils of the Solling are well within the range of those values published in the literature.

Table 2-7. Rates of silicate weathering in the Solling area (keq $\cdot$ ha^{-1} $\cdot$ yr^{-1}).

	Na	K	Ca	Mg	Σ
Spruce ($\bar{x}$ for 1973–1983)					
	0.00	0.08	0.00	0.36	0.44
Beech ($\bar{x}$ for 1969–1983)					
	0.00	0.00	0.02	0.27	0.29

The accuracy of the flux balance approach for determining the rates of silicate weathering obviously allows only a rough estimate of the actual rate (Matzner, 1988). The accuracy may be still high enough to draw major conclusions by comparison with the rate of H$^+$ input to the soil. However, the method seems to be unsuitable to parameterize the process of silicate weathering according to mineral composition, dependence on pH, and other soil factors. Therefore, future work should focus on long-term laboratory studies with soils to overcome these difficulties. In order to reduce the inaccuracy in separating cation exchange and silicate weathering, the removal of exchangeable Na, K, Ca, and Mg by ions such as NH$_4$, Cs, and Ba will then be necessary.

Buffer Reactions of the Mineral Soil. In line 1 of Table 2-8 the protons remaining in the mineral soil and therefore subjected to buffer reactions are 2.5 and 5.2 keq $\cdot$ ha^{-1} $\cdot$ yr^{-1} for beech and spruce, respectively. The most important buffer reaction in both sites is the release of Al ions from hydroxides or from exchange sites. The dominating buffer reaction therefore characterizes the Al buffer range, according to Ulrich (1981c).

The source function of the soils under spruce for sulfate was attributed to the dissolution of Al hydroxosulfates under equivalent H$^+$ consumption. This process released annually 0.7 kmol Al^{3+}, which is equivalent to the buffering of 0.7 kmol H$^+$ (line 5).

The small S accumulation (0.2 kmol ha^{-1}yr^{-1}) found under beech is interpreted as the precipitation of Al hydroxosulfates causing a proton consumption of 0.4 kmol (line 6). The buffering of protons by Ca exchange was found only under spruce (line 7).

The sum of the buffer reactions given in lines 2 to 7 exceeds the protons remaining in the soil by 0.2 keq ha^{-1}yr^{-1}: Calculating the total proton load according to Equation 4, the uptake (source function) of PO$_4^{2-}$ of 0.2 keq ha^{-1}yr^{-1} reduces the excess cation release from the soil. To derive the buffer rates given in Table 2-8, only the sources of cations, the sulfur budget, and indirectly the N budget have been taken into account. The remaining difference of 0.2 keq is therefore equal to the proton consumption due to phosphate uptake. Thus the buffer rates in lines 2 to 7 must be reduced in total by 0.2 keq. In order to keep the approach used distinct and clear and because of the small difference, this was omitted.

Table 2-8. H^+ buffering in the mineral soil (keq $\cdot$ ha^{-1} $\cdot$ yr^{-1}).

	Beech ($\bar{x}$ for 1969–1985)	Spruce ($\bar{x}$ for 1973–1985)
1 H^+ load remaining in the mineral soil from internal and external sources	2.5	5.2
2 Buffering by silicate weathering	0.3	0.4
3 Buffering by release of Mn^{2+} from oxides	0.3	0.4
4 Buffering by release of Al^{3+} from hydroxides or exchange places	1.7	3.5
5 Buffering by release of Al^{3+} from $AlOHSO_4$	0.0	0.7
6 Formation of $AlOHSO_4$	0.4	0.0
7 Buffering by release of Ca from exchange places	0.0	0.4
Σ 2–7	2.7	5.4

e. Transfer of Acidity with Seepage Water

The soil represents an open system continuously exchanging matter and energy with its environment. Underneath the concerned soil compartment, the environment is given by the deeper soil compartment, the geological bedrock, or the hydrosphere. The exchange of matter between the soil and its underlaying systems takes place mainly by transport of seepage water. Changes of soil chemistry, especially soil solution chemistry, may have drastic effects on the quality of ground and springwaters, a chain of events that has already often been demonstrated (for an overview, see Drablos and Tollan, 1980). The transfer of strong acids by seepage water is of major importance in this respect, besides the problems related to the transport of heavy metals and other pollutants.

Recently the problem of water acidification has been discussed also in the Federal Republic of Germany. The data available so far indicate that the acidification has reached groundwater level in many cases, and several surface waters were shown to be acid (Schoen et al., 1984; Feger, 1986; Puhe and Ulrich, 1985; Linkersdörfer and Benecke, 1987).

Quantifying the transfer of strong acids by seepage water having pH less than 4.5, NH_4^+, and cationic acids have to be considered in addition to the flux of protons. NH_4 ions have to be taken into account because of their "physiological" acidification potential in case of plant uptake or nitrification. Cationic acids are defined as those cations that may produce protons in contact with solutions of pH > 5.0 according to Equation 14.

$$Me^{n+} + H_2O \quad Me(OH)^{(n-1)+} + H^+ \qquad (14)$$

In soil solutions, Al, Fe, and Mn ions may act as cation acids. The transfer of cation acids with seepage water is restricted to the presence of strong acid anions such as sulfate, nitrate, chloride, or organic anions. The input of these anions by deposition causes high risks because they allow the transfer of acidity from soils to deeper layers—even in soils that have been acidified by other than acid-deposition processes. So far the negative effects of nitrate deposition in this respect are counteracted by the uptake of nitrate in many ecosystems. However, several examples show increasing nitrate losses from forest ecosystems (Hauhs, 1985; Kreutzer et al., 1986; Schofield et al., 1985; Wright et al., 1986) accompanied by increasing transfer of acidity.

In Table 2-9 the input by deposition and output of H^+ and cation acids is summarized.

The average input of total acidity by deposition was about 3.2 keq $ha^{-1}yr^{-1}$ in the beech site and 5.4 under spruce. The output of protons and cation acids by seepage water from the beech site reached 2.6 keq and was only 0.4 keq less than the average input rate.

Caused by the release of sulfate during the period of investigation, the actual output of total acidity exceeds the input by far in the case of spruce. The acidity stored in the soil previously is partly mobilized and resulted in a strong increase in the acid load of deeper layers.

The percentage of sulfate of the sum of all anions emphasizes the importance of S emission and deposition for soil and water acidification. Thus drastic reduction of S emissions would reduce the proton load by deposition and the acid transfer to deeper layers.

3. Chemical Characteristics of Deeper Soil Layers

In 1986 soil samples were taken and analyzed in both sites in large soil pits up to a depth of 2.2 m (Table 2-10). The acidification of the Solling soils has already

Table 2-9. Input by deposition and output of H^+ and cation acids (keq $\cdot$ ha^{-1} $\cdot$ yr^{-1}).

	Beech ($\bar{x}$ for 1969–1985)	Spruce ($\bar{x}$ for 1973–1985)
Total deposition		
H^+	2.00	3.86
NH_4^+	0.96	1.17
Al^{3+}	0.23	0.29
Mn^{2+}	0.02	0.04
Σ	3.21	5.36
% SO_4^{2-} of all anions	63	68
Output by seepage water (90-cm depth)		
H^+	0.47	0.40
NH_4^+	0.01	0.01
Al^{3+}	1.95	5.85
Mn^{2+}	0.18	0.36
Σ	2.61	6.62
% SO_4^{2-} of all anions	63	68

Table 2-10. Chemical characteristics of deeper oil layers sampled in 1986.

Depth (cm)	pH (CaCl$_2$)	H	Na	K	Ca	Mg	Mn	Fe	Al	CEC (ceq·kg^{-1})
				Exchangeable Cations [%]						
				Beech						
70–90	3.86	0.5	2.0	2.9	0.7	0.5	0.4	0.4	92.6	7.16
90–110	3.85	1.6	1.8	3.3	0.8	0.6	0.7	0.5	90.7	6.86
110–130	3.90	1.2	1.7	3.8	1.3	0.6	1.4	0.7	89.3	4.92
130–150	3.87	1.6	1.5	4.0	1.2	0.7	1.0	0.9	89.1	5.38
150–170	3.90	2.3	1.4	3.8	0.9	0.8	0.6	1.1	89.1	6.72
170–190	3.85	2.1	1.4	4.2	0.9	1.1	0.5	0.7	89.1	7.28
190–210	3.90	2.6	1.7	3.7	1.0	0.9	0.9	0.9	88.3	5.96
210–220	3.83	3.1	1.5	3.9	1.0	0.8	1.1	0.6	88.0	5.40
				Spruce						
80–100	3.70	3.1	1.1	3.0	0.7	0.5	0.5	0.5	90.6	7.59
100–120	3.72	3.8	1.0	3.1	0.9	0.5	0.5	0.5	89.7	7.14
120–140	3.70	3.4	1.1	2.2	0.8	0.5	0.5	0.5	91.0	7.69
140–160	3.70	4.1	0.8	3.2	0.9	0.7	0.9	0.5	88.9	7.20
160–180	3.72	3.8	1.1	3.0	0.9	0.6	0.9	0.5	89.2	6.25
180–200	3.70	3.5	1.1	2.6	0.8	0.5	1.6	0.5	90.0	6.71
200–210	3.70	3.7	0.9	2.5	0.9	0.7	1.8	0.5	89.0	8.33

passed to a depth of more than 2.2 m, as indicated by extremely low saturation of Ca and Mg of the exchange sites and by pH values lower than 3.9. The degree of acidification of the deeper soil equals the one found in the top soil, which is not surprising, regarding the transfer rates of total acidity with seepage water given in Table 2-9. The high degree of acidification of the deeper layers is supported by the reduction of the stone-free soil volume to less than 10% of the total volume. Stone-free soil is found only in the space between large sandstone rocks. Thus the seepage water flow and the related acid load is focused only on a very limited volume of soil, which subsequently acidifies quickly. As these conditions remain constant up to a depth of several meters, one might presume that the acidification front has already proceeded far and is proceeding quickly.

4. Decouplings of the Ion Cycle: Seasonal Acidification Phases

The long-term effects of the proton load from acid deposition and ecosystem internal processes on soil buffer reactions and soil solution quality have been covered above. These long-term effects are enlarged by short-term H$^+$ loads resulting from seasonal "decouplings of the ion cycle," as defined by Ulrich (1981a, 1981b).

The internal ion cycle of a forest ecosystem may be simplified to two central processes: ion uptake and mineralization. In this case the element input by weathering is assumed to be an external input.

Ion uptake and mineralization are related to proton turnover according to their cation-anion balance. Proton turnover by ion uptake and mineralization balance each other as long as their rates are equal (steady-state conditions). If one process exceeds the rate of the other in time or space, a net proton turnover may result within a defined soil compartment, its duration depending on the duration of this decoupling of the ion cycle.

Decouplings of the ion cycle causing proton production result from the excess nitrification (HNO_3 production) or organic N or deposited NH_4. *Excess* means that the rate of nitrification exceeds the rate of nitrate uptake.

According to the hydrological conditions, the excess nitrate may be leached from the rooting zone or may be accumulated at the place of nitrification or in other horizons of the rooting zone, with subsequent uptake by the stand. In the first case, a long-lasting proton load would result; in the second case, the acidification phase is followed by a deacidification phase.

Because nitrification is a microbial process, its rate depends on environmental factors like temperature and moisture. Soil temperature is normally far from optimum for nitrification (Beese, 1986; Lang, 1986), and an increased rate of nitrification is to be expected under warmer than normal climatic conditions. Furthermore, excess nitrification can be expected after rewetting of the soil following periods of drought.

The effects of seasonal acidification phases on the chemistry of the soil solution and thereby on the roots of plants depend on the intensity of the excess nitrification, its duration, and the buffer reactions of the soil. A well-buffered soil with high base saturation of the CEC is able to buffer seasonal proton loads quickly by exchange of Ca, and strongly acidified soils may react by decreasing pH values or by release of Al ions.

The link between seasonal acidification phases and soil buffer properties emphasizes the interaction with acid precipitation. Acid precipitation continuously impacts on and slowly exhausts the buffer capacities of the soil and thus the risk of reaching physiologically disturbing concentrations of H^+ and Al ions during seasonal acidification phases increases with time under acid-deposition conditions.

Ulrich (1981b) attributed the temporal development of actual and past forest declines in Europe, which is characterized by increasing damage following warm and/or dry seasons, to the occurrence of acidification phases as described above. This hypothesis was widely criticized (Rehfuess, 1981), triggering in 1981 measurements of soil solution chemistry in the rooting zone in four forest sites of different degrees of acidification. The results derived from the Solling spruce stand will be discussed here in detail; data for the others will be summarized by comparison.

The development of the concentrations of nitrate, pH, and Al in the soil solution of the spruce stand at depths of 10 and 20 cm is given in Figure 2-8.

The climatic conditions during the growing season were warm and/or dry in 1982 and 1983, but 1981 was extremely wet, with precipitation reaching 1500 mm.

The nitrate concentration of the soil solution shows distinct seasonal peaks already in 1981 and increasing in 1982, with maximum concentrations of about 12 mg NO_3-N/L. Even higher values were recorded in 1983, especially in 20 cm depth, with a maximum of about 30 mg/L.

The seasonal development of the nitrate concentration does not correspond directly to the rate of excess nitrification because of the influence of the changing water content of the soil. This point is discussed later.

The increasing nitrate concentrations caused equivalent changes of the H^+ concentration (pH) of the soil solution in 1981 and 1982. In 1982 the pH at a depth of 10 cm dropped from values of 3.9 to less than 3.3 within a few months, which is equal to an increase in the H^+ concentration of 500%. The direct relationship between increasing nitrate concentrations and decreasing pH indicates the lack of quick-reacting buffer compounds in the top soil.

The picture is somewhat different in 1983: The acidification phase by excess nitrification is buffered by Al release without significant changes in pH. In 1985 the seasonal peak of nitrate is again related to low pH of up to pH 3.2, which is even lower, than for 1982. The differing buffer characteristics throughout the measuring period are difficult to explain. Changing availability of quick-reacting Al compounds must be assumed.

The effect of seasonal acidification phases on the quality of the soil solution can be demonstrated by comparing all element concentrations at dates of low versus high nitrate concentrations (Table 2-11). The concentrations of organic anions were calculated from the cation-anion balance of the solution.

Table 2-11. Comparison of the element concentrations (in meq $\cdot$ 1^{-1}) of the soil solution of the 10-cm depth at different dates.

Month	Jan. 1982	Sept. 1982	Δ	March 1983	Oct. 1983	Δ	Dec. 1984	June 1985	Δ
H	0.12	0.51	+0.47	0.23	0.25	+0.02	0.24	0.69	+0.45
Na	0.10	0.17	+0.07	0.10	0.18	+0.08	0.08	0.12	+0.04
K	0.12	0.05	−0.07	0.03	0.02	−0.01	0.01	0.02	+0.01
NH_4-N	0.00	0.01	+0.01	0.01	0.00	−0.01	0.01	0.01	0.00
1/2 Ca	0.12	0.30	+0.18	0.22	0.37	+0.15	0.16	0.27	+0.11
1/2 Mg	0.07	0.13	+0.06	0.09	0.16	+0.07	0.08	0.12	+0.04
1/3 Fe	0.01	0.02	+0.01	0.01	0.02	+0.01	0.02	0.15	+0.13
1/2 Mn	0.04	0.06	+0.02	0.05	0.10	+0.05	0.04	0.05	+0.01
1/3 Al	0.68	0.90	+0.22	0.54	1.12	+0.58	0.52	0.84	+0.32
1/2 SO_4-S	0.65	0.78	+0.13	0.54	0.60	+0.06	0.72	0.78	+0.06
1/2 PO_4-P	0.00	0.00	0.00	0.00	0.00	0.00	0.00	0.00	0.00
Cl	0.18	0.27	+0.09	0.21	0.30	+0.09	0.12	0.17	+0.05
NO_3-N	0.20	0.76	+0.56	0.48	1.21	+0.73	0.21	0.71	+0.50
org. Anions	0.12	0.41	+0.29	0.05	0.11	+0.06	0.15	0.61	+0.46

The increase of the nitrate concentration of 0.56 meq/L in 1982 is accompanied by an equivalent increase of the H^+ concentration. In 1983 the nitrate level rose by 0.73 meq/L and was almost completely accompanied by increasing Al concentrations (0.58 keq/L).

Besides nitrate, organic anion concentration increased in 1985. The accompanying cations are Al and H^+.

By comparing the change of element concentrations of all ions, the influence of changing water content of the soil versus excess nitrification can be evaluated. Increasing concentrations of Na and Cl are presumably related to hydrological effects because the internal cycling of these elements can be neglected. In 1982, 1983, and 1985, the Na and Cl concentrations changed by a factor of about 1.5, and the nitrate levels were raised by a factor of 2.5 to 3.4. These relationships elucidate the importance of excess nitrification and the related H^+ production for the observed seasonal variations of soil solution chemistry.

Comparison of the results with those derived from other sites (Matzner and Cassens-Sasse, 1984; Matzner, 1988) reveal that seasonal acidification phases are common in all sites. However, the effects on the soil solution chemistry are determined by the degree of soil acidification. In soils of high base saturation (>20%), the peaks of nitrate are balanced completely by increasing Ca concentrations without changes of pH and Al level.

5. Discussion

The effects of acid precipitation on soils are often coined *soil acidification* and are discussed by several authors without being defined precisely in many cases.

The decrease of soil pH with time is commonly used as the criterion for soil acidification. However, soil acidification may occur while the pH remains constant. Reuss and Johnson (1986) have shown that over a large pH range the base saturation of the exchange sites may decrease without significant pH change. From this example, one may proceed to define soil acidification by changes of the base saturation of the CEC, but this approach is again not useful in many cases. Examples are calcareous soils and already strongly acidified soils like those in the Solling. In the former case, protons are buffered by the dissolution of limestone and pH, and base saturation remains constant. In the latter case, further reduction of the base saturation is no longer possible because it has already reached a minimum. The pH of the soil solution is controlled by the input of strong acids and the buffer kinetics of Al compounds and thus may keep constant over long periods. This situation represents the Solling beech site. In order to overcome these problems and to use a more general definition of soil acidification, van Breemen et al. (1983) introduced the concept of acid neutralizing capacity (ANC). A decrease of the ANC is equivalent to soil acidification; the increase corresponds to soil alkalinization. ANC is measured by total soil analysis.

The decrease of the soil storage of cations bound in oxides, hydroxides, silicates, and carbonates in excess of the loss of acid-reacting compounds results in a decrease of the ANC and thus in soil acidification. The concept of ANC allows the quantification of the rate of soil acidification, which is consequently equal to the quantity of protons buffered in the soil, as given in Tables 2-4 and 2-5 for the Solling sites. Van Breemen and associates (1983) compare several ecosystems and their rates of soil acidification. Highest rates were recorded in calcareous soils, reaching 14 keq $ha^{-1}yr^{-1}$ as a result of the dissociation of carbonic acid. Calcareous soils hence may have extremely high rates of soil acidification without accumulating a base neutralization capacity, without reduction of base saturation, and without pH changes. These considerations emphasize the need for an appropriate definition of *soil acidification*. The definition via change of ANC is the most general one. As the example of calcareous soils show, a reduction of the ANC gives no information on the ecological effects of soil acidification and the related risks. Evaluating these requires the inclusion of other parameters like pH, proton and base saturation of CEC, and Ca/Al ratios of the soil solution.

Summing up the results derived from the Solling sites in this respect, a decrease of the ANC and hence soil acidification took place. As the soils were already strongly acidified at the beginning of the measurement, this caused only slight changes of the exchangeable cations (increasing amounts of exchangeable protons in the top soil) and pH values of the soil. In both sites the Ca/Al and Mg/Al ratios of the soil solution decreased to values of very high risk to tree root damage and antagonistic effects during cation uptake. The proton budget of the mineral soil gives information about the driving processes: The quantity of protons buffered in the mineral soil and leading to soil acidification amounts to 70% (beech) and 85% (spruce) from the deposition of air pollutants (H^+ and NH_4).

Nilsson (1985) and Driscoll and Likens (1982) investigated forest ecosystems receiving less deposition than the Solling sites. They concluded that about 50% of the total proton load of the soil was derived from the deposition of air pollutants. The absolute amounts of the internal proton load corresponded well with the one given for the Solling sites.

Evaluating the influence of acid deposition on the soil by establishing H^+ budgets is a useful and necessary tool but cannot describe the resulting ecological effects on organisms, deeper soil layers, and water quality in an appropriate way. Many soils, especially in central Europe, have been acidified largely in the past by mismanagement and without acid deposition impacts. This kind of acidification is characterized by reduction of pH and base saturation of the soil without leading to high acid concentrations of the soil solution because of the lack of mobile strong acid anions. Reuss and Johnson (1986) describe the common situation of large amounts of acid stored in soils and small amounts of actual acid input by deposition of sulfuric or nitric acid. However, they conclude that even low inputs of nitrate and sulfate to acidified soils may change the soil solution drastically by forcing the cation acids bound at the exchange sites into the soil solution.

The major processes causing water acidification may therefore differ from those causing soil acidification. Under specific conditions, the proportions of ecosystem internal processes inducing soil acidification may still be high when compared to acid deposition effects. However, the acidification of deeper soil layers, spring-water, and groundwater must be related to the impact of acid deposition (Reuss and Johnson, 1986; Nilsson, 1985; Seip, 1980).

Only the reduction of sulfate and nitric acid inputs can stop the acidification from proceeding to deeper layers, besides the application of lime to acid soils. Even strong reduction of the acid inputs may not lead to direct mitigation because the acid load of the soil solution might be kept high by the dissolution of Al sulfates stored previously.

As the Solling spruce stand shows, the acid load of the seepage water may change quickly, a finding that makes the prediction of acidification processes and of effects of mitigation measures more difficult.

The data presented here clearly show that the actual composition of the soil solution not only depends on long-term processes like acid deposition but also is strongly influenced by seasonal variations caused by decouplings of the ion cycle. These decouplings may be the reason for the seasonal pH changes in forest soils that have been reported in the past (Runge, 1974a; Ellenberg, 1939).

Nitrification is evidently common in strongly acidified soils. As no autotrophic organisms are detectable (Lang, 1986), heterotrophic organisms must be responsible, being differently adjusted to environmental conditions as compared to autotrophic organisms. Hence, the present knowledge about the effects of environmental factors on nitrification in acid forest soils is very limited (Beese, 1986).

The existence of seasonal acidification and deacidification phases as postulated by Ulrich (1981b) has recently been shown by others (Jörgensen, 1986; Büttner et al., 1986; van Breemen and Jordans, 1983). Results from different sites indicate the variability of the buffer processes. Even the same site (e.g. Solling) may have different buffer properties in subsequent years.

However, one can conclude that the base saturation of the CEC is of major importance. The higher the base saturation, the lower the risk of the occurrence of potential toxic ions like Al and H^+ in the soil solution during acidification phases. The decrease of the soil solution pH found in 1982 and 1985 under spruce represents a severe stress to the fine root system. Rost-Siebert (1985) found damage to roots of beech and spruce seedlings in hydroculture solutions with pH lower than 3.5.

Murach (1984) investigated the fine root dynamics of the Solling spruce stand in 1981 and 1982. He found a drastic decrease of the living fine root biomass in the topsoil following the pH drop in autumn 1982 and concluded that the acid toxicity induced by the acidification phase could be the only reason.

Increasing the base saturation of acid forest soils (e.g. by liming; Murach, 1987) would improve the situation and prevent conditions of high-toxicity risks to roots following excess nitrification. However, acid deposition and the resulting soil acidification obviously would progressively cause more negative effects in natural seasonal acidification phases in the future. Intensified nitrification and acidification phases may result from the accumulation of deposited nitrogen in forest ecosystems, leading to reduced C/N ratios of the organic matter (v. Zezschwitz, 1985; Beese and Matzner, 1986) and thus to increasing nitrification potential.

V. Conclusions

In the Solling area, the rates of deposition of almost all elements in question have reached a magnitude that significantly influences the element budget of the whole ecosystem. The knowledge of the actual deposition rates as well as their development thus becomes a necessity to understand the present situation and to predict future developments. At present, the flux balance method used to quantify the rates of deposition seems to be appropriate to measure long-term rates of deposition to vegetation surfaces for a large number of elements, including H^+. However, there are still uncertainties, especially when determining processes by calculating differences of large fluxes, such as dry deposition, H^+ buffering in the canopy, cation leaching, and N uptake within the canopy. The rates of these processes need critical review specific for each element and must be a subject of future experimental studies.

The long-term measurements of soil solution chemistry and element budgets of the mineral soil in the Solling area have led to better understanding of basic ecosystem internal processes and their causal relationship to acid deposition. The central component is the H^+ budget of the mineral soil, emphasizing the importance of acid precipitation for the total H^+ load of the soil.

The effects of the actual acid load from deposition become obvious when regarding the buffer rates, the quality of the soil solution, and the output of total acidity by seepage water.

The small changes of soil pH and exchangeable cations found during the period of investigation indicate that these soils have already reached an equilibrium with their environmental conditions and acid deposition load in respect to these parameters. The dominant actual process is the release of Al ions by H^+ buffering and the transfer of acidity to deeper soil layers, followed by their acidification.

The S budget of the spruce stand is of special importance in this context. It shows the risks of S accumulation in acid soils and the problems and uncertainties in predicting the dynamic of acidification. The solution of Al sulfates previously stored in the soil may also modify the effects of future reductions of the sulfur load by deposition. Thus, the behavior of sulfate in acid soils needs closer investigation. In respect to the stands, the reduction of the Ca/Al and Mg/Al ratios of the soil solution in time is of major importance and will result in increasing damage to roots, especially in connection with seasonal acidification phases. As a consequence of this development, severe Mg deficiency symptoms have been visible since 1981 in the spruce stand, having extremely shallow fine-root distribution and the gradients evidently controlled by adverse Ca/Al ratios of the soil solution (Murach, 1984).

The Solling case study has shown that acid deposition influences the whole ecosystem, induces measurable soil chemical changes within a few years, and undoubtedly is a major factor in destabilizing forest ecosystems in large areas of Europe and North America.

Appendix

Average ($\bar{x}$ for 1969–1985) fluxes of elements by bulk precipitation, throughfall and seepage water in the SOLLING area.

$(kg \cdot ha^{-1})$

Bulk precipitation

	(mm) H₂O	H	Na	K	Ca	Mg	Fe	Mn	Al	S	P	Cl	NH₄-N	NO₃-N	N₍org₎-N	N₍ges₎
$\bar{x}$	1032	0.82	7.8	3.7	9.8	1.7	0.7	0.4	1.2	23.2	0.4	16.7	11.9	8.7	4.3	24.9
range-values	688–1544	0.61–1.27	4.9–12.1	2.4–5.6	6.5–21.8	1.3–3.9	0.2–1.1	0.1–0.9	0.6–2.1	19.6–27.2	0.1–0.8	10.5–25.5	9.0–15.6	6.1–11.0	1.8–7.5	18.9–31.3

Throughfall (stemflow + canopy drip): Beech

	(mm) H₂O	H	Na	K	Ca	Mg	Fe	Mn	Al	S	P	Cl	NH₄-N	NO₃-N	N₍org₎-N	N₍ges₎
$\bar{x}$	870	1.34	14.1	27.9	24.1	4.0	1.2	3.9	1.6	50.0	0.6	32.5	13.4	11.5	9.8	34.8
range-values	628–1249	0.79–1.72	9.5–18.2	18.1–40.0	18.4–32.1	3.2–4.9	0.3–1.7	2.4–6.5	1.3–2.3	39.1–66.0	0.2–0.9	27.1–40.9	9.0–24.8	8.3–20.9	3.0–18.7	20.3–49.8

(Continued)

$[\text{kg} \cdot \text{ha}^{-1}]$

Throughfall: Spruce

	H₂O (mm)	H	Na	K	Ca	Mg	Fe	Mn	Al	S	P	Cl	NH₄-N	NO₃-N	Nₒᵣ₉-N	N₉ₑₛ
x̄	752	3.15	17.0	28.0	31.4	4.7	1.9	5.3	2.9	83.1	0.5	38.6	15.5	15.7	9.6	40.8
range-values	550– 1132	2.08– 5.03	9.0– 25.4	20.0– 41.5	19.6– 41.6	3.0– 6.1	0.6– 2.8	2.6– 8.1	2.1– 4.6	54.0– 107.6	0.2– 1.0	26.5– 53.3	8.7– 19.4	10.2– 20.4	2.7– 17.1	21.7– 51.4

Seepage water: Beech

	H₂O (mm)	H	Na	K	Ca	Mg	Fe	Mn	Al	S	P	Cl	NH₄-N	NO₃-N	Nₒᵣ₉-N	N₉ₑₛ
x̄	578	0.47	12.0	3.4	9.4	3.1	0.1	4.9	17.6	40.8	0.0	28.5	0.2	1.5	2.2	4.7
range-values	304– 1130	0.18– 1.0	5.5– 22.7	2.0– 7.4	4.0– 34.5	1.7– 5.8	0.0– 0.5	2.6– 10.1	5.6– 33.9	14.8– 79.6	0.0– 0.0	13.7– 63.2	0.0– 0.5	0.1– 4.3	0.4– 4.7	1.2– 15.0

$[\text{kg} \cdot \text{ha}^{-1}]$

Seepage water: Spruce

	H₂O (mm)	H	Na	K	Ca	Mg	Fe	Mn	Al	S	P	Cl	NH₄-N	NO₃-N	Nₒᵣ₉-N	N₉ₑₛ
x̄	424	0.40	19.5	3.7	14.1	5.8	0.2	9.9	52.7	96.6	0.0	42.3	0.2	12.5	2.2	14.9
range-values	240– 893	0.16– 0.84	7.8– 36.6	1.9– 7.7	7.4– 31.1	2.7– 11.2	0.0– 0.4	4.2– 18.4	10.5– 120.3	22.2– 225.0	0.0– 0.0	24.1– 72.8	0.1– 0.3	2.5– 28.7	1.2– 4.4	4.1– 33.2

References

Becker, K. W. 1984. Z Pflanzenernähr Bodenkde 147:476–484.

Beese, F. 1986. Göttinger Bodenkdl Ber 90:1–344.

Beese, F., and E. Matzner. 1986. Ber d Forschungszentrums Waldökosysteme/ Waldsterben d Universität Göttingen, Reihe B, Bd 3.

Benecke, P. 1984. Schriften aus der Forstl Fak d Univ Göttingen, Bd 77:1–150.

Boynton, D. 1954. Ann Rev Plant Physiol 5:31–54.

Bredemeier, M. 1987. Dissertation Univ Göttingen (in press).

Bredemeier, M., E. Matzner, and B. Ulrich. 1988. *In* M.H. Unsworth and D. Fowler (eds): *Acid deposition processes at high elevation sites, Edinburgh Sept.* D. Reidel, 607–619.

van Breemen, N., J. Mulder, and C. T. Driscoll. 1983. Plant and Soil 75:283–308.

van Breemen, N., and E. R. Jordans. 1983. *In* B. Ulrich and J. Pankrath, eds. *Effects of accumulation of air pollutants in forest ecosystems*, 171–182. D. Reidel, Dordrecht, NL.

van Breemen, N., P. H. B. de Visser, and J. J. M. van Grinsven. 1986. J Geol Soc London 143:659–666.

Büttner, G., N. Lamersdorf, R. Schultz, and B. Ulrich. 1986. Ber d Forschungszentr Waldökosysteme/Waldsterben d Universität Göttingen, Reihe B, Bd 1.

Cole, D. W., and D. W. Johnson. 1977. Water Resources Research 13:313–317.

Cronan, C. S., and W. A. Reiners. 1983. Oecologia (Berlin) 59:216–223.

Drablos, D., and A. Tollan. 1980. *Ecological Impact of Acid Precipitation*. SNSF-project, 1432. Ås, Norway, 388 pp.

Driscoll, C. T., and G. E. Likens. 1982. Tellus 34:283–292.

Ellenberg, H. 1939. Mitt Florist Soziol Arb gem Niedersachsen 5:3–135.

Ellenberg, H. sen., R. Mayer, and J. Schauermann, eds. 1986. *Ökosystemforschung: Ergebnisse des Sollingprojekts 1966–1986*. Ulmer Verlag, Stuttgart.

Feger, K. H. 1986. Freiburger Bodenkundliche Abhandlungen, Heft 17.

Fölster, H. 1985. *In* J. I. Drever, ed. *The chemistry of weathering*, 197–209. D. Reidel, Dordrechts, NL.

Förster, R. 1986. Dissertation Univ. Göttingen, 1–146.

Godt, J. 1986. Ber d Forschungszentr Waldökosyst/Waldsterben d Univ Göttingen, Bd 19:1–265.

Hauhs, M. 1985. Ber d Forschungszentrums Waldökosyst/Waldsterben der Univ Göttingen 17:1–206.

Heller, H. 1986. *In* H. Ellenberg, ed. *Ökosystemforschung: Ergebnisse des Sollingprojekts 1966–1986*, 109–126. Ulmer Verlag, Stuttgart.

Horntvedt, R., G. H. Dollard, E. Joranger. 1980. *In* D. Drabløs, and A. Tollan, eds. *Ecological Impact of Acid Precipitation*, SNSF-project Oslo-Ås, p 192.

Hüttermann, A. 1983. Allg Forstzeitschrift, 663–664.

Isermann, K. 1983. VDI-Berichte Nr. 500:307–335.

Johnson, D. W., D. D. Richter, G. M. Lovett, and S. E. Lindberg. 1985. Can J For Res 15:773–782.

Johnson, N. M., C. T. Driscoll, J. S. Eaton, G. E. Likens, and W. H. McDowell, 1981. Geochim Cosmochim Acta 45:1421–1437.

Jörgensen, R. G. 1986. Göttinger Bodenkundl Ber 91:1–409.

Jorns, A., and C. Hecht-Buchholz. 1985. Allg Forstzeitschrift 46:1248–1252.

Kreutzer, K., and J. Bittersohl. 1986. Forstw Cbl 105:357–363.

Kreutzer, K., E. Deschu, and G. Hösl. 1986. Forstw Cbl 105:357–363.

Krug, E. C., and C. R. Frink. 1983. Bull 811. The Connecticut Agric. Exp. Stat., New Haven.

Lang, E. 1986. Gött Bodenkdl Ber 89:1–119.

Lang, G. E., W. A. Reiners, and R. K. Heier. 1976. Oecologia (Berlin) 25:229–241.

Linkersdörfer, S., and P. Benecke. 1987. Materialien des Umweltbundesamtes 4/87, Berlin.

Lovett, G. M., and S. E. Lindberg. 1984. J Appl Ecol 21:1013–1027.

Marschner, H., and V. Römfeld. 1983. Z Pflanzenphysiol Bd III, 241–251.

Matzner, E. 1984. In Agren, ed. State and change of forest ecosystems. Swed. Univ. Agric. Sci., Dept. Ecology and Environmental Research. Report 13, 303–311.

Matzner, E. 1985. Allg Forstzeitschr 43:1143–1147.

Matzner, E. 1986. In H. W. Georgii, ed. Atmospheric pollutants in forest areas, 247–262. D. Reidel, Dordrecht, NL.

Matzner, E. 1988. Ber d Forschungszentrums Waldökosysteme der Universität Göttingen, Reihe A, Bd. 40, 1–217.

Matzner, E., and E. Cassens-Sasse. 1984. Ber d Forschungszentrums Waldökosysteme/ Waldsterben der Univ. Göttingen, Bd 2, 50–60.

Matzner, E., P. K. Khanna, K. J. Meiwes, E. Cassens-Sasse, M. Bredemeier, and B. Ulrich. 1984. Ber d Forschungszentrums Waldökosysteme/Waldsterben der Univ. Göttingen, Bd 2, 29–49.

Matzner, E., P. K. Khanna, K. J. Meiwes, M. Lindheim, J. Prenzel, and B. Ulrich. 1982. Gött Bodenkdl Ber 71:1–267.

Mayer, R. 1981. Gött Bodenkdl Ber 70:1–292.

Mayer, R., E. Matzner, and B. Ulrich. 1986. In H. Ellenberg, ed. Ökosystemforschung: Ergebnisse des Sollingprojekts 1966–1986, 375–417. Ulmer Verlag, Stuttgart.

Meiwes, K. J. 1979. Gött Bodenkundl Ber Bd 60:1–108.

Meiwes, K. J., P. K. Khanna, and B. Ulrich. 1986. Forest Ecology and Management 15:161–179.

Meyer, F. H. 1984. Allg Forstzeitschrift, 212–228.

Miller, H. G., and J. D. Miller. 1980. In B. Ulrich and J. Pankrath, eds. Effects of accumulation of air pollutants in forest ecosystems, 33–40. D. Reidel, Dordrecht, NL.

Murach, D. 1984. Göttinger Bodenkundl Ber 77:1–126.

Murach, D. 1987. Symposium of the Comm. of Europ. Communities: Effects of air pollution on terrestrial and aquatic ecosystems, Grenoble, May 1987 (in press).

Nilsson, S. I., H. G. Miller, and J. D. Miller. 1982. Oikos 39:40–49.

Nilsson, S. I. 1985. Ecological Bull. (Stockholm) 37:311–318.

Nilsson, S. I. 1983. In B. Ulrich and J. Pankrath, eds. Effects of accumulation of air pollutants in forest ecosystems, 105–112. D. Reidel, Dordrecht, NL.

Paces, T. 1986. J Geol Soc 143:306–311.

Prenzel, J. 1982. Gött Bodenkdl Ber Bd 72:1–113.

Puhe, J., and B. Ulrich. 1985. Arch Hydrobiol 102:331–342.

Rehfuess, K. E. 1981. Forstwiss Cbl 100:363–381.

Reid, J. M., D. A. Macleod, and M. S. Cresser. 1981. Earth Surfaces Processes and Landforms 6:447–457.

Reuss, J. O., and D. W. Johnson. 1986. Ecol Studies 59:1–119.

Rost-Siebert, K. 1983. Allg Forstzeitschrift 686–689.

Rost-Siebert, K. 1985. Ber d Forschungszentrums Waldökosysteme/Waldsterben der Univ Göttingen, Bd 12, 1–219.

Runge, M. 1970. Flora 159:233–257.

Runge, M. 1974a. Oecol Plant 9:201–218.

Runge, M. 1974b. Oecol Plant 9:219–230.

Salihi, O. O. A. 1984. Dissertation Univ Göttingen.

Schaller, G., and W. R. Fischer. 1985. Z Pflanzenernähr Bodenkde 148:306–320.

Schoen, R., R. Wright, and M. Krieter. 1984. Naturwissenschaften 71:95–97.

Schofield, C., J. Galloway, and G. Hendry. 1985. Water, Air and Soil Pollution 26:403–423.

Schrimpff, E. 1983. Staub-Reinh Luft 43:240.

Seibt, G. 1981. Schriften Forstl Fak Univ Göttingen, Bd 72, Sauerländer-Verlag, 1–109.

Seip, H. M. 1980. *In* D. Drablos and A. Tollan, eds. *Ecological Impacts of Acid Precipitation*, SNSF-Project, Oslo, 358–366.

Sollins, P., C. C. Grier, F. M. McCorison, K. Cromack Jr., R. Fogel, and R. L. Fredriksen. 1980. Ecol Monographs 50:261–283.

Ulrich, B. 1972. Forstarchiv 43:41–43.

Ulrich, B. 1980. *In* T. C. Hutchinson and M. Havas, eds. *Effects of acid precipitation on terrestrial ecosystems*, 255–282. Plenum Press, New York.

Ulrich, B. 1981a. Z Pflanzenernähr Bodenkde 144:647–659.

Ulrich, B. 1981b. Forstwiss Cbl 100:228–236.

Ulrich, B. 1981c. Z Pflanzenernähr Bodenkde 144:289–305.

Ulrich, B. 1983a. *In* B. Ulrich and J. Pankrath, eds. *Effects of accumulation of air pollutants in forest ecosystems*, 1–29. D. Reidel, Dordrecht, NL.

Ulrich, B. 1983b. *In* B. Ulrich and J. Pankrath, eds. *Effects of accumulation of air pollutants in forest ecosystems*, 33–45. D. Reidel, Dordrecht, NL.

Ulrich, B. 1986. Z Pflanzenernähr Bodenkd 149:702–717.

Ulrich, B., and E. Matzner. 1983a. Forst- und Holzwirt 18:468–474.

Ulrich, B., and E. Matzner. 1983b. Forschungsbericht 104-02-615, Umweltbundesamt, Berlin.

Ulrich, B., R. Mayer, and P. K. Khanna. 1979. Schriften aus der Forstl Fak d Univ Göttingen Bd 58, Sauerländer-Verlag.

Ulrich, B., U. Steinhardt, and A. Müller-Suhr. 1973. Göttinger Bodenkdl Ber 29:133–192.

Wiedey, G., and M. Gerriets. 1986. Berichte des Forschungszentrums Waldökosyst/Waldsterben d Univ Göttingen, Bd 2, Reihe B, 26–55.

Wisniewski, J. 1982. Water, Air and Soil Pollution 17:361–377.

Wright, R. F., E. Gjessing, N. Christophersen, E. Lotse, H. M. Seip, and B. Sletaune. 1986. Water, Air and Soil Pollution.

v. Zezschwitz, E. 1985. Forstwiss Cbl 104:205–220.

Zielinski, J. 1984. *In* Grodzinski et al., eds. *Forest ecosystems in industrial regions*. Ecol Studies 49:149–166.

Zöttl, H. W. 1958. Z Pflanzenernährung Bodenkde 81:149–166.

Zöttl, H. W. 1960. Forstwiss Cbl 79:16–32.

Decline of Red Spruce in High-Elevation Forests of New York and New England

A.H. Johnson,* T.G. Siccama,† W.L. Silver,†
and J.J. Battles‡

Abstract

A high rate of red spruce (*Picea rubens* Sarg.) mortality has been noted in the Adirondack, Green, and White Mountains of the northeastern United States over the past two decades. Because spruce mortality is greatest at high elevations where the forest is frequently immersed in cloudwater, and because cloudwater contains high levels of acids and other dissolved chemicals, several hypotheses regarding the role of acid deposition in the spruce decline have been formulated. Although several possibilities remain to be tested, at present no clear mechanistic evidence links acid deposition to spruce decline. Several natural stress factors are associated with the timing and severity of spruce mortality and will likely be judged causal agents. Winter damage (desiccation and/or freezing injury) appears to play a major role as an initiating and synchronizing influence. Experiments carried out with red spruce seedlings showed that realistic levels of ozone had an unfavorable effect on winter hardiness and suggested a way in which air pollution might play a role in the decline.

I. Introduction

Between the early 1960s and mid-1980s, red spruce (*Picea rubens* Sarg.) died at an unusually rapid rate on the Adirondack, Green, and White Mountains in the United States. The early reports of spruce mortality (1962 to 1965) caused some concern locally, but the ongoing and regionwide nature of this phenomenon did not become known until much later (e.g., Johnson and Siccama, 1983).

When large changes in spruce populations were quantitatively documented in the early 1980s (Siccama et al., 1982; Johnson and Siccama, 1983; Scott et al., 1984), and when it became apparent that the most severe mortality was occurring at high elevations where exposure to acidified clouds and rainwater was greatest, considerable attention was given to air pollution as a possible contributor to the mortality. Because an apparently similar decline of conifers in the Federal Republic of Germany was being studied in the early 1980s, and because acid

*Department of Geology, University of Pennsylvania, Philadelphia, PA 19104, USA.

†Yale School of Forestry and Environmental Studies, New Haven, CT 06511, USA.

‡Department of Natural Resources, Cornell University, Ithaca, NY 14850, USA.

deposition was the initial focus in those efforts, many hypotheses regarding how acid deposition and other pollutants might contribute to spruce mortality were proposed. Insects and biotic diseases have, since the earliest reports (e.g., Tegethoff, 1964), been regarded as secondary causal agents in most areas, thus increasing interest and speculation about air pollution effects.

When ideas of climatic extremes (drought and cold winters) are factored in as possible triggering stresses (e.g. Johnson et al., 1986, 1988), the concept of a decline (a complex disease that has biotic and abiotic components) seems to fit. Thus, attention has shifted to trying to determine if air pollution can alter spruce in ways that might increase the effects of natural stresses.

Ideas about air pollution have met with skepticism because there are no obvious symptoms of air pollution damage, the subtle responses of red spruce to ambient levels of air pollution are unknown, and declines of several species have occurred at times and in places where air pollution could not have been a factor. Additionally, data regarding the many natural causes of tree mortality have not been collected and analyzed.

A Spruce-Fir Research Cooperative has been formed to address the many aspects of this change in the forest (Hertel et al., 1987), including the potential for acid deposition effects. Thus, over the next few years, there will be much new information to use in judging whether or not the assimilation of airborne chemicals was an important contributor to the rather large change that has occurred in the high-elevation spruce-fir forests of the Appalachians.

In this review, we have summarized the data that define the nature and timing of the spruce decline, the data relevant to known natural contributors, and data addressing some of the hypotheses implicating acid deposition and other airborne chemicals. As adapted from Manion (1981), we have approached the recent episode of red spruce mortality as a complex disease with predisposing factors, initiating or synchronizing stresses, and contributing or sustaining stress factors. Below, we discuss the factors we judge will be most likely agreed upon as serving in each role and suggest where acid deposition and other airborne chemicals are being considered as possible causal agents.

II. The Occurrence of Spruce Decline

Figure 3-1 shows a compilation of reports of red spruce mortality in the northeastern states. About 1959 and especially by 1962 there were several reports of dead and dying red spruce across the region, particularly at high elevations. By 1984, another set of reports indicated a widespread decline of spruce in the mountain forests. This is the second severe, regionwide episode of spruce mortality in the past 150 years. The first lasted from about 1871 to 1890 (Hopkins, 1901). In the Adirondacks (NY) it was estimated that 50% of the mature spruce died between 1870 and 1885 (Seventh Report of the New York Forest Commission). Many smaller-scale incidents of spruce mortality have occurred in the last two centuries as reported by Hopkins (1891, 1901) and Weiss and associates (1985).

Figure 3-1. Years when widespread or severe red spruce mortality was reported. Confirmed reports are those written by trained scientists (after Johnson et al., 1986).

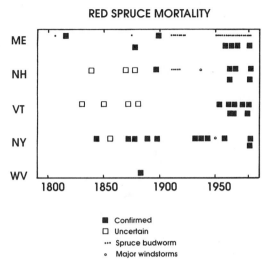

RED SPRUCE MORTALITY

■ Confirmed
□ Uncertain
⋯ Spruce budworm
○ Major windstorms

It is difficult to determine at this time if the most recent decline has the same set of characteristics as the decline of the late 1800s, or if the characteristics are substantially different, but clearly red spruce have undergone periods of extensive mortality in the past.

Hertel and co-workers (1987) reported that a consensus of scientists who have studied the recent decline of high-elevation red spruce in the northern Appalachians indicates that the high degree of mortality and pronounced growth reduction of spruce growing above 900 m are not readily explained by existing models of usual stand dynamics and forest growth, and that there are no consistent or unusual changes in the vigor or abundance of the major co-occurring species, white birch (*Betula papyrifera* var. *cordifolia* Marsh Regel) and balsam fir (*Abies balsamea* Mill.).

The symptoms of declining spruce vary from place to place, depending in many cases on the presence of insects like the spruce beetle (*Dendroctonus rufipennis* Kirby) and other diseases (McCreery et al., 1987). One of the most prominent symptoms at higher elevations is dieback of the terminal and lateral apices. Declining trees often have slightly chlorotic older needles and relatively few year classes of needles retained (e.g., 4 to 6).

III. Patterns and Rates of Mortality

The gradients in species composition and environmental conditions on the Green Mountains (VT), White Mountains (NH), Adirondacks (NY), and Catskills have been summarized by several investigators, most recently by Siccama (1974), Foster and Reiners (1983), Holway and associates (1969), and McIntosh and Hurley (1964). Figure 3-2 shows the gradients in major species at Whiteface Mountain, New York, which is fairly representative of montane forests of the

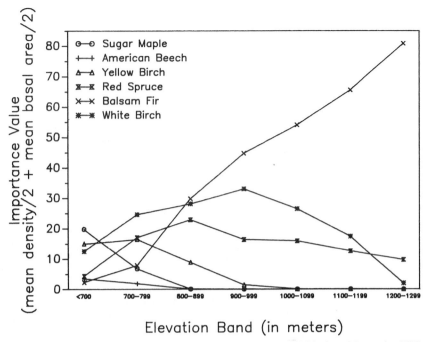

Figure 3-2. Elevational gradients in species importance at Whiteface Mountain (NY) (after Battles et al., 1988).

Northeast. Red spruce tends to peak in its importance on the midslopes (800 to 900 m) and is a minor component of the hardwood forest (below about 750 m) and the fir-dominated forest (above about 1,300 m). The natural conditions and mechanisms controlling the upper limit of red spruce are not well defined at present.

At a smaller scale, patterns of vegetation on the northern mountains are complex mosaics that result from the gradients in climate and soils and from the effects of natural and human disturbances, including fir, windthrow, landslides, disease and logging (Foster and Reiners, 1983).

In the recent period of mortality, severe dieback and tree death are most pronounced at higher elevations in most cases (Johnson and Siccama, 1983; Johnson and McLaughlin, 1986), and, on average, there is an increase in the percentage of standing dead trees in larger size classes (Johnson and McLaughlin, 1986).

The condition of red spruce in the high-elevation forests changed substantially between the mid-1960s and early 1980s, when at some sites live basal areas decreased by 40 to 70% (Scott et al., 1984; Siccama et al., 1982). Roughly comparable reductions occurred in larger and smaller size classes. In 1982, we established sampling transects to determine the status of mature spruce in the canopy and >10 cm dbh (Johnson and Siccama, 1983; Johnson and McLaughlin,

1986), and we resurveyed those in 1987 to assess changes in the 5-year interval (W. L. Silver, T. G. Siccama, and A. H. Johnson, unpublished data).

Figure 3-3 shows the percent of standing dead spruce in 1987 at the three most intensively surveyed sites (Mt. Washington, NH; Mt. Mansfield, Vt.; and Whiteface Mountain, NY) and Figure 3-4 shows the progress of spruce decline between 1982 and 1987 at those sites. Above 900 m, the majority of trees that were severely declining (crown class 3, with >50% loss of foliage from the upper part of the live crown) died between 1982 and 1987, and the percentage of spruce in crown classes 1 and 2 (0 to 10 and 11 to 50% loss of foliage from the top of the live crown) was unchanged. Below 900 m, the pattern was different. Severely declining trees (crown class 3) were rare at lower elevations in both surveys. A decrease was noted in crown class 1 trees with a corresponding increase in dead spruce (class 4 and 5). Within the sample, the greatest change between 1982 and 1987 occurred at Whiteface Mountain, where spruce died at most elevations at a rate in excess of 4% per year (Figure 3-5).

The different patterns in Figures 3-3 through 3-5 indicate that there are differences between high- and low-elevation sites and differences among the

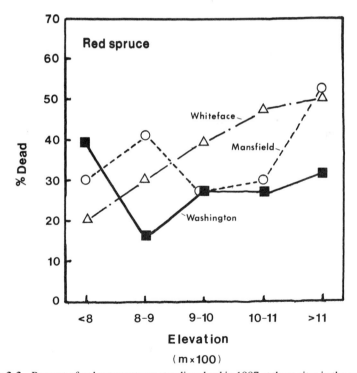

Figure 3-3. Percent of red spruce stems standing dead in 1987 at three sites in the northern Appalachians. Each site had approximately 350 to 400 spruce trees observed along systematically located sampling transects (W. L. Silver, T. G. Siccama, and A. H. Johnson, unpublished data). Methods are given in Johnson and McLaughlin, 1986.

Northern Sites Pooled − $\geq$ 900 m

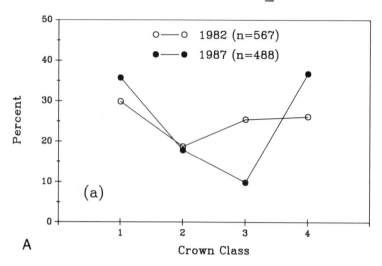

Northern Sites Pooled − $<$ 900 m

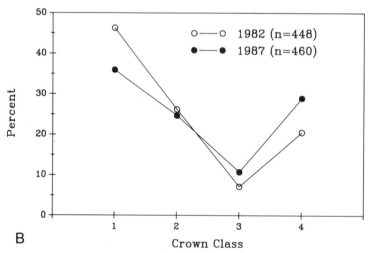

Figure 3-4. Change in crown condition between 1982 and 1987 at five sampling sites in the northern Appalachians. The data are from 56 100-m transects on Mt. Mansfield (VT), Mt. Washington (NH), and Whiteface Mountain (NY). Crown class 1 is 0–10% loss of foliage from the upper portion of the live crown, crown class 2 is 11–50%, class 3 is 50–99%, class 4 is dead/intact, and 5 is dead, broken above dbh (W. L. Silver, T. G. Siccama, and A. H. Johnson, unpublished data). Methods are given by Johnson and McLaughlin, 1986.

Figure 3-5. Percentage of the spruce stems at Whiteface Mountain that were dead in 1982 and 1987 as a function of elevation. Data are from 20 100-m transects on SW and SE slopes (W. L. Silver, A. H. Johnson, and T. G. Siccama, unpublished data).

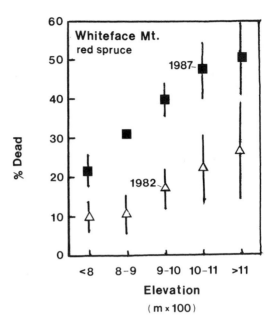

mountains surveyed. Based on reports of spruce beetle damage in some of our low-elevation stands (Crawford Notch near Mt. Washington, NH, personal communication from T. C. Weidensaul, Ohio State University), we suspect that the pattern of mortality and crown vigor at low elevation (<900 m) is a result, at least in part, of insect damage. Many years of observations of red spruce at our research sites suggest to us that the high-elevation mortality generally occurs gradually, in that individuals decline for several years before they die. Based on Figure 3-4, it appears that most of the trees currently in crown class 3 will die, but that at least a temporary stabilization of the canopy at high elevations will occur, as vigorous-looking trees did not begin to decline during the mid-1980s.

IV. Relationships between Mortality, Age, Aspects, and Elevation

As a means of refining some of the spatial relationships between site factors and spruce mortality, a detailed vegetation survey was done at Whiteface Mountain in 1986 to 1987 (Battles et al., 1988). A systematic sampling was accomplished by establishing 331 ten-m diameter permanent plots at 120-m intervals on 21 transects that covered all aspects and elevations between 700 and approximately 1,350 m.

Figure 3-6 shows the relationship between elevation and the percentage of dead spruce for the mountain as a whole. The upper three elevations are about the same

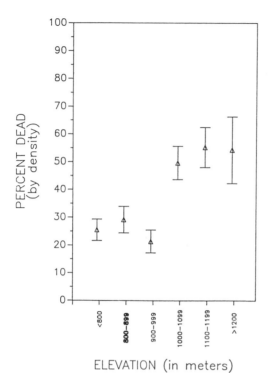

Figure 3-6. Percentage of spruce that were dead in 1986–1987 in 331 permanent plots at Whiteface Mountain (NY) (after Battles et al., 1988). Calculations are based on the number of live and dead stems >5 cm dbh per unit area. All aspects are pooled.

(55% of the spruce dead) and significantly different from the three lower elevational bands (about 25% of the spruce were dead). Figure 3-7 shows that the elevational difference persists when tree diameters are taken into account. The diameter distributions are important because mortality appears to have been greater in larger size classes (assuming that the higher proportion of dead stems in larger size classes does not reflect only the tendency for larger trees to remain standing longer than smaller trees) and because there is a higher proportion of larger trees above 1,000 m (data not shown).

The maximum mortality on Whiteface has occurred above 1,000 m on the northwest aspect (Figure 3-8) and there is generally a higher proportion of dead spruce in all size classes compared to other aspects at high elevation (Figure 3-9). The northwest aspect has the greatest development of fir waves (Sprugel, 1976), due most likely to the more severe impact of winter winds, clouds, and storms that cause mechanical defoliation and death of ranks of exposed, mature balsam fir. We suspect that these mechanical stresses are at least partially responsible for the increased spruce mortality on the northwest aspect, as they are for the increased fir mortality. At Whiteface Mountain, the most polluted clouds impact the southwest side of the mountain, and the cleanest air comes from the northwest (National Research Council, 1983).

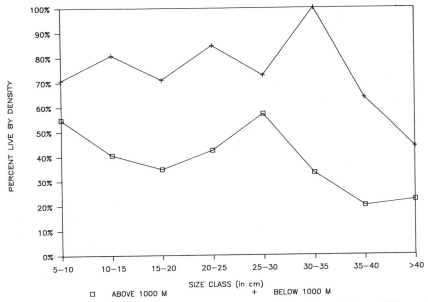

Figure 3-7. Distribution of live spruce by diameter classes above and below 1000 m at Whiteface Mountain (NY) (after Battles et al., 1988).

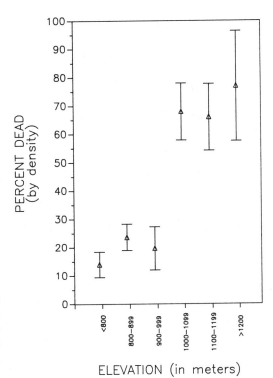

Figure 3-8. Percentage spruce >5 cm dbh that were dead as a function of elevation on the northwest-facing side of Whiteface Mountain (after Battles et al., 1988).

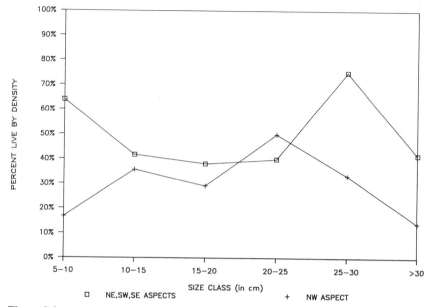

Figure 3-9. Percentage of live spruce by size class on the northwest-facing slope compared to the percent dead on the other three major aspects. Data are for all plots over 1000-m elevation (after Battles et al., 1988).

A. Age and Poor Site Conditions as Predisposing Factors

Age and poor sites are factors that have been associated with other declines, where they appear to be determinants of the spatial pattern of severity (Houston, 1981; Manion, 1981; Weiss and Rizzo, 1987). In the mountain forests of the Northeast, the upper slopes have thinner, more acidic soils, considerably shorter growing seasons, and greater wind stress and are poorer sites (Siccama, 1974). The current evidence suggests that the spatial pattern of spruce mortality is related to site and tree age across the New England mountains (Johnson and Siccama, 1983; Johnson and McLaughlin, 1986) as well as at Whiteface Mountain.

V. Climatic Extremes as Initiating Stresses

Climate stress in the form of drought or frost (freezing) injury has been considered a cause of other declines in the Northeast (Weiss and Rizzo, 1987; Manion 1981), and climate stress traditionally has been regarded as a likely contributor to forest diseases. Hepting (1963) discussed some of the indirect ways in which climate change (particularly long-term warming) can alter important life-supporting factors in forests and lead to disease, but, except in cases of severe climatic events, it is difficult to show convincing proof of climatic influence on mature trees.

Johnson and associates (1986), Cook and co-workers (1987), and the Johnson group (1988) used a combination of empirical studies to infer climate involvement in the recent spruce decline. Foremost is the inferred effect of winter damage to foliage and buds. Figure 3-10 shows that winter damage that resulted in red needles observed in the springtime was a common occurrence from 1959 to 1965 across New England. According to foresters' reports, red spruce suffered repeated and severe damage during the late 1950s and early 1960s. The foliar damage observed was usually attributed to *winter injury,* the desiccation of tissues that is thought to occur on warm, sunny days when the transpiration demand cannot be met because of frozen ground and/or stems. Freezing injury (called *frost damage*) could also cause the observed symptoms (Wardle, 1981).

The degree of needle and bud death was sufficient for some of the observers in the 1960s to conclude that winter conditions had killed spruce trees outright (Tegethoff, 1964), and the survey of Curry and Church (1952) suggests severely winter-damaged red spruce died soon after the injurious winter and that moderately damaged spruce suffered repeated winter damage in subsequent years, sometimes leading to death. We suggest that the type and degree of damage observed in the early 1960s was sufficient to initiate a decline. This is supported by the fact that sightings of winter damage in the early 1960s are temporally consistent with the beginning of reports of dead and dying spruce in the high-elevation forests of New England shown on Figure 3-1.

Tree rings also provide evidence that places the initiation of the period of decline in the late 1950s to mid-1960s. Johnson and McLaughlin (1986), the Cook group (1987), McLaughlin and associates (1987), and Johnson and co-workers (1988) showed that a highly synchronized regionwide decrease in ring width began in the

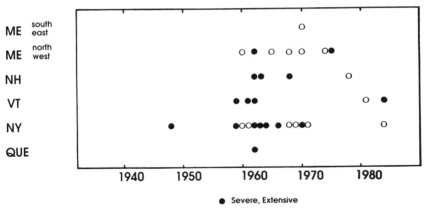

Figure 3-10. Reports of winter damage to red spruce in the northeastern states and Quebec. Reports are generally of red (dead) needles in the spring, bud and twig death, and tree death attributed by the observers to "winterkill." Solid circles designate events the observer noted as severe (after Johnson et al., 1986).

1960s, was not related to tree age, stand age, basal area, or disturbance history, and was thus not readily explained by usual stand dynamics.

More important than the reduction in ring width, the relationship between tree rings and climate changed abruptly sometime between 1959 and 1965. Across the mountains of New England and New York, standardized tree ring widths in red spruce growing at elevations over 900 m are highly correlated and show an essentially uniform relationship with later summer and early winter temperatures in the year prior to ring formation from about 1860 to 1960.

Warm, late summers or cold, early winters were associated with narrower-than-expected rings in the following growing season during this period. After 1960, the ring-width–climate relationships changed abruptly at all of the 14 high-elevation sites sampled. Figure 3-11 shows that a climate model based on August temperature (year prior to ring formation) and December temperature (year prior to ring formation), calibrated for the 1885 to 1940 period, accurately predicts standardized ring widths from 1941 to about 1960, but not after about 1960 or so. Although the physiological basis for the relationships observed in Figure 3-11 is not known, there appears to have been an abrupt, regionwide shift in the way spruce trees responded to their environment. Interestingly, the August–December temperature model appears to begin to predict accurately ring widths after the mid-1970s. Perhaps this is an indication that the trees were shocked in the 1960s but then began to recover a decade or so later.

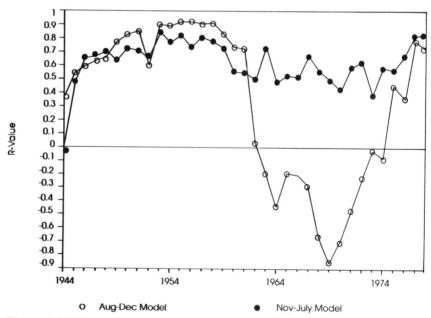

Figure 3-11. Seven-year running correlation coefficients for actual versus climate-predicted standardized ring widths for 11 sites in the Adirondacks and northern Appalachians (after Johnson et al., 1988).

The shift in tree-ring widths and the tree-ring–climate relationships corresponds with the reports of mortality and winter damage, and it is temporally consistent with the beginning of a decade of unusually cold winters (Namias, 1970; Johnson and McLaughlin, 1986; Johnson et al., 1988), suggesting a link between cold winters, winter damage to foliage, and the recent spruce decline.

In summary, the temporal consistency of cold winters, winter damage, reports of dead and dying spruce, the beginning of unusually small rings, and the change in tree-ring–climate relationships, coupled with the apparent adverse effect of cold winters on growth, suggests that winter damage is a major initiating and synchronizing factor in the decline. We also have found that historical episodes of red spruce mortality have occurred during or shortly after periods of unfavorable temperatures (Johnson et al., 1986; Johnson et al., 1988). Figure 3-12 shows a representation of the periods of climatic stress as positive index values (see Johnson et al., 1988, for methods and rationale). When compared to Figure 3-1, the regionwide periods of spruce mortality in the 1870s and 1880s, 1930s and 1940s, and in the 1960s to 1980s appears to have occurred during or shortly after periods when August temperatures were unusually warm and/or when December temperatures were unusually cold.

A. Drought

During the mid-1960s widespread drought in the Northeast was prolonged and particularly severe (Cook and Jacoby, 1977) and the potential for drought as an initiating stress in other declines has been noted (Manion, 1981; Houston, 1981; Burgess et al., 1984). Information relating to the importance of the mid-1960s

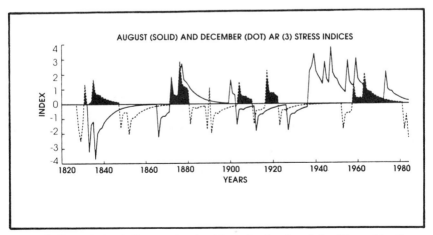

Figure 3-12. Periods of climate stress (positive values) and periods of favorable climate (negative values) for red spruce based on August (solid) and December (dashed) temperatures. Methods are explained by Johnson et al., 1988. The anomalously warm August of 1871 is obscured.

drought in the high-elevation forests is conflicting. Spatial patterns do not strongly suggest drought as a major factor, as precipitation during the drought increased with elevation (Siccama, 1974; Johnson and McLaughlin, 1986), where mortality has been the greatest. However, the high-elevation soils are thinner and hold less water. Temporally, the available evidence shows clearly that major episodes of high-elevation spruce mortality were in progress by 1962 or 1963, two years prior to severe drought (Tegethoff, 1964); the episodes had the same symptoms and characteristics that have been reported more recently.

Although drought index is not correlated with the standardized ring widths or with residuals from the temperature-regression models, nevertheless, the tree-ring analyses often show strong negative residuals from the temperature-based predictions during 1965 to 1967 (Johnson et al., 1988), suggesting that drought may be a stress adding to the decline syndrome. Additionally, the timing of the breakdown in the climate–tree-ring relationships in the Adirondacks (1967 or so, according to Cook, 1987) corresponds more with the drought period than with the earlier episodes of winter damage.

VI. The Influence of Insects and Diseases

Very commonly, insects and diseases contribute to declines as factors responsible for the death of trees that have been subject to other types of stress. In some cases, defoliating insects can initiate a period of decline (Weiss and Rizzo, 1987). Currently, many symptoms are recognized as related to known insects and diseases, and many symptoms are of unknown origin (Hertel et al., 1987). Hopkins (1901) and Weiss and associates (1985) reviewed the occurrence of insects and diseases in episodes of spruce mortality that occurred during the past 100 years or so. The spruce beetle *Dendroctonus rufipennis* Kirby has been implicated in many cases of spruce mortality and is present in some, but not all, locations in the current episode of decline (McCreery et al., 1987; T. C. Weidensaul, personal communication).

Armillaria mellea (Vahl:Fr.), the shoestring root rot fungus, is relatively abundant in declining spruce at lower elevations but rare at higher elevations (in about 10% of the severely declining trees in a study done by Carey et al., 1984). Other fungi (e.g. *Fomes pini* and *Cytospora* canker), which are considered to be secondary in their action, are present and have been identified in declining spruce in some places over the past two decades (Hadfield, 1968). At lower elevations, dwarf mistletoe (*Arceuthobium pusillum* Peck) has been identified as a factor associated with growth loss and mortality (McCreery et al., 1987).

Early observers of spruce mortality at high elevations in New Hampshire and western Maine (Kelso, 1965; Wheeler, 1965; Tegethoff, 1964; D. Stark, unpublished field notes, 1962) found no evidence of insects or fungal disease as a cause of the spruce mortality they studied through repeated visits in 1963 and 1964. They attributed the mortality and the visible decline in spruce to the effect of severe winter conditions. Overall, the current evidence suggests a variety of insects and diseases as contributors to decline, with different combinations present in different areas.

VII. Summary of Natural Factors

To date, many of the naturally occurring factors that have been implicated in declines of other species are shown to be temporally and/or spatially associated with the current red spruce decline. Tree age, elevation and aspect (indicators of more stressful sites), winter damage, drought, insects, and fungal diseases have been identified thus far. Using the scheme of Manion (1981), age and site quality might be logically assumed as predisposing factors, winter damage and drought might be considered as inciting or triggering factors, and the fungal diseases might be considered as contributing factors. Spruce beetles, where present, might be contributing factors or, in some cases, primary factors in tree death. If the forthcoming findings of research from the several intensively studied sites are consistent with the findings presented above, those natural factors will likely be designated as causes of the decline.

VIII. The Possibility of Air Pollution Involvement

Although the available information suggests that the spruce decline fits the pattern of other declines where air pollution has not been suspected as a contributor, the presence of such evidence does not rule out air pollution involvement. Many possible roles for acid deposition and other airborne chemicals have been identified and summarized by other authors (McLaughlin, 1986; Klein and Perkins, 1987; Johnson and Siccama, 1983; Friedland et al., 1984a). At present, no symptoms are attributable solely to air pollution, so it is logical to ask questions about the interaction of airborne chemicals and site conditions, winter damage, drought, insects, and disease. Could air pollution at present and past levels have changed site conditions or plant function in a way that would make spruce more susceptible to the natural stresses associated with decline? Could the recovery period after acute climatic stress require enough energy for repair that a background of air pollution stress, for which the plants could ordinarily compensate, becomes significant? A few empirical studies, and a few experimental studies in which realistic levels of pollutants were used, provide a basis for determining the direction of future investigations.

A. Airborne Chemicals and Site Quality

Acids and heavy metals are constituents of cloudwater and precipitation that have been regarded as chemicals that might have adverse effects on the quality of forest soils. They are deposited at relatively high rates in the upper montane forests largely because of the interception of cloudwater, which has much higher levels of dissolved chemicals than does rain. Table 3-1 summarizes some data sets on deposition of chemicals in cloudwater and precipitation to forests of the Northeast.

1. Trace Metals

Reiners and associates (1975), the Johnson group (1982), and Friedland and co-workers (1984b, 1984c) have documented a rather rapid accumulation of lead, copper, zinc, and organic matter in the forest floor in montane forests of New England. The forest floor in the subalpine coniferous forests is usually deep (>10 cm), and because red spruce are particularly shallow rooted, changes in the chemistry or processes regulating nutrient availability and turnover in that ecosystem compartment could be important.

Friedland and associates (1984b) showed that trace metal accumulations in the forest floor of declining spruce-fir stands on Camels Hump (VT) are not presently in the range where alterations of biological activity have been noted in controlled experiments but lie in the range where experimental results indicate no significant changes in the measured indicators of biological activity. Camels Hump metal levels are among the highest measured values (Friedland et al., 1984b, 1984c) in the Northeast; although those levels could increase, there is no indication in these or other data available now that trace metal deposition is a likely factor in the current spruce decline. However, effects on special-purpose organisms that are crucial to nutrient cycles (ammonifiers, nitrifiers, mycorrhizae) cannot be judged adequately with the data available, and this leaves open an area for further study.

2. Soil Acidification

At sufficient input levels, the leaching of base cations by mobile anions resulting from atmospheric inputs can probably increase soil acidity in some types of soils (Ulrich et al., 1980; van Breemen et al., 1984). Soil acidification would be expected to result in lower soil fertility, increased availability of potentially toxic elements, particularly aluminum, and, as such, additional site-related stress. A recent review by Binkley and Richter (1986) summarized the processes that govern soil acidification and the factors that make soils susceptible or resistant to acidification by atmospheric sources.

Owing to the high rates of sulfuric and nitric acid deposition to subalpine forests, an understanding of long-term changes in soil pH and base status is helpful in understanding the impact to date of acid deposition. In 1930 to 1932, C. C. Heimburger (1934) measured horizons in soil profiles throughout the Adirondack Mountains (NY) and determined the pH of the organic horizons and dilute acid-extractable calcium in organic and mineral horizons. Those measurements were repeated in 1984 by S. B. Andersen (University of Pennsylvania, Ph.D. Thesis) and shown in Figures 3-13 to 3-16. As expected based on theoretical considerations, there was no systematic change in the pH of extremely acid organic horizons (original pH <3.5) over five decades. There was no systematic change in the extractable Ca in B horizons of any soils, although E horizons acidified. Higher pH organic horizons tended to show a decrease in pH and

Table 3-1. Deposition of chemicals in precipitation and clouds to high-elevation forests in the Northeast compared to deposition in a lower elevation forest at Hubbard Brook, New Hampshire (kg ha^{-1} yr^{-1}).

Ion	Hubbard Brook (NH)[1] (1963–1974)	Mt. Moosilauke (NH)[2] (1220 m)	Whiteface Mountain (NY)[3] (1060 m)	Ion concentrations in stratiform cloudwater (μeq/1) at Whiteface Mountain (NY)[4]
H$^+$	1.0	3.9	1.03	280 (pH 3.6)
NH$_4^+$	2.9	20.5	9.1	89
SO$_4^{2-}$	38.4	202.7	60.2	140
NO$_3^-$	19.7	124.9	27.4	110

[1] 1963–1974 data from Likens et al., 1977.

[2] 1980–81 data from Lovett et al., 1982.

[3] E. K. Miller, A. J. Friedland, and A. H. Johnson, unpublished data for 1985–1986.

[4] Mean of 28 samples collected at Whiteface Mountain summit, Castillo et al., 1983.

Lovett, G.H., W.R. Reiners and R.K. Olsen. 1982. Cloud droplet deposition subalpine balsam fir forests: hydrological and chemical inputs. Science 218:1303–1304.

Likens, G.E., F.H. Bormann, R.S. Pierce, J.S. Eaton, N.M. Johnson. 1977. Biogeochemistry of a Forested Ecosystem. Springer Verlag, New York, 147 p.

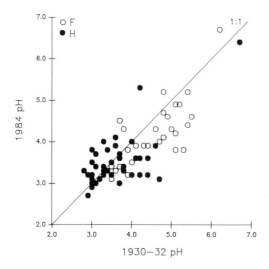

Figure 3-13. Changes in pH of organic horizons of Adirondack forest soils between 1930–1932 and 1984. Values are shown for 32 sites that were relocated with very high confidence (S. B. Andersen, Univ. of Pennsylvania, Ph.D. Thesis). The original study was carried out by C. C. Heimburger (1934), and the 1984 measurements were made using the same collection and analytical procedures.

extractable Ca, which is probably consistent with normal forest aging and Ca accumulation in biomass over 50 years. For the most part, high-elevation sites do not appear to be particularly susceptible to acidification at present and past rates of acid deposition. In light of those data, the likelihood that acid deposition has significantly increased stress related to long-term acidification of high-elevation forest soils is small. An exception to this generalization might stem from the effects that highly acidic precipitation or snowmelt events might have on soil solution chemistry, roots, mycorrhizae, and the like.

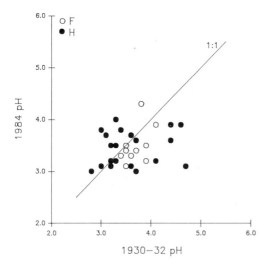

Figure 3-14. Changes in pH in organic horizons in high-elevation Adirondack forest stands with appreciable amounts of red spruce and balsam fir (S. B. Andersen, unpublished data).

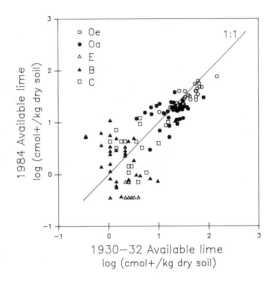

Figure 3-15. Changes in dilute acid-extractable calcium (0.2N HCl) in organic and mineral horizons between 1930–1932 and 1984 at sites relocated with the most confidence (S. B. Andersen, unpublished data).

B. Foliar Leaching and Foliar Nutrition

Related to the issue of site quality and soil acidification is the ability of plants to obtain adequate base cations for normal functions. While there is no empirical evidence that soil-available Ca, Mg, or K levels have decreased due to atmospheric deposition, experiments with several different species have shown that acidic precipitation increases the leaching of cations from foliage. Given the

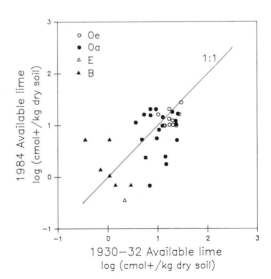

Figure 3-16. Changes in dilute acid-extractable Ca at high-elevation Adirondack sites (S. B. Andersen, unpublished data).

association of Mg and K deficiency with declining Norway spruce (Heuttl and Wisniewski, 1987) in Germany and the high potential for foliar leaching in the cloud-immersed forests, the levels of cations in red spruce foliage are of interest.

Friedland and associates (1987) sampled healthy, mature red spruce at high and low elevations in the Green and Adirondack Mountains and reported that foliar Mg levels were lower in their high-elevation sites than all other reported values for red spruce, and in the range of moderate deficiency according to Swan (1971).

Studies of foliar Ca, Mg, and K carried out using standard sampling procedures at Whiteface Mountain (T. N. Schwartzman and R. G. Miller, University of Pennsylvania, unpublished data) show some interesting patterns. Of the three elements measured, only potassium was related to crown class (Figure 3-17). Interestingly, levels appear to be sufficient, even in severely declining trees

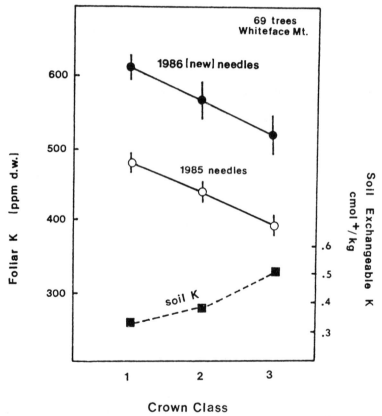

Crown Class

Figure 3-17. Foliar and soil potassium as a function of crown class at Whiteface Mountain (NY). Foliage was sampled from the south side of the 4th–7th whorl in October 1986. Soil K is 2N NH_4Cl-extractable K. At each tree, soils were sampled at 0–15 cm (n = 4 per tree) and 15–30 cm (n = 4 per tree) depths and bulked into one sample for each depth. Values reported are the arithmetic mean of the two depths (T. N. Schwartzman and R. G. Miller, University of Pennsylvania, unpublished data). Bars represent standard errors.

(Tables 3-2 and 3-3), and the lower foliar levels are not related to low levels of available K in the soil. This contrasts with the case of Norway spruce, where K and Mg deficiencies are associated with soils having very low available levels (Heuttl and Wisniewski, 1987). The lower levels of foliar K in declining trees may be related to increased foliar leaching (due to poorer foliar integrity?) or decreased uptake capability (due to deteriorating root systems?).

Although Mg is only weakly associated with crown class, Figure 3-18 shows that at Whiteface Mountain foliage is not in general well supplied with Mg. According to the values of Swan (1971), red spruce at high elevations show a moderate Mg deficiency. Foliar and soil levels both decrease with increasing elevation, so the roles of foliar leaching (which is expected to increase at higher elevations due to greater precipitation and cloudwater acidity) and soil supply are obscured.

Foliar Ca values measured at Whiteface Mountain are not correlated with elevation or crown class and even in severely declining trees (crown class 3) are present at levels required for good growth (Table 3-3). Available Ca in the soils at Whiteface is quite high relative to most high-elevation forest soils due to the rapidly weathering Ca feldspar that is a major component of the anorthosite that dominates the local bedrock and till.

At present, no conclusions can be drawn regarding the role of foliar nutrient status in the spruce decline or regarding the effect of acid deposition on foliar nutrient status. However, those questions deserve continued study, particularly as foliar nutrition relates to the natural stresses contributing to the spruce decline.

C. Airborne Chemicals and Climate Stress

Some recent research has been designed to test the effect of realistic levels of pollutants on the resistance of red spruce seedlings to climatic stresses. In a growth

Table 3-2. Values for foliar Ca and NH_4 Cl-extractable Ca in the soil at 0–15 and 15–30 cm depths at Whiteface Mountain.

Crown class	n	Foliage		Soil	
		Ca (g per 100 g dry wt)		Ca (cmol$^+$/kg)	
		1986 needles	1985 needles	0–15 cm	15–30 cm
1	45	.22 + .01	.30 + .01	8.1 + 1.2	5.7 + 1.5
2	14	.21 + .01	.31 + .03	6.2 + 1.7	5.1 + 1.4
3	10	.22 + .02	.30 + .03	10.8 + 3.6	9.3 + 3.1

T. N. Schwartzman, R. G. Miller, and A. H. Johnson, unpublished data.

Foliage was collected in October 1986 and dried immediately upon return from the field. Digestion in HCl/HF was carried out as in Friedland et al. (1987).

Four soil samples were taken from the base of each tree at each depth and combined for analysis. Extraction was in 1 N_4Cl.

Crown class descriptions are given in the text.

Trees were sampled based on random selection from permanent plots covering 700–1300 m elevation on the northwest face.

Table 3-3. Suggested provisional standards for the evaluation of the results of foliar analyses for red spruce (foliar concentration expressed as percent dry matter).

Element	Range of acute deficiency	Range of moderate deficiency	Transition zone from deficiency to sufficiency	Range of sufficiency for good to very good growth	Range of luxury to excess consumption
K	Below 0.19	0.19–0.30	0.30–0.40	0.40–1.10	1.10 and up
Mg	Below 0.04	0.04–0.06	0.06–0.08	0.08–0.17	0.17 and up
Ca	Below 0.05	0.05–0.08	0.08–0.12	0.12–0.30	0.30 and up

Modified from Swan, 1971.

These suggested standards are essentially judgments; they are based both on the results of greenhouse studies and on experience gained from the use of foliar analysis in field studies.

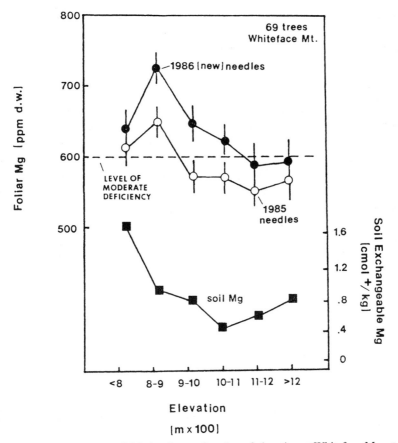

Figure 3-18. Foliar and soil Mg levels as a function of elevation at Whiteface Mountain (NY). Methods are the same as for Figure 3-17 (T. N. Schwartzman and R. G. Miller, unpublished data).

chamber experiment, Norby and associates (1987) used potted red spruce seedlings from Maine planted in soil from Camels Hump (VT) and Acadia National Forest (ME), coupled with ozone and acid mist and/or acid rain treatments. In that experiment, mist and rain acidified to realistic values (pH 3.6 mist and pH 4.1 rain) and applied at realistic rates was associated with a large and statistically significant increase in drought stress. The authors attributed the effect to an increase in shoot-to-root ratio brought about by the application of mist and rain and speculated that this might be the result of the increased nitrogen applied.

As explained previously, the tree-ring analyses for intervals prior to 1960 and from 1961 to 1980 and the timing of reports of mortality in 1962 or 1963 did not strongly suggest drought as the primary factor initiating the current spruce decline, but the tree-ring analyses cannot account for threshold responses to acute stress, so a particularly severe drought (and the mid-1960s drought was extreme; Johnson and Siccama, 1983) could have been an important factor. As is the case for all experimental work to date, it is not known if experiments done with seedlings are applicable to the mature trees in the field.

D. Winter Damage and Airborne Chemicals

As explained above, there is substantial evidence that winter damage played a key role in initiating the spruce decline. Thus, the effects of airborne chemicals on resistance of red spruce to winter stress is an important area for investigation. Many factors, some reasonably well understood and others much less so, contribute to a plant's ability to resist damage in winter (Weiser, 1970).

Davison and associates (1987) reviewed the types of injury plants can sustain in winter and the ways in which air pollution might alter resistance to those types of damage. In winter, conifers must contend with photo-oxidation of chlorophyll (resulting in bleaching of the needles), desiccation (generally thought to be caused by water loss through cuticles on bright, sunny days when water in the soil and/or conducting tissues is frozen), and freezing injury (usually called *frost damage*). The last two stresses result in the death of needles, which turn reddish brown in spring.

As for evidence that air pollution might affect resistance to winter damage, electron microscopy in Finland done in polluted areas and unpolluted areas suggested that conifers exposed to pollution were more susceptible to winter damage (Davison et al., 1987), but the exact reasons remain unknown. Because several pollutants attack constituents of cell membranes and changes in cell membranes occur during hardening against freezing injury, Davison and co-workers (1987) suggest that there is a sound theoretical basis for suspecting that air pollution might interfere with the development of freezing resistance. Their experiments with ozone fumigations (120 ppb O_3, 6 hours day^{-1}, for 70 days) indicated that ozone at those dose rates was associated with increased cuticular water loss from the previous year's needles in one of eight Norway spruce clones tested. Four clones showed an increase in freezing-related damage when exposed to the ozone treatments.

Weinstein and associates (1987) used red spruce seedlings and exposed them to filtered air, the ambient levels of ozone in Ithaca, New York, and twice ambient levels. Ozone dose was associated with increased respiration, decreased pigment concentrations, and continued high photosynthesis rates during the hardening period (October), whereas plants exposed to charcoal-filtered air showed the expected decrease in photosynthesis rates in the fall. Related work by Alscher and co-workers (R. Alscher, VPI, personal communication) showed a delay in starch removal from red spruce needles and changes in the ultrastructural characteristics of foliage, both of which suggest a delay in hardening. When compared to the experimental conditions of the Weinstein group (1987), the summertime ozone doses received by the trees at Whiteface Mountain have probably been in the ambient to twice-ambient range of the Weinstein group's experiments (Burgess et al., 1984; J. Panek, Atmospheric Sciences Research Center, State University of

New York, Albany, unpublished data). Again, the application of the results to mature trees on the mountains is suspect, but ozone stress clearly becomes a plausible candidate for more detailed research as an important factor leading to spruce decline.

Weinstein and associates (1987) also tested the effect of nitric acid additions in artificial mist and rain on red spruce seedlings. Using reasonable misting periods and reasonable water application rates, they observed no increase in winter damage at even unrealistically high rates of N input (pH 2.5 nitric acid), thus suggesting that increased N inputs via cloudwater deposition do not increase the likelihood of winter damage, as suggested by Friedland and co-workers (1984a). Jacobson and Lassoie (1987) showed similar results with spruce seedlings, but noted an increase in foliar lesions with decreasing mist pH, which was associated with sulfate concentrations. The greatest effects were noted at pH values below 3.0, which are very infrequently observed in ambient cloudwater. On the other hand, Cape et al. (1988) have shown that freezing resistance in red spruce seedlings is decreased by exposure to acidic mist. Treatment with mist at pH 3.5 (equimolar ratio of $SO_4:NO_3$) increased the freezing temperature during the autumn by 3-5°C compared to pH 5.0 mist. More study is warranted here to clarify the relationships between acidic mist, ozone and winter hardiness.

IX. Summary

Experimental work now in progress and future research may show that realistic levels of acids, ozone, or other airborne chemicals affect spruce in ways that may have contributed to declining vigor, but the record to date does not clearly implicate acid deposition as a contributor to the recent spruce decline. The record of experimental results from which to choose the most appropriate targets for research suggests that acidic cloudwater ozone should be given a high priority. A clear record has been established with seedlings and saplings of many species that suggests ozone at ambient levels can reduce net photosynthesis and growth (Wang et al., 1986; Skelly et al., 1983; Reich and Amundson, 1985; also see Bormann, 1985), and recent results with red spruce (Weinstein et al., 1987) suggest a possible link to increased winter damage.

References

Battles, J., A. H. Johnson, T. G. Siccama, and W. L. Silver. 1988. *In* G. D. Hertel, ed. *Effects of air pollutants in spruce fir forests of the U.S. and Federal Republic of Germany.* U.S. Forest Service, Broomall, PA (in press).

Binkley, D., and D. D. Richter. 1987. *In* Advances in ecological research, vol. 16, 1–51, Academic Press.

Bormann, F. H. 1985. Bioscience 35(7):434–441.

Burgess, R. L., M. B. David, P. D. Manion, M. J. Mitchell, V. A. Mohnen, D. J. Raynal, M. Schaedle, and E. H. White. 1984. *Effects of acidic deposition on forest ecosystems in the northeastern United States: an evaluation of current evidence.* New York State College of Environmental Science and Forestry, Syracuse, NY.

Cape J. N., L. J. Sheppard, I. D. Leith, M. B. Murray, J. D. Deans and D. Fowler 1988. Aspects of Applied Biology (in press).

Carey, A. C., E. A. Miller, G. T. Geballe, P. M. Wargo, W. H. Smith, and T. G. Siccama. 1984. Plant Dis 68:794–795.

Castillo, R. A., J. E. Jiusto, and E. M. McLaren. 1983. Atmospheric Environ 17:1497–1505.

Cook, E. R. 1987. *In* T. C. Hutchinson and K. M. Memmea, eds. *Effects of atmospheric pollutants on forests, wetlands and agricultural ecosystems,* 277–290. Springer Verlag, Berlin.

Cook, E. R., and G. C. Jacoby. 1977. Science 198:399–401.

Cook, E. R., A. H. Johnson, and T. J. Blasing. 1987. Tree Phys 3:27–40.

Curry, J. R., and T. W. Church. 1952. J. Forestry 50:114–116.

Davison, A. W., J. D. Barnes, and C. J. Renner. *In Interactions between air pollution and cold stress.* Proc. 2d int. symposium on air pollution and plant metabolism April 6–9, 1987. Neuhrenburg, FRG. (in press).

Foster, J. R., and W. A. Reiners. 1983. Bull Torrey Bot Club 110:141–153.

Friedland, A. J., G. J. Hawley, and R. A. Gregory. Plant and Soil (in press).

Friedland, A. J., A. H. Johnson, and T. G. Siccama. 1984a. Can J For Res 14:963–965.

Friedland, A. J., A. H. Johnson, and T. G. Siccama. 1984b. Water Air Soil Pollut 21:161–170.

Friedland, A. J., A. H. Johnson, T. G. Siccama and D. L. Mador. 1984c. Soil Sci. Soc. America J. 49:422–425.

Hadfield, J. S. 1968. File Report A-68-8 5230. Amherst, MA: USDA-Forest Service Northeastern Area, State and Private Forestry, Amherst FPC Field Office.

Heimburger, C. C. 1934. Forest Type Studies in the Adirondacks. Ph.D. Thesis, Cornell University, Ithaca, NY.

Hepting, G. H. 1963. Ann Review Phytopathology 1:31–50.

Hertel, G., S. J. Zarnoch, T. Arre, C. Eager, V. A. Mohnen, and S. Medlarz. 1987. Status of the Spruce-Fir Research Cooperative research program. Proc. 80th Ann. Meeting of the Air Pollut. Control Assn, New York, NY, June 21–26, 1987.

Holway, J. G., J. T. Scott, and S. Nicholson. 1969. *In* J. G. Holway and J. T. Scott, eds. *Vegetation-environment relations at Whiteface Mt., NY.* Rept. no. 92, Atmospheric Sciences Research Center, State University of New York at Albany, 1–44.

Hopkins, A. D. 1891. Forest and shade tree insects. II. West Virginia Experiment Station Third Annual Rep, 171–180.

Hopkins, A. D. 1901. Insect enemies of the spruce in the Northeast. U.S. Dept. of Agriculture, Div. of Entomology Bull 28, new series, 15–29.

Houston, D. B. 1981. Stress triggered tree diseases, the diebacks and declines. USDA Forest Service NE-INF-41-81. Washington, D.C.

Huettl, R., and J. Wisniewski. 1987. Water Air Soil Pollut 33:265–276.

Jacobson, J. S., and J. P. Lassoie. *In* G. D. Hertel, ed. *Proc. of a conference on the effects of atmospheric pollutants on the spruce-fir forests of the Eastern United States and the Federal Republic of Germany.* U.S. Forest Service, Broomall, PA (in press).

Johnson, A. H., E. R. Cook, and T. G. Siccama. 1988. Relationships between climate and red spruce growth and decline. Proc Nat Academy Sci (in press).

Johnson, A. H., A. J. Friedland, and J. Dushoff. 1986. Water Air Soil Pollut 30:319–330.

Johnson, A. H., and S. B. McLaughlin. 1986. *In Report of the committee on monitoring and trends in acidic deposition,* 200–230. National Research Council. National Academy Press, Washington, D.C.

Johnson, A. H., and T. G. Siccama. 1983. Environ Sci Technol 17:294a–305a.

Johnson, A. H., T. G. Siccama, and A. J. Friedland. 1982. J Env Qual 11:577–580.

Kelso, E. G. 1965. Memorandum 5220,2480, July 23, 1965. Amherst, Mass., U.S. Forest Service Northern FPC Zone.

Kline, R. M., and T. D. Perkins. 1987. Ambio 16:86–93.

Manion, P. D. 1981. *Tree Disease Concepts.* Prentice-Hall, Englewood Cliffs, N.J.

McCreery, L. R., M. N. Weeks, M. J. Weiss, and I. Millers. 1987. *In Proc. integrated pest management symposium for northern forests, March 24–27, 1986.* Cooperative Extension Service, University of Wisconsin, Madison.

McIntosh, R. P., and R. T. Hurley. 1964. Ecology 45:314–326.

McLaughlin, S. B. 1986. J Air Pollut Control Assn 35:512–534.

McLaughlin, S. B., D. J. Downing, T. J. Blasing, E. R. Cook, and H. S. Adams. 1987. Oecologia 7:487–501.

Namias, J. 1970. Science 170:741–743.

National Research Council. 1983. *Acid deposition: atmospheric processes in Eastern North America.* National Academy Press, Washington, D.C. 375 pp.

New York State. 1891. Seventh Report of the Forest Commission. Albany, NY.

Norby, R. J., G. E. Taylor, S. B. McLaughlin, and C. A. Gunderson. 1987. *In* C. G. Tauer and T. C. Hennessey, eds. *Proc. 9th North American forest biology workshop,* 34–41. Dept. Forestry, Oklahoma State U., Stillwater.

Reich, P. B., and R. G. Amundson. 1985. Science 230:566–570.

Reiners, W. A., R. H. Marks, and P. M. Vitousek. 1975. Oikos 26:264–275.

Scott, J. T., and J. G. Holway. 1969. *In* J. G. Holway and J. T. Scott, eds. *Vegetation-environment relations at Whiteface Mt. in the Adirondacks,* 44–88. Rep. No. 92, Atmospheric Sciences Research Center, State University of New York, Albany, NY.

Scott, J. T., T. G. Siccama, A. H. Johnson, and A. R. Breisch. 1984. Bull Torrey Bot Club 111:438–444.

Siccama, T. G. 1974. Ecol Monogr 44:325–349.

Siccama, T. G., M. Bliss, and H. W. Vogelmann. 1982. Bull Torrey Bot Club 109:163–168.

Skelly, J. M., Y. Yang, B. Chevone, S. J. Long, J. E. Nellessen, and W. E. Winner. *In Air Pollution and Productivity of the Forest,* 143–159. Izaak Walton League, Washington, D.C.

Sprugel, D. 1976. J Ecol 64:889–911.

Stark, D. Unpublished notes on red spruce disease, mortality and winter injury, 1957–1977. State of Maine Department of Conservation, Entomology Laboratory, Augusta.

Swan, H. S. D. 1971. *Relationships between nutrient supply, growth and nutrient concentrations in the foliage of white and red spruce.* Woodlands Rep. WR/34, Feb., 1971. Pulp and Paper Research Institute of Canada. 27 p.

Tegethoff, A. C. 1964. Memorandum 5220, September 25, 1964. Amherst, Mass. U.S. Forest Service Northern FPC Zone.

Ulrich, B., R. Mayer, and P. K. Khanna. 1980. Soil Sci 130:193–199.

van Breemen, N., J. Mulder, and C. T. Driscoll. 1983. Plant and Soil 75:283–308.

Wang, D., D. F. Karnosky, and F. H. Bormann. 1986. Can J Forest Research 16:47–55.

Wardle, P. 1981. Arctic and Alpine Res 13:419–423.

Weinstein, L., R. J. Kohut, and J. S. Jacobson. 1987. Research at Boyce Thompson Institute on the effects of ozone and acidic precipitation on red spruce. Proc. 80th Ann. Meeting, Air Pollut. Control Assn. June 21–26, 1987, New York, NY.

Weiser, C. J. 1970. Science 169:1269–1278.

Weiss, M. J., and D. M. Rizzo. 1987. Forest declines in major forest types of the eastern United States. Proc. of workshop on Forest Decline and Reproduction: Regional and Global Consequences. Krakow, Poland, March 23–28, 1987.

Weiss, M. J., L. R. McCreery, I. R. Millers, J. T. O'Brien, and M. M. Weeks. 1985. Cooperative survey of red spruce and balsam fir decline in New York, New Hampshire and Vermont—1984. Interim Report. USDA Forest Service, Forest Pest Management. Durham, N.H.

Wheeler, G. S. 1965. Memorandum 2400, 5100. July 1, 1965. Laconia, N.H. U.S. Forest Service Northern FPC Zone.

ALBIOS: A Comparison of Aluminum Biogeochemistry in Forested Watersheds Exposed to Acidic Deposition

C.S. Cronan* and R.A. Goldstein†

Abstract

This chapter presents a case study of the broad interregional patterns of aluminum biogeochemistry and aluminum toxicity in the forest landscapes of North America and northern Europe. Sulfur deposition at the 14 ALBIOS study catchments ranged 20-fold from approximately 4 kg S ha^{-1}yr^{-1} at the Experimental Lakes Area, Ontario, to >80 kg S ha^{-1}yr^{-1} at Solling, West Germany. On a regional basis, the lowest total monomeric aluminum (MAL) concentrations (0–10 μM) were found in watersheds of the southeastern and midwestern USA, which contain soils characterized by high soil percent base saturation and/or a large sulfate adsorption capacity. Intermediate MAL concentrations (15–80 μM) were found in soil solutions of northeastern North America and northern Europe, where soils are characterized by low percent base saturation and low sulfate adsorption capacity. The highest MAL concentrations (up to 240 μM) were observed in the soil solutions of a West German spruce stand. Headwater streams in the study catchments contained MAL concentrations that ranged from less than 1 μM to peak values of approximately 55 μM in one West German stream.

Two dominant geochemical patterns were observed in most watersheds: (1) upper soil horizon and wetland zones characterized by aluminum adsorption-desorption reactions on solid-phase humic materials; and (2) mineral soil horizon and groundwater zones dominated by aluminum solubility relationships with some form of Al(OH)$_3$. Much of the overall variation in aquo aluminum ion activity could be explained on the basis of relatively simple equilibrium pH-solubility and adsorption models.

The ALBIOS evidence indicated that the relationship between watershed inputs of H$_2$SO$_4$ or HNO$_3$ and outputs of soluble aluminum is not necessarily simple and straightforward. However, for those watersheds characterized by aluminum-saturated soils and low retention of strong acid anions, increased concentrations and fluxes of sulfate and nitrate in soil water were accompanied by increased concentrations and fluxes of soluble aluminum, both on a broad geographic basis and on a single catchment basis.

Experimental plant response studies showed that honey locust (*Gleditsia*

---

*Department of Botany and Plant Pathology, University of Maine, Orono, ME 04469, USA.

†Environmental Science Department, Electric Power Research Institute, 3412 Hillview Avenue, Palo Alto, CA 94303, USA.

triacanthos), red spruce (*Picea rubens*), sugar maple (*Acer saccharum*), American beech (*Fagus grandifolia*), red oak (*Quercus rubra*), and loblolly pine (*Pinus taeda*) could be grouped into sensitive, moderately sensitive, and insensitive classes on the basis of growth and nutritional responses to soluble aluminum. The toxicity thresholds for the sensitive and moderately sensitive tree species were within range of the peak concentrations of soluble aluminum observed in soil solutions at some of the northern and European watersheds. Likewise, soluble aluminum concentrations in many headwater streams were in excess of toxicity thresholds for fish species like brown trout (*Salmo trutta* L.) and brook trout (*Salvelinus fontinalis*). As such, it is likely that aluminum toxicity serves as a contributing stress factor in these kinds of northern watersheds.

I. Introduction

Much has been written about the biogeochemical linkages between acidic deposition and aluminum chemistry in natural systems. Initially, aluminum was identified as an important buffer system in terrestrial catchments exposed to acidic deposition (Norton, 1976; Johnson, 1979; Cronan and Schofield, 1979; Driscoll and Bisogni, 1984). Then, aluminum in acidified surface waters was noted to cause toxicity in some fish populations (Baker and Schofield, 1980). Finally, Ulrich and associates (1980) hypothesized that forest dieback in German spruce stands could be attributed to soil aluminum toxicity associated with acidic deposition. Because of this interest in the potential toxicity of aluminum in catchments exposed to acidic deposition, there have been renewed efforts to understand aluminum biogeochemistry in natural ecosystems.

Unfortunately, much remains to be learned about the patterns of occurrence, mobility, and toxicity of aluminum in the environment. Although the chemistry of aluminum has been studied intensively over the past several decades, no substantive attempts have been made to synthesize available data and theory into a general model that would allow one to predict the effects of acidic deposition on aluminum solubility and transport in a range of contrasting environments. Moreover, there are few complete sets of data that would allow one to test for relationships between atmospheric deposition of strong acids, solid-phase aluminum chemistry, and aqueous aluminum chemistry in any given watershed ecosystem.

Here we present an interregional case study that was designed specifically to examine the general patterns of aluminum biogeochemistry and aluminum toxicity in forested ecosystems exposed to acidic deposition. The overall study objectives were: (1) to quantify the regional patterns of aluminum mobilization, chemistry, and transport in areas receiving acidic deposition; and (2) to evaluate the potential for aluminum toxicity in the forests of eastern North America and northern Europe. The investigation was conducted by an interdisciplinary team of 18 scientists.

The geochemical and plant toxicity components of ALBIOS (aluminum in the biosphere) were organized into complementary field and laboratory studies to address two hypotheses: (1) acidic deposition increases the concentrations and transport of soluble aluminum in soils and surface waters of forested watersheds; and (2) in forest ecosystems, acidic deposition may increase bioavailable Al to levels that are toxic to trees and aquatic biota, causing growth reductions, nutritional deficiencies, or mortality.

II. Study Sites

The ALBIOS investigation involved a comparison of aluminum biogeochemistry in ten North American and four European forested catchments (Table 4-1 and Figure 4-1). These study sites were selected to provide a broad range of contrasts in key environmental parameters, including atmospheric deposition, climate, soil properties, bedrock and surficial geology, and forest cover. Sulfur deposition at these watersheds ranged 20-fold from approximately 4 kg S ha^{-1}yr^{-1} at the Experimental Lakes Area, Ontario, to >80 kg S ha^{-1}yr^{-1} at Solling, West Germany. There were similar ranges of variation for such factors as climate (Table 4-1), soil texture (Table 4-2), soil mineralogy (Table 4-3), soil permeability (Figure 4-2), soil pH (Figure 4-3), soil cation exchange capacity, and soil percent base saturation (Figure 4-3).

Table 4-1. The 14 ALBIOS study watersheds.

Watershed	Location	Mean annual T (°C)	Mean annual rainfall (cm)	Estimated S deposition[b] (kg ha^{-1})
ELA[a]	Ontario	3	60	4
PL	Ontario	4	110	11
HF	New York	5	105	8
BM	New York	5	125	18
RL	Wisconsin	5	85	6
UMF	Missouri	15	115	8
OX	Mississippi	17	140	8
CB	Tennessee	13	140	31
CO	North Carolina	12	200	9
OR	Virginia	13	105	12
BI	Norway	5	140	23
GR	Sweden	5	126	28
LB	West Germany	6	130	46
SL	West Germany	6.5	110	88

[a] ELA = Experimental Lakes Area; PL = Plastic Lake; HF = Huntington Forest; BM = Big Moose/Pancake-Hall; RL = Round Lake; UMF = University Forest; OX = Oxford Pine; CB = Camp Branch; CO = Coweeta WS40; OR = Old Rag Mt.; BI = Birkenes; GR = Lake Gardsjon F1; LB = Lange Bramke; SL = Solling spruce stand.

[b] Deposition estimate includes wet + dry.

Figure 4-1. Location of ALBIOS study sites. The site codes are explained in Table 4-1.

Table 4-2. Average soil particle size distributions for the ALBIOS watersheds.

Watershed	% Gravel	% Sand	% Silt	% Clay	Texture
Glaciated Sites[a]					
ELA	24	43	27	6	Sandy loam
Plastic Lake	11	50	37	2	Sandy loam
Huntington Forest	15	67	15	2	Loamy sand
Big Moose	8	61	28	3	Sandy loam
Round Lake	1	93	4	2	Sand
Birkenes	17	66	15	2	Loamy sand
Gardsjon	5	44	47	4	Sandy loam
Unglaciated Sites					
University Forest	22	40	18	20	Sandy clay loam
Oxford Pine	0	41	42	18	Loam
Camp Branch	6	23	48	22	Silt loam
Coweeta	3	50	31	16	Loam
Old Rag Mt.	9	39	29	23	Loam
Lange Bramke	53	23	19	5	Loam
Solling	5	21	54	20	Silt loam

[a] Data from R. Newton, Smith College.

III. Methods

At each catchment, four to seven plots were established along topographic and/or drainage gradients, and soils were sampled by horizon for detailed physical-chemical analysis. Solution chemical samplers and collection points were established for collecting wet deposition, canopy throughfall, stream water, and soil solutions in the O, B, and BC or C horizons. Replicated plastic tension and tension-free lysimeters were used for soil water collections. Most watersheds were sampled on a storm event basis during the spring or fall between 1983 to 1986. After each collection, solution samples were immediately processed for aluminum speciation (Figure 4-4) and were then shipped on ice to the University of Maine for analysis of inorganic solutes by ion chromatography and flame atomic absorption spectroscopy and for dissolved organic C by persulfate oxidation and IR analysis (Cronan, 1985). Aluminum species in solution were separated using the method of Driscoll (1984) into the following operationally defined categories: total monomeric aluminum (MAL), organic monomeric aluminum (MOAL), total organic aluminum (TOAL), and total acid-soluble aluminum (TAL). Free and complexed inorganic aluminum species were calculated using the solution chemistry data and a chemical equilibrium model calibrated with thermodynamic constants from Nordstrom and associates (1984).

   Soil chemical samples were analyzed using the selective extractants illustrated in Figure 4-5. Soil mineralogy was determined using x-ray diffraction analysis; bulk chemistry was determined by x-ray fluorescence. Tree foliage and root tissues were also sampled by species at most sites and were analyzed, following

Table 4-3. Comparison of bulk soil mineralogy in the ALBIOS forest soils.

	ELA	PL	GR	BI	SL	LB	BM	HF	RL	OR	OX	UMF	CB	CO
Major Minerals														
Quartz	X	X	X	X	X	X	X	X	X	X	X	X	X	X
Plagioclase	X	X	X	X			X	X						X
K-feldspar	X	X	X	X	X	X	X	X	X	X				
Mica					X	X						X	X	X
Cristobalite											X			
Chert												X		
Kaolinite													X	
Accessory Minerals														
Hornblende	X	X	X	X			X	X						
Mica			X	X										
Kaolinite													X	
Ilmenite							X	X						
Magnetite							X	X						
Amorphous Fe_2O_3									X				X	X
Gibbsite														X
Ferrihydrite														X
Lepidocrocite													X	
Trace Amounts														
Plagioclase					X	X			X	X	X			
Hornblende										X				

R. April, Colgate University unpublished data.

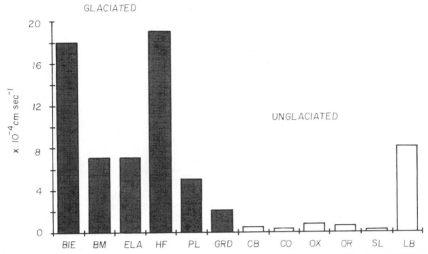

Figure 4-2. Comparative soil permeability for the upper 50 cm of soil in the ALBIOS watersheds (R. Newton, unpublished data).

nitric-perchloric acid digestion, for element concentrations of Al, P, Ca, and Mg, using ICP-AES spectroscopy.

Controlled laboratory experiments and field studies were used to examine aluminum toxicity responses for a number of major tree species. These studies included hydroponic studies of aluminum toxicity thresholds (Thornton et al., 1986a, 1986b, 1987), greenhouse potted seedling experiments (Joslin, 1987), and field root ingrowth core experiments (Joslin, 1987).

IV. Results and Discussion

A. Interregional Differences in Aluminum Geochemistry

The field results showed broad interregional differences in the aqueous aluminum concentrations of drainage waters in forested catchments exposed to acidic deposition. Table 4-4 illustrates that peak MAL concentrations in soil waters varied significantly across the ALBIOS geographic gradient, ranging from <1 μM in parts of the southeastern USA to ≥ 240 μM in West German spruce stands. Peak MAL values are given as an index of the highest short-term exposure levels that might be encountered by a tree root in each system. Compared to soil waters, first-order streams typically contained lower MAL concentrations, ranging from <5 μM in several northern and southern streams to peak values of approximately 55 μM in one West German headwater stream. Table 4-4 also contains estimates for Al^{3+} ion activity in each soil and stream water sample. It has been suggested that this chemical species is particularly toxic to organisms (Pavan

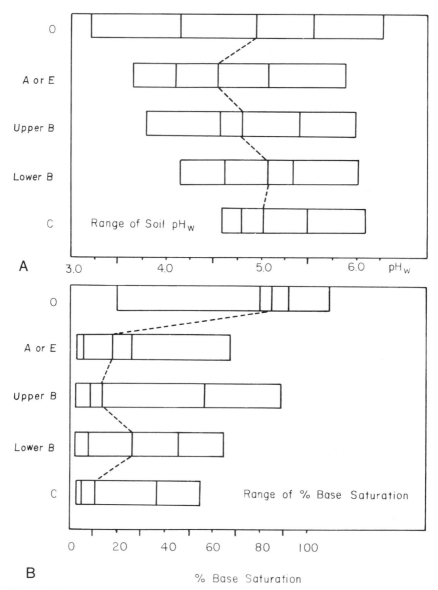

Figure 4-3. Ranges of soil:water pH values and percent base saturation by horizon in the ALBIOS catchments. Quartiles are indicated and medians are connected by a dashed line.

and Bingham, 1982). As indicated in the table, Al^{3+} activities were generally much lower than MAL or TAL concentrations, ranging from 0 to 104 μM in soil solutions and from 0 to 34 μM in headwater stream samples.

One can note a number of other important points in Table 4-4. First, the trend of increasing MAL in soil water samples generally correlated with a gradient of increasing soil solution sulfate and ionic strength across the sites. In contrast, the

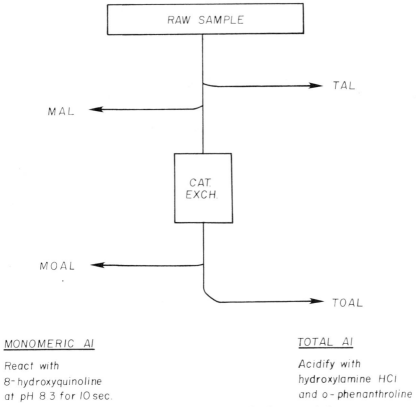

Figure 4-4. Flow sheet for aqueous aluminum speciation.

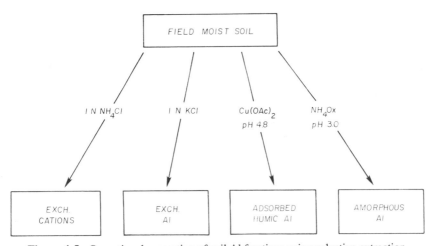

Figure 4-5. Operational separation of soil Al fractions using selective extraction.

Table 4-4. Comparison of solution chemistries for the ALBIOS watersheds. Examples show chemistry of individual soil solution and stream samples exhibiting the maximum MAL concentration for the given catchment.

Site	Max. MAL[a] (μmol/L)	Max. TAL[b] (μmol/L)	$[Al^{3+}]$[c] (μmol/L)	pH	SO_4^{2-} (μmol/L)	Ca^{2+} (μmol/L)	Ionic strength (M)
			Soil Water[d]				
CO	0.1	0.7	0	5.6	45	22	.0002
OR	3.0	4.8	0.4	5.7	20	32	.0002
OX	4.4	23.7	0.3	5.6	50	37	.0004
UMF	5.9	8.5	1.4	5.3	65	25	.0003
CB	10.4	27.8	0	4.3	50	22	.0003
ELA	16.7	25.2	11.0	4.6	55	30	.0003
PL	33.3	49.3	17.0	4.6	105	30	.0004
BIE	42.4	44.7	27.2	4.6	135	14	.0005
HF	52.6	64.1	25.0	4.7	125	40	.0006
BM	60.4	102.2	33.0	4.7	135	25	.0006
GRD	68.1	74.4	31.0	4.6	230	20	.0013
LB	81.9	170.7	45.0	4.2	395	27	.0015
SL	240.0	368.9	104.0	4.2	555	65	.0026
			Stream Water				
CO	nd	0.7	nd	6.60	23	16	.0002
HF	2.1	3.3	0.2	6.10	67	101	.0005
OR	2.3	6.1	0	7.20	16	23	.0002
UMF	3.7	10.7	0	5.24	72	27	.0004
PL	6.5	12.4	4.1	4.32	95	70	.0005
CB	6.6	10.8	1.6	4.65	28	11	.0002
BIE	19.0	33.1	11.5	4.35	72	19	.0004
GR	21.3	27.0	6.7	4.08	129	29	.0009
BM	25.3	29.9	15.9	4.74	72	35	.0004
LB	55.1	72.7	33.8	4.48	166	60	.0009

[a] Maximum monomeric Al.

[b] Maximum total acid-soluble Al.

[c] Al^{3+} activity calculated using temperature, solution chemistry data, and thermodynamic computations.

[d] There was no soil water collection at RL.

stream samples in Table 4-4 showed no apparent consistent relationship between MAL and sulfate concentration or ionic strength. The data also illustrate that soil water calcium concentrations were surprisingly similar across watersheds and that stream water calcium concentrations were quite variable. From the perspective of root growth in the soil zone, this means that tree roots are exposed to decreasing solution Ca/Al ratios across the geographic gradient.

Table 4-5 contains data comparing the observed differences in peak MAL

Table 4-5. Interregional variations in peak monomeric aluminum concentrations by ecosystem strata (concentrations are in μmol/L).

Ecosystem level	Southeastern ultisols and alfisols	Northern spodosols	German inceptisols
Throughfall	3	4	16
O Horizon	10	32	138
B Horizon	9	60	136
BC/C Horizon	5	76	240
Streams	7	25	55

concentrations by region and ecosystem strata. In general, peak MAL concentrations were very low in southeastern catchments, intermediate in north temperate catchments, and high in the West German catchments. Maximum MAL concentrations decreased from O to B to C horizons in southeastern Ultisols, Inceptisols, and Alfisols, and were generally ≤ 10 μM for all soil solutions. An exception to this was found in a high-elevation spruce site in the Smoky Mountains, Tennessee. Soil solutions in that Inceptisol contained up to 40 to 50 μM MAL. In Spodosols of eastern North America and Scandinavia, maximum MAL concentrations peaked in the Bs horizon, and mineral soil solutions typically contained ≤ 70 μM MAL. It is important to emphasize that although Bs horizon solutions were typically much less acidic than O horizon solutions in the northern Spodosols, the Bs horizon samples usually contained much higher MAL concentrations. Overall, the highest concentrations of monomeric aluminum were found at the West German sites. For those catchments, peak MAL concentrations occurred in the Bs3 or BC horizons and generally ranged from 100 to 240 μM.

Several important differences were noted in the chemical speciation of aqueous aluminum between sites and ecosystem strata. Some of the differences in the operationally defined aqueous aluminum species—MAL, MOAL, TOAL, and TAL—are presented in Figure 4-6 for samples from the southeastern, northern, and German regions. At the Camp Branch catchment, solution samples contained low concentrations of aluminum and were dominated by the acid-soluble (TAL-MAL) and organic monomeric (MOAL) fractions. Samples from the Big Moose watershed contained much higher concentrations of aluminum, and monomeric aluminum was a major component of the total aluminum in all ecosystem strata. Finally, the samples from Lange Bramke exhibited the greatest extremes, with some of the lowest aluminum concentrations in the O horizon and downstream samples and the highest aluminum concentrations in the B horizon and headwater samples. All Lange Bramke samples were characterized by a low percentage of organic monomeric aluminum (MOAL).

The graphs also illustrate that (1) the MOAL/MAL ratio tended to decrease between O and B horizons; (2) concentrations of inorganic monomeric aluminum (MAL-MOAL) were highest in the O and A horizons of southern soils and the B horizons of northern and West German soils; and (3) aluminum speciation in

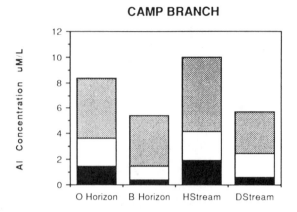

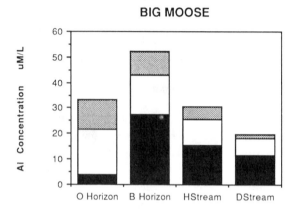

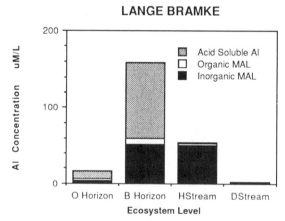

Figure 4-6. Comparison of operationally defined aluminum fractions for O/A horizon leachates, B horizon leachates, headwater stream samples, and downstream samples from southeastern, northern, and German ALBIOS catchments.

headwater streams tended to vary between sites because of different hydrologic conditions. Using thermodynamic calculations, it was also determined that the major aqueous species in all B horizons were Al^{3+} and the hydrolyzed inorganic aluminum species.

1. Solid-Phase Aluminum Chemistry

One of the concerns in ALBIOS was to determine the interregional variations in solid-phase soil aluminum chemistry and the extent to which variations in aqueous aluminum chemistry are controlled by differences in soil chemistry. Table 4-6 presents a comparison of the major aluminosilicate secondary minerals for the watersheds in each of the ALBIOS study regions. Both vermiculite and kaolinite were common to all regions. At several northern sites, the vermiculite contained hydroxy-Al interlayers that may have exerted solubility control on aqueous Al^{3+}. The southern sites were distinguished by large amounts of total clay-sized particles (>16–20% by weight), relatively larger amounts of gibbsite, and, in some cases, abundant illite. In comparison, the German sites contained larger amounts of mixed-layer illite-vermiculite and illite.

Selective extractions were used to quantify and to compare several operationally defined, partially overlapping soil aluminum pools in the ALBIOS study soils (Figure 4-5). Results showed that median values for soil-exchangeable aluminum, humic aluminum, and amorphous hydroxy-aluminum were highest in the northern and German forest soils (Table 4-7). Although the selective extractions did not provide direct predictions for MAL or Al^{3+} in the different soils, the percent aluminum saturation of organic or mineral horizons could be used to help predict solution aluminum in many soils. For organic horizons, the percent aluminum saturation or bound aluminum ratio (Cronan et al., 1986) was calculated as the ratio of Cu-extractable Al to titratable carboxyl content. For mineral horizons, the

Table 4-6. Comparison of major secondary minerals in soils of the ALBIOS watersheds. Data are summarized by region.

Region	Major secondary minerals[a]
Southeastern USA	Vermiculite
	Kaolinite
	Gibbsite
	Illite
Upper Midwest	Vermiculite
	Kaolinite
Northeastern North America	Vermiculite
and Scandinavia	Kaolinite
Germany	Vermiculite
	Illite
	Mixed layer illite/vermiculite
	Kaolinite

[a] Data from R. April, Colgate University.

Table 4-7. Interregional variations in soil-extractable aluminum fractions. Median values are given in meq Al^{3+} 100 g^{-1}.

Horizon[a,b]	Southern	Midwest	Northern	German
Exchangeable Al				
O	2.7 (0.3)	1.7 (0.8)	3.6 (1.0)	6.6 (0.9)
B2	2.4 (0.2)	0.8 (0.2)	3.6 (0.4)	6.4 (0.9)
C	2.1 (0.3)	0.1 (0.002)	0.8 (0.2)	1.2 (0.002)
Humic Al				
O	3.3 (0.7)	2.6 (0.6)	5.1 (4.4)	5.4 (1.0)
B2	1.6 (0.2)	1.2 (0.3)	16.9 (12.6)	4.2 (0.8)
C	1.3 (0.2)	0.2 (0.003)	3.7 (1.5)	2.1 (0.0)
Amorphous Al				
O	16.7 (7.3)	11.6 (1.6)	15.0 (8.8)	14.7 (2.4)
B2	43.9 (4.8)	28.2 (6.5)	131.9 (72.9)	56.8 (14.7)
C	17.3 (6.9)	8.9 (0.1)	56.5 (12.6)	29.5 (0.0)

[a] Data from R. Bartlett, University of Vermont, and first author.

[b] For reference, standard errors of the mean are provided in parentheses.

ratio was calculated as KCl-exchangeable Al divided by the effective CEC (cation exchange capacity).

B. Influence of Acidic Deposition and Other Factors on Aluminum Chemistry

One of the principal challenges for the ALBIOS investigation was to elucidate the environmental and geochemical factors underlying the interregional variations in solution aluminum chemistry. It was hypothesized initially that aluminum mobilization and transport in forest soils are strongly influenced by acidic deposition and that aluminum leaching increases with increasing soil inputs of strong acid. However, field evidence indicated that the relationship between inputs of H_2SO_4 or HNO_3 and outputs of aluminum is not necessarily simple and straightforward. As in the earlier Integrated Lake-Watershed Acidification Study (Goldstein et al., 1985), the ALBIOS results demonstrated that aluminum leaching is a function of several interacting factors. Hence, there was considerable variability between watersheds in terms of the sensitivity and magnitude of the aluminum leaching response to atmospheric sulfate. In general, it was shown that the impact of variable strong acid inputs on aluminum mobilization and transport depends upon the interplay of two major factors: (1) the mechanisms and rates of alkalinity generation in a given watershed (through base cation exchange, aluminum desorption from clays, aluminum desorption from humic material, dissolution of $Al(OH)_3$ or similar solid, mineral weathering, sulfate adsorption, nitrate immobilization, denitrification, or sulfate reduction); and (2) the hydrologic characteristics of the catchment. For those soils dominated by aluminum buffering and characterized by low retention of strong acid anions, there was a strong correlation between increased fluxes of sulfate and/or nitrate from atmospheric deposition and increased aluminum leaching (Figure 4-7). Typically, these soils had pH_w values below 4.9 and soil base saturation levels $\leq 15\%$. In contrast, there was little or no aluminum leaching response from the ALBIOS forest soils characterized by high base saturation (Coweeta, Round Lake, Oxford, University Forest, and Old Rag Mt.) or pronounced sulfate adsorption (Camp Branch).

As shown in Figure 4-8, there was a clear relationship between solution pH and Al^{3+} activity across the diverse field sites. Thus, more acidic solutions tended to have higher MAL concentrations and Al^{3+} activities than less acidic solutions. In many respects, the samples followed the kind of pH-solubility relationship that one would expect for solutions in equilibrium with an aluminum trihydroxide solid phase. The major exceptions to this pattern were the acidic solutions from surface organic horizons and wetlands that contained much lower Al concentrations than respective mineral soil horizons.

Samples from the ALBIOS catchments also exhibited a strong positive relationship between solution ionic strength and Al^{3+} activity (Figure 4-9). In general, as one followed a geographic gradient across the southeastern, northern, and German watersheds, there was a trend of increasing strong anion flux through the soil, increasing soil solution ionic strength, and increasing Al^{3+} activity in soil water. This evidence is consistent with the hypothesis of Reuss (1983), James and Riha (1986), and others suggesting that much of the aluminum leaching response in acidic forest soils results from the "salt effect" or ionic strength effect of acidic deposition. In mechanistic terms, this suggests that the short-term release of aluminum to mineral horizon soil solutions occurs through ion exchange rather than dissolution processes.

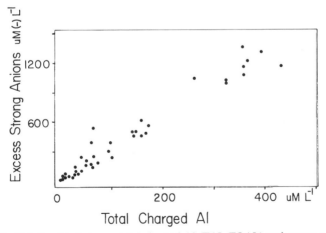

Figure 4-7. Relationship between total charged Al (TAL-TOAL) and excess strong acid anions $(C_A - C_B - H^+)$ for mineral horizon soil solutions. The relationship can be described by the following equation:

$$y = 4.17 + 3.22x \quad (r^2 = 0.96).$$

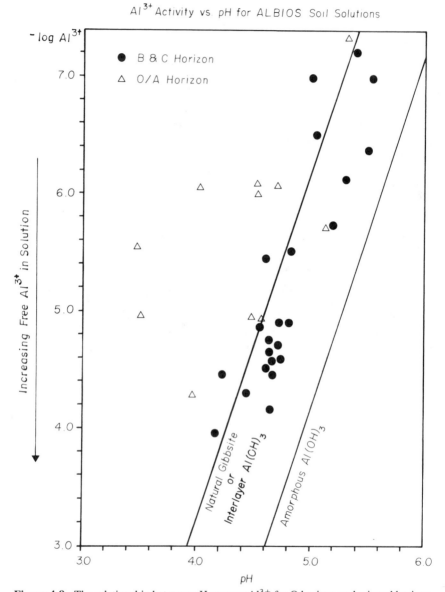

Figure 4-8. The relationship between pH versus pAl^{3+} for O horizon and mineral horizon soil solutions.

In organic surface horizons and wetland source areas, Al^{3+} activity was apparently controlled by adsorption-desorption reactions with humic carboxyl groups (Cronan et al., 1986). By combining titration estimates of carboxyl content in the soil horizon with estimates of copper-extractable aluminum in soil organic matter, it was possible to calculate a carboxyl-bound aluminum ratio that was

Figure 4-9. Relationship between −log ionic strength and
pAl^{3+} for ALBIOS mineral horizon soil solutions.

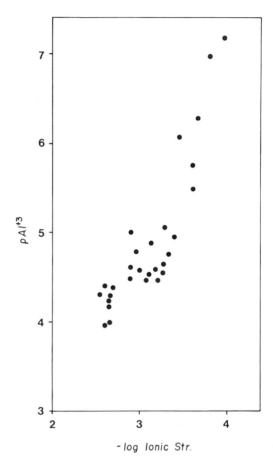

positively correlated with aqueous Al^{3+} activity. Thus, forest floors with larger
bound-aluminum ratios contained higher Al^{3+} activities in solution.

C. Conceptual and Predictive Models

The final objective of the geochemical portion of the ALBIOS investigation was to
develop improved models for predicting aqueous aluminum chemistry in forested
watersheds (Gherini et al., 1985; Cronan et al., 1986; Schecher and Driscoll,
1987). Although it was originally suspected that there might be important kinetic
constraints on some of the aqueous complexation or solution-solid phase reactions
involving aluminum, the ALBIOS experimental kinetic studies generally supported the validity of applying an equilibrium modeling approach to the chemistry
and transport of dissolved aluminum (Plankey et al., 1986; Plankey and Patterson,
1987; Walker et al., 1988). Overall, it was concluded that much of the variation in
solution aluminum chemistry could be explained on the basis of one or more

simple chemical equilibrium models, coupled with appropriate hydrologic and general soil chemical models.

At a conceptual level, two dominant geochemical patterns emerged in the ALBIOS study catchments: (1) zones characterized by aluminum adsorption-desorption reactions with solid-phase soil humic materials, and (2) zones dominated by aluminum solubility relationships with some form of $Al(OH)_3$. These zones roughly correspond to surface organic soil horizons and peaty hydrologic source areas versus lower mineral soil horizons and groundwater reservoirs. Figure 4-8 shows the pH-pAl^{3+} relationships for samples from these two different types of source zones and illustrates that many of the data points from mineral horizons and groundwater-dominated streams fell near or between the solubility line for gibbsite (or interlayer Al-hydroxide) and amorphous $Al(OH)_3$. This general pattern implies that $[Al^{+3}]$ in these waters may be largely controlled by an aluminum trihydroxide solid phase. It is important to note, however, that this pattern does not necessarily mean that dissolution reactions are the immediate source of dissolved inorganic aluminum in this zone. Given the evidence presented in Figure 4-9 for the relationship between $[Al^{3+}]$ and ionic strength and the kinetic differences between $Al(OH)_3$ dissolution (Bloom, 1983) and Al^{3+} exchange on clays (Walker et al., 1988), it seems likely that exchangeable aluminum is the important short-term source of aqueous aluminum in the mineral soil. The aluminum trihydroxide phases probably serve as a longer-term resupply reservoir for exchangeable aluminum and as the key solid phase controlling Al^{3+} solubility. For the organic zone, Cronan and associates (1986) proposed that the pH-pAl^{3+} solubility relationship is controlled by equilibrium partitioning of Al^{3+} between the solution phase and solid phase soil organic matter. This results in a different solubility relationship than that described for the aluminum trihydroxide system (Figure 4-10).

In general, the ALBIOS results indicated that much of the spatial and temporal variability in aqueous aluminum chemistry could be accounted for using an equilibrium model that combines a solid-phase humic adsorption reaction and an aluminum trihydroxide solubility relationship. For organic surface horizons and peaty hydrologic source areas, aquo aluminum ion activity could be modeled using the following expression:

$$pAl^{3+} = b(pH) + c \tag{1}$$

where b is the slope $= 1.05 \times$ (bound Al ratio) $+ 0.345$, pH refers to the solution pH, and C is the intercept $= -(5.47 \times$ bound Al ratio) $+ 3.879$. In order to model aquo aluminum ion activity in mineral horizons and groundwater zones, the following aluminum trihydroxide solubility expression was used:

$$\log [Al^{3+}] = 8.77 - 3pH \tag{2}$$

where 8.77 corresponds to the average solubility product, log K, for natural gibbsite or interlayer aluminum hydroxide. Finally, streams were modeled by combining these chemical models with accurate hydrologic flux estimates derived from the ILWAS model (Gherini et al., 1985).

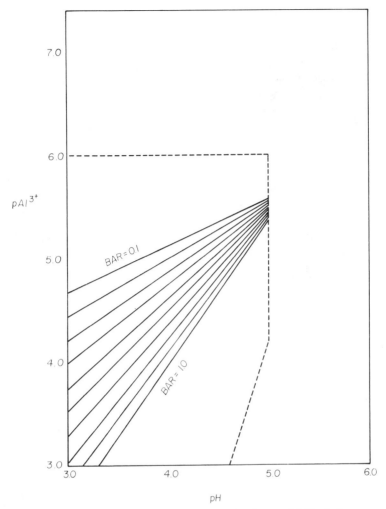

Figure 4-10. Theoretical stability field for apparent aluminum solubility in the presence of soil humic material in the pH range 3.0 to 5.0. The diagonal lines show the predicted pAl^{3+} at bound aluminum ratios (BAR) ranging from 0.1 to 1.0. The BAR is calculated as the copper-extractable Al divided by the carboxyl content of the organic matter.

D. Relationships between Aluminum Chemistry and Tree Toxicity

One of the goals of the ALBIOS investigation was to link the geochemical results with an analysis of tree sensitivity to aluminum. Aluminum toxicity has been recognized as a significant factor limiting the growth of many plant species (Foy, 1984). The toxicity of aluminum to trees varies with such environmental factors as solution pH, the speciation of aluminum, the concentration of Ca and Mg in solution, the overall ionic strength of the medium, the form of inorganic nitrogen in the soil solution, mycorrhizal interactions, temperature, soil moisture, the nutrient status of the plant, and the species and genetic stock of the plant. In general, studies have shown that the monomeric species of inorganic aluminum (Al^{3+}, $AlOH^{2+}$, and $Al(OH)_2^+$) are the major phytotoxic forms of aluminum. Polymeric aluminum may be equally as or more toxic than the monomeric species; however, it is not clear how common polymers are in the rhizosphere.

Aluminum toxicity may occur as a result of a number of specific cellular or biochemical mechanisms (Haug, 1984). In vitro studies suggest that the major mechanisms of aluminum toxicity are related to the strong interactions of aluminum with orthophosphate groups of organophosphates and the interference of aluminum with calcium metabolism. These interactions may have the following consequences: (1) aluminum can bind to nucleotides and nucleic acids, inhibiting cell division; (2) aluminum can bind to ATP, ADP, or membrane-bound ATPases, interfering with energy transfer; (3) aluminum may interfere with enzyme systems like acid phosphatases; (4) aluminum may bind to calmodulin and other calcium peptides, seriously disrupting cellular control and contractile processes; (5) aluminum can bind to phospholipid groups and alter membrane permeability; and (6) aluminum can interfere with membrane selectivity for nutrient ions. Thus, aluminum toxicity can potentially affect energy transformations, cell division, membrane transport, and nutrient accumulation, along with additional activities regulated by calmodulin.

The visible or measurable symptoms of aluminum toxicity in laboratory or field trees include reduced root elongation and branching, a generalized thickening of the roots, and necrosis of cells near the root and shoot meristems. Other possible symptoms of aluminum toxicity are reductions in aboveground growth and reduced uptake of Ca, Mg, P, and other essential nutrients.

Experiments were conducted by the ALBIOS investigators in order to develop toxicity response surfaces relating the concentration of soluble aluminum to the growth and mineral status of representative tree species from the ALBIOS watersheds (Thornton et al., 1986; Thornton et al., 1987; Joslin, 1987; Zhao et al., 1987). In designing these studies, efforts were made to address two important concerns: (1) the influence of experimental conditions on the quantification of aluminum sensitivity in a given species; and (2) the need to be able to extrapolate from the laboratory to the natural field system. By including solution culture, soil culture, and field ingrowth core manipulations, the ALBIOS experiments were intended to bridge the gap between laboratory and field conditions and to provide cross-comparisons between toxicity thresholds determined with different experimental approaches.

Results from the plant response studies are plotted in Figure 4-11, along with data illustrating the range of soluble aluminum concentrations observed among the ALBIOS watersheds. Several points are evident from the data. First, tree species differed significantly with respect to aluminum sensitivity, ranging from sensitive to tolerant. As noted in the figure, nutritional effects from aluminum exposure generally occurred at lower soluble aluminum concentrations than those associated with growth reductions. Comparisons between studies indicated that there

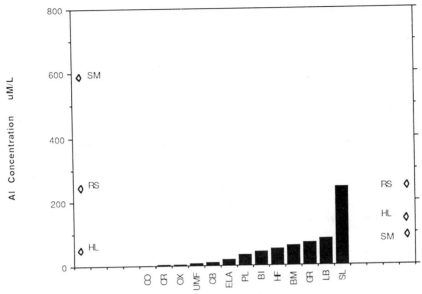

Figure 4-11. Al toxicity thresholds in relation to maximum soluble aluminum concentrations observed in soil solutions at the ALBIOS watersheds. The plant toxicity thresholds on the left axis represent the treatment conditions at which root biomass or total plant biomass decreased significantly compared to a control; the thresholds on the right axis indicate the aluminum concentration where there was a significant effect on plant tissue nutrient concentrations. Tree species include sugar maple (SM), red spruce (RS), and honey locust (HL). The thresholds for loblolly pine and red oak were above 1000 μmol/L.

were variable differences in aluminum toxicity thresholds determined by separate experimental approaches. For example, most of the solution and soil culture experiments demonstrated consistently that significant growth reductions in red spruce occurred at a soluble aluminum concentration of 200 to 300 μM. By contrast, there was considerable variability in the different estimates of the toxicity threshold for red oak.

Overall, we would expect an increased probability of forest impacts from aluminum toxicity under the following conditions: in forests containing sensitive or moderately sensitive species (e.g. in mixed hardwood forests that include red spruce or sugar maple); in situations where much of the fine root biomass is concentrated in the aluminum-rich mineral horizons; in ecosystems characterized by significant soil aluminum buffering and high inputs of acidic deposition; and in forest systems experiencing drought stress and increased dependence on deeper rooting. The consequences of aluminum stress and toxicity might vary with the situation. In some cases, exposure to elevated concentrations of soluble aluminum might serve to exclude roots from certain soil layers or might effectively cause root pruning and increased fine root turnover. In other cases, the chronic effects of aluminum toxicity might result in decreased forest growth or gradual selection of more aluminum-tolerant species or clones.

V. Conclusions

In general, MAL concentrations and solution Al^{3+} activities were greatest in soil solutions and surface waters at sites with higher sulfur deposition and soil sulfate fluxes. On a regional basis, the lowest MAL concentrations were found in the southeastern and midwestern watersheds characterized by high soil percent base saturation and/or significant sulfate adsorption. Intermediate MAL concentrations were found in the Spodosols of northeastern North America and northern Europe characterized by low percent base saturation and low sulfate adsorption capacity. Highest MAL concentrations were found in a West German Inceptisol. Much of the variation in aquo aluminum ion activity could be explained on the basis of relatively simple equilibrium pH-solubility and adsorption models.

The ALBIOS results have indicated that watersheds with certain soil chemical and hydrologic characteristics exhibit a positive relationship between solution fluxes of strong acid anions and soluble aluminum. For these sites, there is a clear link between acidic deposition and aluminum mobilization. In addition, the soil culture studies showed that root aluminum concentrations and plant toxicity responses were most strongly correlated with soluble aluminum in soil water or soil paste extracts. This, then, suggests that aluminum availability to plants is more directly a function of the concentration or activity of aluminum in solution than of the concentration of solid-phase exchangeable aluminum. Based upon this evidence, the following conclusion emerges. There are certain types of watersheds (notably the catchments containing soils with <10–15% base saturation) where soluble aluminum increases with increasing acidic deposition. This then increases the availability of aluminum to plants and aquatic life and presumably increases the incidence of aluminum toxicity to these organisms.

Acknowledgments

This paper is a contribution from the ALBIOS project, an interdisciplinary research investigation supported by contract RP2365-01 from the Electric Power Research Institute. The authors wish to express their gratitude to the following co-investigators who contributed to the success of ALBIOS: Richard April, Richmond Bartlett, Paul Bloom, Charles Driscoll, Susan Erich, Steve Gherini, Gray Henderson, Dev Joslin, Mike Kelly, Robert Newton, Rod Parnell, Howard Patterson, Dudley Raynal, Michail Schaedle, Carl Schofield, Edward Sucoff, Herbert Tepper, Frank Thornton, and William Walker. Betty Lee kindly handled the graphics and word processing. The authors also thank all of the site cooperators who made the field studies possible.

References

Baker, J. P., and C. L. Schofield. 1980. Water, Air, Soil Pollut 18:289–309.

Bloom, P. R. 1983. Soil Sci Soc Amer J 47:164–168.

Cronan, C. S. 1985. Water, Air, Soil Pollut 26:355–371.

Cronan, C. S., W. J. Walker, and P. R. Bloom. 1986. Nature 324:140–143.

Cronan, C. S., and C. L. Schofield. 1979. Science 204:304–306.

Driscoll, C. T. 1984. Int. J. Envir. Anal. Chem. 16:267–284.

Driscoll, C. T., and J. J. Bisogni. 1984. *In* J. L. Schnoor, ed. *Modelling of total acid precipitation impacts,* Butterworth, Boston, MA.

Foy, C. D. 1984. *In* Adams, F., ed. *Soil Acidity and Liming,* 57. American Society of Agronomy, Madison, WI.

Gherini, S., L. Mok, R. J. M. Hudson, G. F. Davis, C. W. Chen, and R. A. Goldstein. 1985. Water, Air, Soil Pollut 26:425–459.

Goldstein, R. A., C. W. Chen, and S. A. Gherini. 1985. Water, Air, Soil Pollut 26:327–337.

Haug, A. 1984. CRC Crit. Rev. Pl. Sci. 1:345–373.

James, B. R., and S. J. Riha. 1986. J Envir Qual 15:229–234.

Johnson, N. M. 1979. Science 204:497–499.

Joslin, J. D. 1987. *In* R. Perry, et al., eds. *Acid rain: scientific and technical advances.* Selper Ltd., London.

Nordstrom, P. K., S. D. Valentine, J. W. Ball, L. N. Plummer, and B. F. Jones. 1984. Water Resources Investigations Report 84-4186.

Norton, S. A. 1976. *In* L. S. Dochinger and T. A. Seliga, eds. U.S. Forest Service General Tech. Rpt. NE-23, Upper Darby, PA.

Pavan, M. A., and F. T. Bingham. 1982. Soil Sci Soc Amer J 46:993–997.

Plankey, B. J., H. H. Patterson, and C. S. Cronan. 1986. Envir Sci Tech 20:160–165.

Plankey, B. J., and H. H. Patterson. 1987. Envir Sci Tech 21:595–601.

Reuss, J. O. 1983. J Envir Qual 12:591–595.

Schecher, W. D. and C. T. Driscoll. 1987. Water Resour Res 23:525–534.

Thornton, F. C., M. Schaedle, D. J. Raynal, and C. Zipperer. 1986a. J Exp Bot 37:775–785.

Thornton, F. C., M. Schaedle, and D. J. Raynal. 1986b. Can J For Res 16:892–896.

Thornton, F. C., M. Schaedle, and D. J. Raynal. 1987. Envir Exp Bot 27:489–498.

Ulrich, B., R. Mayer, and P. K. Khanna. 1980. Soil Sci 130:193–199.

Walker, W. J., C. S. Cronan, and H. H. Patterson. 1988. Geochim Cosmochim Acta 52:55–62.

Zhao, X. J., E. Sucoff, and E. J. Stadelmann. 1987. Plant Phys 83:159–162.

Long-Term Acidic Precipitation Studies in Norway

G. Abrahamsen,* H.M. Seip,† and A. Semb‡

Abstract

Long-range transported air pollutants have probably been deposited in Norway for more than 100 years but with a significant increase in the 1950s and 1960s. The deposition of S and N has been fairly constant during the last 15 to 20 years.

The atmospheric deposition of acid is likely to increase leaching of cations from soils. This may result in water acidification if H^+ and aluminum ions are leached and in soil acidification if base cations are leached. Sulfate is the most important vehicle for increased leaching from soils. Because N, in the form of NO_3^- and NH_4^+, is largely absorbed by vegetation and soil, it does not contribute substantially to the leaching.

Increase in soil acidity is a slow process, and dramatic immediate effects on the terrestrial ecosystems are not expected. However, in the long run increased leaching of base cations is likely to influence the forest negatively. The deposition of N will probably increase tree growth. But it is unknown if and when the forest becomes saturated with N and leaching of NO_3^- increases. The combination of increased growth due to the N deposition and the increased leaching of base cations could accelerate possible nutrient deficiencies in the future. In the long run also other terrestrial organisms (e.g. soil organisms, ground flora, etc.) may suffer from the atmospheric deposition.

Dramatic declines in the populations of fish, particularly trout (*Salmo trutta*) and salmon (*Salmo salar*), have been taking place during the last century. Declines have been especially dramatic during the last three to four decades. The decline in fish populations is due to increased water acidity combined with increased concentrations of toxic Al species. Other freshwater organisms, including crustacea, snails, insect larvae, and various zooplankton species, also suffer from the acidification.

The possibility that increased uptake of aluminum may cause dementia in humans gives cause for concern.

*Institute of Soil Sciences, Agricultural University of Norway, N-1432 Aas-NLH, Norway.

†Center for Industrial Research, P.O. Box 124, Blindern, N-0314 Oslo 3, Norway.

‡Norwegian Institute for Air Research, P.O. Box 64, N-2001 Lillestrøm, Norway.

I. Introduction

In Norway the population density is low (i.e. 12 km^{-2}), and the demand for electricity is met by hydroelectric power plants. Consequently, damage to plants, soil, and water by nationwide emissions of air pollutants has been limited. However, smelter industries, especially for copper, nickel, and aluminum, have locally damaged vegetation and soil in the past. Today obvious effects occur in relatively small areas exposed to emissions of hydrogen fluoride from aluminum smelters.

Due to seemingly limited damage by air pollution, little attention has been given to research on pollution effects in the past. But research significantly intensified around 1970 when it became evident that Norway was exposed to air pollutants originating from abroad.

Scientific research on acidification of rivers and lakes started in the 1920s. Trout (*Salmo trutta*) populations were declining in relatively large areas (Dahl, 1926), concomitant with declining populations of salmon (*Salmo salar*) in rivers. Experiments indicated that the depleted populations of trout were a consequence of low pH in water. As early as 1934 a connection between the acidity of freshwater and precipitation was suggested (Torgersen, 1934).

After this period characterized by initial significant scientific curiosity in water acidification, interest declined until the 1950s. In 1959, a freshwater biologist claimed that the acidity of freshwater depended mainly on chemical compounds from the precipitation or compounds that were produced in the soil (Dannevig, 1959). Shortly thereafter, the discussion of long-distance transport of air pollutants to Norway escalated. Also effects on forest productivity were taken into consideration (Royal Ministry for Foreign Affairs, 1971; Dahl and Skre, 1971), and it was argued that increased leaching of calcium due to increased input of sulfuric acid would reduce the growth of forest in southern Scandinavia by 0.5 to 1.5% year^{-1}.

The potential effects on freshwater and especially forest were the stimuli for establishing a large scientific program "Acid precipitation—effects on forest and fish" (the SNSF project) in 1972. Since then about 300 publications have been published in internal series and in national and international scientific journals (See Overrein et al., 1980; Tollan, 1981).

After the conclusion of the SNSF project in 1980, the research activity was reduced. But reports on the dramatic decline of West German forests stimulated the scientific activity again. Research is now divided among several institutes and universities, coordinated by a committee appointed by the Norwegian Research Council for Science and Humanities. The most active participants in the SNSF project and in the present research are the Norwegian Institute for Air Research (NILU), the Norwegian Institute for Water Research (NIVA), the Norwegian Institute for Forest Research (NISK), and Center for Industrial Research (SI). Important contributions also come from the universities in Oslo, Bergen, and Trondheim and the Agricultural University of Norway.

The present paper summarizes the Norwegian research on long-distance imported air pollutants and their effects on soil, water, vegetation, and humans.

II. Atmospheric Transport and Deposition

The earliest records of precipitation chemistry in Norway were due to the collaboration between the initiators of the European Atmospheric Chemistry Network (EACN) at the Swedish Agricultural University and their Norwegian colleagues. The measurements were primarily intended to quantify the supply of plant nutrients by precipitation (Låg, 1963). Interest in the meteorological process as a link between emissions and deposition of air pollutants in Norway was triggered, however, by Odèn's (1968) interpretation of EACN data and the subsequent United Nations meeting in Stockholm in 1972 (Royal Ministry for Foreign Affairs, 1971). Observations of soot-stained snow had been made in Norway as early as 1881 (Br., 1881). Elgmork and associates (1973) showed that soot-stained layers in snow were associated with low pH and high conductivity in melted snow. Acidity in rainwater was also shown to be correlated with wind direction (Førland, 1973).

The international ramifications of acid precipitation were sensed by the OECD Air Management Group, which started the planning of a cooperative investigation program in 1968. This preliminary work included preparation of methods for sampling and chemical analysis of air and precipitation, and plans for the collection of meteorological data and interpretation of results. This planning work was supported by NORDFORSK, a cooperative organization of Nordic Scientific Research Councils (Ottar, 1978a). On the basis of these preparations, the OECD cooperative "Technical Programme to Study the Long-Range Transport of Air Pollutants" (LRTAP) was launched in 1972 with active participation from 11 countries in northern and western Europe. A central coordinating unit was established at the Norwegian Institute for Air Research, which was to coordinate measurements and, in cooperation with the Norwegian Meteorological Institute, evaluate the data using meteorological dispersion models.

A. Long-Range Transport Models and Results

Early work on diagnosis of episodes by air trajectories and simple advection models was reported by Nordø and co-workers (1973) and by Nordø (1974). A statistical study of the decay of SO_2 and the transformation of SO_2 to SO_4^{2-} aerosol under transport over long distances by Eliassen and Saltbones (1975) showed that the observations of airborne SO_2 and SO_4^{2-} aerosol were consistent with reasonable assumptions of mixing height, SO_2 dry deposition rate, and a transformation rate for the oxidation of SO_2 to SO_4^{2-} at around 0.7% hour^{-1}. This simple trajectory model with fixed mixing height and simple parameterization of the transformation process was particularly useful in calculating export and import budgets for the various countries in Europe. The final report of the OECD program (OECD, 1977) described the emissions of SO_2, ground station measurements, interpretation of case studies and "episodes," and model evaluations and results. It was shown that the air and precipitation chemistry in any one European country that is measurably affected by emissions can be quantified by simple advection models. Aspects of the OECD study were also reported by Semb (1978a), Ottar (1978b), and Eliassen (1978).

The OECD LRTAP study also provided the scientific basis for the 1978 ECE Convention on Long-Range Transboundary Air Pollution, and the monitoring and evaluation of the long-range transmission of air pollutants is continued in the EMEP program, which includes 91 stations in 24 European countries.

The evaluation of the data was also built on previous modeling experience at the Norwegian Meteorological Institute. Improvement of the model was made by a series of model experiments (Eliassen and Saltbones, 1983) introducing seasonally variable deposition and transformation rates for SO_2, variable mixing height, and a mass-consistent treatment of the deposition of SO_2 and SO_4^{2-} with precipitation. In the OECD study, use was made of a statistical relationship between calculated airborne SO_4^{2-} concentrations and the concentration of SO_4^{2-} in precipitation to calculate SO_4^{2-} deposition. The new model uses a removal rate for SO_2 and SO_4^{2-}, which is related to precipitation intensity. However, a remaining problem is that the space and time resolution of the model (150 × 150 km and 6 h) is less than the actual temporal and spatial variability for precipitation.

The model also incorporates a seasonal variable transformation rate of SO_2 to SO_4^{2-}, a large part of which is due to seasonal variability in the atmospheric concentration of OH radicals and a reduced dry deposition velocity for SO_2 under winter conditions. The importance of the seasonal variation in atmospheric chemistry and dry deposition is illustrated by the occurrence of surprisingly high concentrations of air pollutants in the Arctic in late winter and spring (Larssen and Hanssen, 1980; Rahn et al., 1980). This phenomenon is obviously linked to long-range transport of air masses from Eurasia and can be explained only on the basis of very little deposition of pollutants en route and virtually no oxidation of SO_2 under Arctic winter conditions. These increased winter-spring concentrations of pollutants are observed also in northern Greenland, Canada, and Alaska (Barrie, 1986) and have been extensively studied over the last 10 years. Results of a 3-year study in the Norwegian Arctic have been summarized by Ottar and associates (1986). One of the main conclusions from this study was that high winter concentrations of air pollutants in the Arctic occurred when large emission source areas in Eastern Europe and USSR were within the stable continental arctic air mass. Transport from south to north under these conditions is connected to the large-scale planetary wave system. However, aircraft sampling also reveals distinct haze layers containing pollutants at higher altitudes (2,000 to 3,000 m above sea level, a.s.l.), which must have originated from source regions with higher ambient temperatures.

B. Deposition Studies

An extensive network of daily precipitation sampling stations was established in Norway in connection with the OECD Program and the SNSF project (Overrein et al., 1980). Results from this network (Dovland et al., 1976) show a general pattern of excess (nonmarine) SO_4^{2-} and strong acid in precipitation, with annual deposition values exceeding $1 \text{ g S m}^{-2} \text{ year}^{-1}$ in the southernmost part of Norway. Deposition patterns of individual events could also be shown, demonstrating the effect of different origins of air masses on the precipitation chemistry. Much of the precipitation in Norway is due to orographic effects (Nordø and Hjortnæs, 1967), and the resulting pattern of precipitation and deposition of pollutants reflects the topographic relief (Eliassen and Saltbones, 1976). In the southernmost part of Norway, which is most affected by acid precipitation, the largest amounts of precipitation and the highest deposition of air pollutants both occur within 100 km from the coast, at elevations of 200 to 300 m a.s.l.

Skartveit (1982) found an inverse relationship between precipitation rate and concentration of various soluble ions in precipitation collected at four stations located in a transect from the coast to 20 km inland in western Norway. This was relatively more marked for NO_3^- than for NH_4^+ and SO_4^{2-} ions. Joranger and associates (1980) reported concentrations of all major soluble ions in precipitation at eight background stations in Norway (Figure 5-1). The ratios between the concentrations of excess SO_4^{2-}, NO_3^-, NH_4^+, and H^+ were found to be similar at all stations, and both the concentrations of these ions in precipitation and the concentrations of SO_4^{2-} and SO_2 in air depend strongly on the origin of the air masses, as described by 850 mb trajectories. It was also shown for southern Norway that the concentrations of SO_4^{2-} in air and SO_4^{2-} in precipitation were correlated, and that the ratio corresponded to a scavenging ratio of about 10^6 (liter air/liter precipitation). This is consistent with the SO_4^{2-} particles acting as cloud condensation nuclei in advected warm and humid air masses. In contrast, SO_4^{2-} concentrations in snow samples from Spitsbergen, where SO_4^{2-} aerosols are associated with cold and relatively dry air, yield scavenging ratios nearly an order of magnitude lower (Semb et al., 1984).

A constant problem with precipitation sampling is the collection efficiency of sampling gauges. For simple gauges at exposed sites, this may be as low as 90% for rain, and the collection efficiency for dry snow is as low as 55% at a wind speed of 5 m s^{-1} (Førland and Joranger, 1980). This precipitation collection deficiency, and inputs from fog and dry deposition, must be taken into account when calculating hydrological and chemical mass balances for small catchments.

Because of the large precipitation amounts and the distance from major source areas, dry deposition represents only a relatively minor contribution to the total atmospheric deposition in Norway. However, the dry deposition also depends on surface characteristics and micrometeorological conditions. Dovland and Eliassen (1976) resampled the surface snow layer at a field near Lillestrøm over a period of 14 days without precipitation and measured the increase in the concentration of lead and SO_4^{2-}, which was compared with ambient concentrations of lead, SO_4^{2-} aerosol, and SO_2 in air. Sulfur dioxide is probably insoluble in snow at ambient temperatures below 0°C, but the results showed that the deposition velocities were less than 2×10^{-3} m per second for all three species. This low deposition velocity could partly be ascribed to the stable thermal stratification that occurs over snow surfaces.

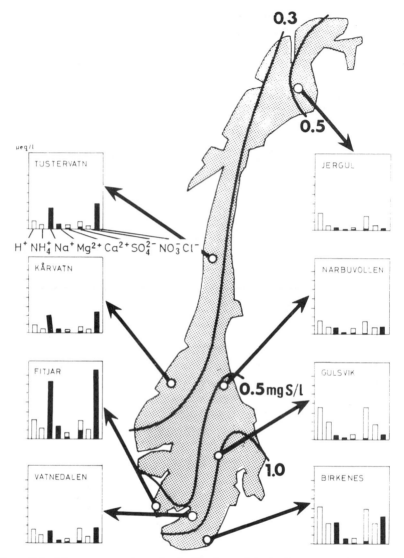

Figure 5-1. Mean concentrations of major ions in precipitation in Norway, 1978–79, given in histograms and with isolines for SO_4^{2-}. Black bars represent the sea salt components (Joranger et al., 1980).

Dollard and Vitols (1980) investigated the uptake of radioactively labeled SO_2 and sulfuric acid aerosols by spruce, pine, and birch seedlings in a specially designed wind tunnel. The uptake of SO_2 by spruce was suppressed during darkness and was not sensitive to variations in wind velocity. In contrast, deposition of aerosols increased markedly with wind speed in the range 0.5 to 2.5 m per second. The deposition velocity also increased with particle size, but

difficulties in obtaining reproducible particle size distributions prevented evaluation of a size-dependent deposition velocity. However, using conservative
estimates of wind speed, and the results obtained for the test aerosol with the
lowest aerodynamic mass median diameter, the authors estimated that the
deposition velocities for SO_2 and SO_4^{2-} aerosols to a typical Norwegian mixed
forest would be nearly equal, about 0.007 m per second. Applied to measured air
concentrations in Norway, this implies that dry deposition accounts for about 25%
of the total deposition of excess SO_4^{2-} in forested catchments. This result seems to
be in accordance with long-term runoff from experimental watersheds (see the
catchment studies section).

Measurements of ambient air quality and precipitation chemistry are being
continued under the auspices of the Norwegian State Pollution Control Authority,
and results are reported annually (e.g. SFT, 1986a, 1986b).

During the 15-year period of measurements since 1972, the concentrations of
SO_4^{2-}, NO_3^-, NH_4^+, and H^+ have changed very little in southern Norway (Figure
5-2). During this period, marked reduction of SO_2 emissions have occurred in
western Europe, mainly as a result of reduced fuel oil consumption after 1979.
This reduction has been partly offset by increased solid fuel consumption in
eastern Europe (Semb and Dovland, 1986).

C. Trace Elements

Apparently, SO_2 and SO_4^{2-} are not the only air pollutants capable of being
advected over long distances. Other components include trace elements emitted
from fossil fuel combustion and other high-temperature processes. Measurements
of various trace elements in air and precipitation, reviewed by Semb (1978b),

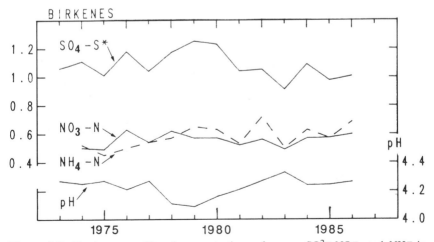

Figure 5-2. Yearly mean pH and concentrations of excess SO_4^{2-}, NO_3^-, and NH_4^- in
precipitation at Birkenes in southern Norway, 1973–1986 (SFT, 1986a).

showed a general correlation with other long-range transported air pollutants. Steinnes and Rambæk (1980) determined trace elements in mosses and lichens from more than 500 localities in Norway and were able to show a general pattern of deposition due to long-range transport that was very similar to the deposition pattern for excess SO_4^{2-}. For some elements the effect of metallurgical industry in Norway was also apparent.

Measurements of trace elements in air and precipitation at four stations over an 11-month period (Hanssen et al., 1980) made it possible to relate the concentrations in the moss samples directly to deposition by precipitation. By using emission estimates for anthropogenic trace elements in Europe (Pacyna, 1984) together with a modification of Eliassen and Saltbones's (1975) trajectory model, Pacyna and associates (1984) were able to demonstrate substantial agreement between the model estimates and measured concentrations.

D. Photochemical Oxidants and Nitrogen Oxides

Early in the 1970s it was realized that photochemical oxidation processes involving nitrogen oxides and hydrocarbons from human activities were not confined to areas producing photochemical smog, but occurred over large areas of Europe and North America. Episodes of high ozone concentrations were also observed in southern Scandinavia (Grennfelt and Schjoldager, 1984; Schjoldager et al., 1984) and are clearly related to situations with long-range transport of polluted air masses under anticyclonic conditions in summer.

Modeling of the formation of photochemical oxidants requires knowledge of emissions of nitrogen oxides and reactive hydrocarbons and incorporation of a complex set of interactive chemical reactions. Episodes with high ozone concentrations in southern Scandinavia have been interpreted using a Lagrangian model with photochemistry (Eliassen et al., 1982; Hov et al., 1985). The photochemical oxidants interact with the long-range transport of sulfur and nitrogen compounds by providing the reactants (ozone, hydroxyl, and peroxy radicals) that oxidize SO_2 and NO_x to sulfuric and nitric acid. Ammonia, which is emitted from animal manure, combined with these acids to form ammonium sulfate and ammonium nitrate aerosols, which have longer atmospheric lifetimes than gaseous ammonia and nitric acid. Work is therefore in progress to link these processes together in a comprehensive model, which will also provide the first model description of long-range transport and deposition of nitrogen oxides and ammonia in Europe. The atmospheric chemistry, transport, and deposition processes of NO_x have been reviewed for the Nordic Council of Ministers by Grennfelt and co-workers (1986).

III. Acidification of Soil and Water

The earliest warning that something was happening in natural ecosystems in Norway was the episodic fish kills. As mentioned above, these fish kills were associated with the acidity of the water, and a relationship among the acidity of precipitation, soil, and fresh water was suggested. The early papers on possible forest effects concerned the processes mediated via the soil. It was therefore early understood that effects of acid precipitation on natural ecosystems were strongly connected with soil chemistry.

A. Regional Soil and Water Acidification

1. Soil

Approximately 80% of the land area of Norway is covered by young, nutrient-poor, acidic soils like Lithosols, Regosols, Rankers, Podsols, and Arenosols (corresponding orders in Soil Taxonomy are Entisols, Inceptisols, and Spodosols). They have developed after the last glaciation about 10,000 years ago. Typically, soil pH (H_2O) varies from 3.5 to 4.5 in the topsoil to 4.5 to more than 5 in the subsoil. Observations of soil acidity in forest soils in the 1920s indicated pH values of the topsoil ranged between 3.5 and 5.3, increasing to about 4.3 to 5.7 in the lower part of the B horizon (Gl∅mme, 1928).

Acid precipitation is likely to affect soil acidity. Most soils have been significantly acidified by natural processes since the last glaciation. Therefore, the natural production of acid must have exceeded the natural consumption of acid. With addition of strong acid from the atmosphere, such soils have to become more acidic. Studies from various European countries indicate significant changes in soil pH during the last three to six decades. But little Norwegian data are available for estimating changes in soil chemistry over the last four to five decades. The only study of this kind is from Rondane Mountains in south Norway, where soil pH was measured in a number of places from 1942 to 1947 (Dahl, 1987a). The area is dominated by quartzite with extremely poor soil. In 1984 the soil was resampled and pH measurements showed significantly lower pH values for all soil horizons than previously found (Table 5-1). The differences increased with soil depth. It is not clear how the changes in soil pH are related to the atmospheric deposition,

Table 5-1. Changes in pH from 1942–1949 to 1984 in soil in the Rondane Mountain area, Norway.

Soil horizon	1942–1949		1984		pH change
	No. of samples	pH	No. of samples	pH	
O	15	4.37	20	4.04	−0.33[a]
E	19	4.92	16	4.34	−0.58[c]
Bh	11	5.14	14	4.61	−0.53[b]
Bf	11	5.49	13	4.75	−0.74[c]
C	13	5.69	19	4.84	−0.85[c]

Dahl, 1987a.

[a] Significant level of 5%.

[b] Significant level of 1%.

[c] Significant level of 0.1%.

which in this area amounts to only about 2 to 3 kg ha^{-1} of both N and S (Joranger et al., 1980; SFT, 1986b). Furthermore, weathering rates for the extremely poor mineral soils of this area have not been determined.

2. Water

Extensive losses of fish in rivers and lakes led to more detailed investigations of some of the affected areas. Monitoring of river chemistry was initiated by Snekvik and associates (1972). The comprehensive lake survey performed by the same group is mentioned in the section about damage to fish. During the SNSF project a number of regional surveys of water chemistry in small lakes were carried out (Gjessing et al., 1976; Wright et al., 1977; Wright and Henriksen, 1978). The pH values measured for 1974 are shown in Figure 5-3. Similar results were obtained for other years. The pattern is similar to that for S in precipitation shown in Figure 5-1. The most acid lakes are found in those parts of southernmost Norway that are dominated by slowly weathering minerals and shallow, often peaty, acidic soils. The acid lakes have in general high SO_4^{2-} and aluminum concentrations. However, the NO_3^- concentrations are generally low.

A national survey of about 1,000 lakes in the autumn of 1986 confirmed the regional pattern observed in previous studies. Of these lakes, 305 had also been sampled in the autumn of 1974 or 1975. In the intervening years, the pH of those lakes had changed little, but maximum SO_4^{2-} concentrations were significantly lower than the earlier figures, although still high. Unfortunately, the NO_3^- concentrations in the lakes in the southernmost areas showed increase. However, the values are still generally much lower than the SO_4^{2-} concentrations (SFT, 1987).

Although thousands of lakes in southern Norway have become more acid in this century, the exact quantification is difficult. Old pH measurements compiled by Wright (1977) indicate considerable acidification in sensitive areas. However, comparison of recent and old pH measurements must be done with great caution. Methodological differences may obscure the pH shifts (Blakar and Digernes, 1984). However, for some rivers there are reliable series of measurements from 1966 that indicate a slightly decreasing trend in pH until 1978 (SFT, 1986b).

The changes in lake acidity over time may be obtained by studying sediment cores. Most often pH is inferred from the diatom species found in the sediments (e.g. Davis and Berge, 1980). Davis and associates (1985) studied two lakes in southern Norway, Hovvatn and Holmevatn, and found that both were somewhat acidic in preindustrial times (pH about 5) and had relatively high concentrations of humic compounds (total organic carbon, TOC, 6 to 9 ppm). Recent acidification apparently started about 1920 in Hovvatn and in the 1940s in Holmevatn. This acidification was accompanied by TOC decrease. By the end of the 1970s, Hovvatn had a pH of 4.4 and TOC of 3.2 ppm, and Holmevatn had a pH of 4.7 and TOC of 2.2 ppm. Crustacean remains have also been used to infer acidification history, but this method is less well established (Nilssen and Sandøy, 1986).

In general, paleoecological studies confirm that substantial acidification of surface waters has taken place in this century. Rosenqvist (1987) claims, however, that acidification periods have occurred earlier, in particular after the Black Death

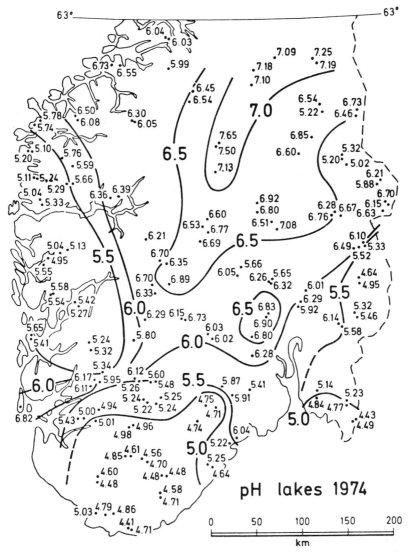

Figure 5-3. Surface water pH levels in 155 lakes in southern Norway (Wright et al., 1977).

when large areas in Norway became depopulated and vegetation changes followed. He finds support for this suggestion in the metal content in the sediments of a lake in southernmost Norway as well as in diatom (Stabel, 1987) and pollen analysis (Høeg, 1987) of these sediments.

B. Mechanisms for Soil and Water Acidification

The first studies on the effect of acid precipitation on freshwater quality seemed to indicate that water chemistry responded rapidly to the acidity of precipitation

(Dale et al., 1974; Johannessen, 1974). In a catchment study (Birkenes, south Norway) the pH of brook water decreased rapidly with increased input of acid rain. Because the rain in this area is almost always acid, it was assumed that the increased acidity was caused by high input of H^+. However, statistical analyses revealed higher correlation between the pH of the brook water and the quantity of rain than with the total input of H^+ (Rosenqvist, 1977; Nordø, 1977).

The effects of acid precipitation on soil and water chemistry are strongly related to the mobility of the deposited anions in soil (Abrahamsen and Dollard, 1979; Seip, 1980). Another factor is the cation exchange occurring in soil-plant systems. Among the cations deposited by acid precipitation, H^+ and NH_4^+ are the most important; H^+ may replace base cations adsorbed to soil particles. In periods with surplus water, the replaced base cations can be leached from the soil, and both soil pH and base saturation percentage may decrease. Then again, high input of sea salts may result in temporarily increased acidity in streams in areas with acid soils due to ion exchange (Skartveit, 1981). Increased loss of base cations may gradually be compensated by base cations mobilized by weathering. Therefore, the effects of acid deposition on soil and water chemistry depend mainly on four factors: the mobility of the deposited anions; the efficiency of H^+ to replace base cations; the weathering rate; and the amount and pathways of water percolating the soil. Increased deposition of N, which is likely to increase tree growth, may also increase soil acidity due to increased uptake of base cations and possibly also in the longer term by increased nitrification.

1. Catchment Studies

a. Input-Output Budgets

Catchment studies have played an important role in acidification research in Norway. During the SNSF project, a total of 37 catchments were studied for 1 to 6 years (Overrein et al., 1980). With one exception these catchments are in southern Norway.

Some catchments have now been studied for nearly a decade. Table 5-2 gives mean input-output budgets for 7 to 9 years (SFT, 1985, 1986a). Kårvatn is in a nearly unaffected area on the western coast, the three other catchments receive precipitation with an average pH of 4.2 to 4.3. There are several difficulties in estimating the total input of water and ions to a catchment (see the section on deposition studies). We have preferred to give wet inputs for all ions and include estimates only for dry deposition of sulfur for Birkenes and Kårvatn (SFT, 1986b). For Storgama and Langtjern the dry deposition of sulfur is probably somewhat less than at Birkenes. In the three most acid catchments, considerable amounts of aluminum are found in stream water. In all four catchments, there is a large retention of NO_3^-. At Birkenes the output of sulfur seems to exceed the input for the period studied, but input and output are in balance within the rather large uncertainties for the three other catchments. Approximate sulfur balances are found for many other catchments studied for shorter periods. During the 2-year period 1974 to 1976, a small net retention seemed to occur in the Tovdal catchment

Table 5-2. Input-output budgets of four catchments in south Norway as mean of 7 or 9 years of measurements.

Period	Birkenes (73–78) + (81–83)		Storgama (75–78) + (80–84)		Langtjern 75–83		Kårvatn 78–84	
	In	Out	In	Out	In	Out	In	Out
Water mm	1370	1067	978	923	669	555	1245	1789
H^+	77	32	52	30	33	11	16	2
Na^+	75	128	15	28	5	14	71	89
K^+	6	6	2	5	3	2	3	6
Ca^{2+}	13	64	7	34	6	32	11	38
Mg^{2+}	19	42	4	13	2	10	16	24
Al^a	pn	74	pn	18	pn	13	pn	6
NH_4^+	54	pn	27	pn	21	pn	9	pn
S-dry	10–30		nm		nm		5–15	
SO_4^{2-}	97	160	55	73	39	45	23	29
Cl^-	81	141	19	28	7	10	83	97
NO_3^-	53	7	28	12	19	1	6	1
HCO_3^-	pn	pn	pn	pn	pn	pn	pn	35

Inputs are generally wet deposition, but for Birkenes and Kårvatn estimates of dry deposition of sulfur are also given (Units: keq km^{-2} yr^{-1}) (SFT, 1986b).

pn: Not measured regularly, but probably negligible.

nm: Concentrations of SO_2 and SO_4^{2-} in air not measured. The dry deposition is probably somewhat less than at Birkenes.

[a] Total Al considered as Al^{3+}.

(Dovland and Semb, 1978), but due to large annual variations budgets for such short periods appear uncertain. It should be stressed also that the annual deposition of SO_4^{2-} has been rather constant since 1970 (Figure 5-2), so that the budgets give little information on the actual residence time for deposited S in natural catchments. In general, SO_4^{2-} is the dominant anion in surface waters in areas affected by acid deposition except in areas close to the coast where Cl^- may be dominant.

b. Episode Studies

Detailed studies of runoff chemistry during rain or snowmelt events have been complementary to the long-term studies described above. High flow generally results in low pH values. During snowmelt most of the ionic pollutants are found in the first meltwater (Johannessen and Henriksen, 1978; Overrein et al., 1980). This may affect the runoff quality to some extent. Even during snowmelt, however, most of the runoff is "old" water that has been stored in the catchment (Christophersen et al., 1985).

Episode studies were carried out from the start of the SNSF project (Dale et al., 1974; Overrein et al., 1980). More recent studies have been conducted mainly at Birkenes (southernmost Norway), Atna (central-eastern Norway) and Høylandet (Nord-Trøndelag county) (For a summary, see Seip, 1987). Atna is a mountain area with moderate deposition (less than 0.5 g S m^{-2}yr^{-1}). Høylandet is a nearly pristine area, designated as a reference site for studies of water, soil, and vegetation in an unpolluted area (Dahl, 1987b). At all these sites pH tends to decrease with flow. However, the lowest pH in stream water at Atna and Høylandet is around 4.8, that is, much higher than at Birkenes, where pH may drop to about 4.2. The Birkenes study has refuted the assumption that aluminum concentrations in stream water are controlled by equilibrium with gibbsite, Al(OH)$_3$. The water chemistry seems to be largely influenced by water pathways in the soil. At high flow more water passes through upper soil horizons with lower pH and less easily available aluminum than in deeper soils (Sullivan et al., 1986, 1987).

2. Experimental Studies

Experimental studies on the effects of acid deposition have been carried out in relation to soil chemical properties, soil organisms, soil biological processes, forest growth, and freshwater quality.

a. Studies on Soil Acidification

To examine the effects of acid deposition on the acid-base status of the soil and the growth of trees, five relatively large field plot experiments were established (Abrahamsen et al., 1976a). The size of the plots in these trials varied from 4 m^2 in the smallest experiment to 625 m^2 in the largest. Artificial rain of different acidities produced by adding H$_2$SO$_4$ to groundwater was given in addition to natural precipitation. The irrigation treatments started between 1972 and 1975 and lasted until 1978 to 1983. Fifty mm of artificial rain were applied five times year^{-1}. Four tree species have been studied: Norway spruce (*Picea abies*), Scots pine (*Pinus sylvestris*), lodgepole pine (*Pinus contorta*), and silver birch (*Betula pendula*). The soil of the field experiments are all acid and relatively poor in nutrients. Detailed description of the soils is given by Stuanes and Sveistrup (1979).

Series of experiments with mineral soil monolith lysimeters have also been carried out in the field (Teigen et al., 1976; Abrahamsen, 1985; Abrahamsen and Stuanes, 1986). These experiments showed that SO$_4^{2-}$ is slightly adsorbed onto the soil particles (Teigen et al., 1976; Abrahamsen, 1980a, 1980b; Bjor and Teigen, 1980; Abrahamsen, 1985; Abrahamsen and Stuanes, 1986). When applied in higher concentration than "natural" levels, SO$_4^{2-}$ is adsorbed in the Bs horizon. This is to be expected when applying an ion in increased concentration (Singh, 1980, 1984a, 1984b; Singh et al., 1980).

These experiments also showed that N is retained by the soil and vegetation (Abrahamsen et al., 1976b, 1977; Abrahamsen, 1980a; Abrahamsen and Stuanes, 1986; Horntvedt et al., 1980). The leaching of N from forest soil in Norway appears therefore at present to be very small. Also watershed studies have revealed an efficient retention of N. In south Norway only 5 to 43% of the incoming N were leached (cf. Table 5-2). The highest figure is from a watershed with a high percentage of barren rock and shallow soils (Gjessing et al., 1976; Overrein et al.,

1980). The retention of N is consistent with the high biological demand for N in forest ecosystems, as demonstrated by a number of fertilizer experiments (Brantseg et al., 1970; Sture, 1984).

Even though large amounts of N are deposited in many terrestrial ecosystems, an accumulation in the soil has not been found. For example, below slopes of barren rock where large amounts of runoff drain into the soil, the total N content was not higher than further away from the rock surface (Stuanes et al., 1987). Compared to the input of N from the atmosphere, the total pool of N in the soil is very large. A small addition will therefore be obscured to observation. Denitrification may also have prevented an accumulation.

The studies mentioned show that the deposition of SO_4^{2-} has a much larger effect on the leaching of cations than that of NO_3^-. The leaching of cations varies between soils. In a Gleyic to Dystric Cambisol (Umbric Dystrochrept, USDA) where the pH of the leachate varied between 4.2 and 5.2, the leaching decreased in the following order (molar basis): Al > Na > Ca > Mg > K > H (Abrahamsen and Stuanes, 1986). In a Cambic Arenosol (Typic Udipsamment, USDA) (pH of leachate 5.0 to 6.0), the leaching decreased in this order: Ca > Na > Mg ≥ K > Al > H (Teigen et al., 1976; Abrahamsen, 1983a). In a more fertile soil (Haplic Phaeozem FAO; Aquic Haploboroll, USDA), where the pH of the leachate varied between 6 and 7, the leaching of cations followed this sequence: Ca > Mg > Na > K > H > Al (Abrahamsen, 1985). When the acidity of the "rain" increased from pH 6 to pH 3 no significant change in the ranking of the different ions was found. At pH 2 no change in the ionic dominance took place in the fertile soil (Haplic Phaeozem), but Al dominated in the two other soil types.

Calcium and Mg were leached in increasing amounts as the "rain" became more acidic. Figure 5-4 shows the effects on the exchangeable amount of the sum of Na^+, K^+, Mg^{2+}, and Ca^{2+} of the various treatments in the lysimeters with the Gleyic to Dystric Cambisol. The exchangeable base cations were mostly affected in the top soil, but as long as the pH of the "rain" was above 3, relatively small leaching occurred. Soil pH was more affected than the base saturation, especially in deeper soil layers.

Similar results were found in a field plot experiment that was established in a 17-year-old Scots pine forest. Application of "rain" of pH 3 and lower has significantly reduced the base saturation and soil pH (Figure 5-5). The effect on the base saturation is more evident in the upper part of the soil than in the deeper soil layers. Soil pH is more affected than base saturation in deeper soil layers. Soil exposed to the most acidic treatment (pH 2) appears to have recovered slightly 3 years after the last application. Overall, a significant reduction in soil pH and base saturation took place, especially in the upper soil horizons, between 1975 and 1978. Depletion of base cations in the top layer appears to be due to uptake by the vegetation. In this soil type, most of the N taken up by plants is in the form of NH_4^+. Therefore, the uptake of cations is much larger than of anions. The larger cation uptake is compensated by excretion of H^+ from the plant roots.

In another experiment stocked with 31-year-old Norway spruce, similar

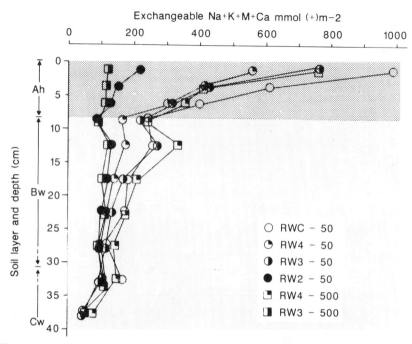

Figure 5-4. Exchangeable base cations in lysimeters with a Gleyic to Dystric Cambisol exposed to a total of 1,250 mm (*circles*) or 12,500 mm (*squares*) of rainwater of different acidities over a period of five years. RWC-50 means that RainWater Control-50 mm month[-1] (control = pH 4.5) was added; RW4-500 means RainWater pH 4 500 mm month[-1]; etc. (Abrahamsen and Stuanes, 1986).

changes in soil chemistry have been found (Figure 5-6). Significant effects of the treatments are obvious in 1975, 1978, and 1981, but in 1984 very small effects of the acid application can be seen. The explanation of the improvement appears to be increased weathering and replenishment of base cations previously leached. Also in this experiment all plots were significantly acidified in the period from 1975 to 1978.

The acidification of the soil increased the mobilization of Al, but when pH of the "rain" was 3 or above, the concentration of total Al did not exceed 0.2 mM. However, for the pH 2 treatment, Al concentrations increased to 1.2 to 2 mM (Abrahamsen, 1983a; Abrahamsen and Stuanes, 1986).

b. Minicatchment Studies

This work, involving natural catchments with areas from a few m^2 to 1,000 m^2 or more, started within the SNSF project (Seip, 1980; Overrein et al., 1980). Barren rock surfaces covered only by various microlichen species changed the pH of artificially added "rain" with pH 3.5 only very slightly. For "rain" with pH 5, pH in runoff depended critically on the salt content of the rain but was generally below

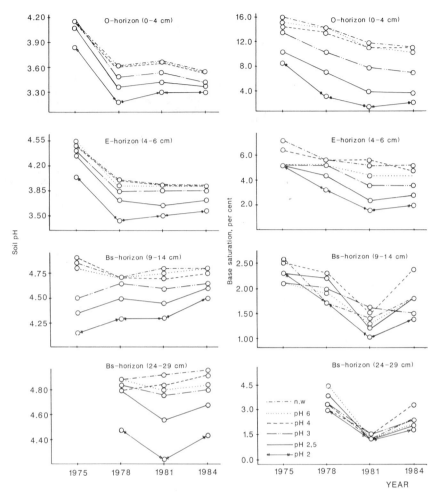

Figure 5-5. Effects of acid application on soil pH(H$_2$O) and base saturation in a field plot experiment with Scots pine. The artificial rain is in addition to the natural precipitation. The bulk pHs of artificial plus natural precipitation of the various treatments are: n.w. (not watered): pH 4.18; pH 6: pH 4.3; pH 4: pH 4.14; pH 3: pH 3.58; pH 2.5: pH 3.14 and pH 2: pH 2.66 (Abrahamsen et al., in press).

5 (Abrahamsen et al., 1979). Experiments in minicatchments with thin, acid soil at Storgama confirmed that the runoff may be more acid than the precipitation (Seip et al., 1978). This was attributed to be partially due to leaching of SO$_4^{2-}$ from the soil during such periods. During snowmelt the SO$_4^{2-}$ concentrations in runoff from the minicatchments decreased approximately by a factor of 4, and pH increased from about 4 to 4.5 or slightly more (Seip et al., 1980b).

The snow of one minicatchment was neutralized with NaOH. Compared to the runoff chemistry from the previous year, the average pH was about the same,

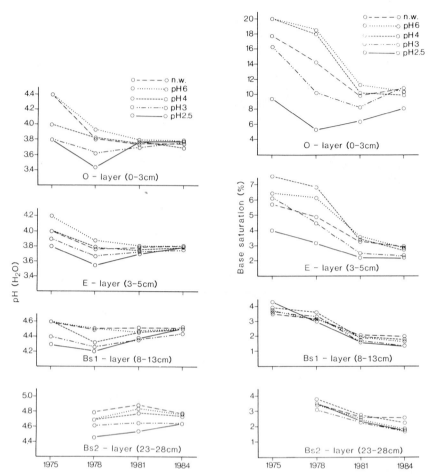

Figure 5-6. Effects of acid application on soil pH(H$_2$O) and base saturation in a field plot experiment with Norway spruce. The artificial rain is in addition to the natural precipitation. The bulk pHs of artificial plus natural precipitation of the various treatments are: n.w. (not watered); pH 4.36; pH 6: pH 4.46; pH 4: pH 4.26; pH 3: pH 3.61 and pH 2.5: pH 3.16 (Stuanes et al., 1988).

although the neutralization may have caused somewhat higher pH values in the beginning of the snowmelt (Seip et al., 1980a). Experiments in which radioactive tracers (HTO, $^{45}Ca^{2+}$) were added to the snowpack showed that the contact between water and soil was sufficient to remove most of the calcium from the meltwater. In another experiment, radioactive SO_4^{2-} ($^{35}SO_4^{2-}$) was added to a minicatchment during the snow-free season. Less than 40% was found in runoff during the first 3 months (Dahl et al., 1979; Seip, 1980).

Project RAIN (Reversing Acidification in Norway) is an international effort for investigating the effects on water and soil by altering the deposition to whole catchments (Wright et al., 1986a). Two large-scale manipulations are carried out: artificial acidification at Sogndal, western Norway, and deacidification of the acid rain at Risdalsheia, southernmost Norway. At Sogndal sulfuric acid has been added to one catchment ($7,220$ m^2) and a mixture of sulfuric and nitric acid to another ($1,940$ m^2). The response to acid treatment has been substantial and rapid (Wright et al., 1986a; Wright, 1986). Results for the snowmelt period 1986 may serve as an example. About 0.45 kmol H^+ ha^{-1} had been added as H_2SO_4 to the snowpack of one catchment. During the first part of the snowmelt, pH in runoff from this catchment was 0.5 to 1 unit lower than in the control. Acidification of runoff is accompanied by increased concentrations of SO_4^{2-} and inorganic aluminum. At Risdalsheia two roofs have been constructed to cover about $1,165$ m^2 and 650 m^2. Under one of these roofs irrigation is carried out with deacidified precipitation; the other serves as control and gets ambient precipitation. Results so far show considerable reductions in concentrations of NO_3^- and SO_4^{2-} in runoff from the catchment receiving deacidified water compared to the control; however, pH shows only a small increase.

In general, these results are in agreement with earlier lysimeter experiments (Abrahamsen and Stuanes, 1980) and the minicatchment studies mentioned above. One reason for the small difference in pH may be the high concentrations of organic compounds in the runoff. Model calculations indicate that larger shifts in pH with varying deposition may occur when the runoff is not quite as acid as at Risdalsheia (where pH is about 4) and in particular when runoff can be considered as a mixture of acid water (from upper soils) and less acid water (from deeper soils) (Seip et al., 1986).

3. Models for Acidification of Soil and Water

As previously mentioned, the acidity in the precipitation is not strongly correlated with the acidity in the runoff during the same event. Rosenqvist (1977, 1978, 1985) pointed out the importance of precipitation amounts and of soil-water interactions. He hypothesized that changes in soil properties, for example, due to vegetation changes, play a dominant role in water acidification. Seip (1980) based a conceptual model on the importance of the concentrations of mobile anions but considered also soil acidification and that a fraction of the precipitation and/or meltwater may reach the watercourses relatively unchanged.

Henriksen (1979, 1980) described acidification as a large-scale titration. He developed an empirical model for the change in alkalinity with varying deposition. The key factor is simply that there must be a balance between cations and anions. In the first model it was assumed that the concentrations of Ca and Mg in runoff did not change with changed deposition. This part has later been modified, and the model has been used to estimate restoration of lakes by reduction in sulfur deposition (Wright and Henriksen, 1983).

Simulation of water chemistry in small streams has been an important part of acidification research in Norway. One model was first developed for the Birkenes catchment (Christophersen and Wright, 1981; Christophersen et al., 1982b) and is therefore known as the Birkenes model, although it has with various modifications been applied to other catchments. The model is based on the mobile anion concept. The hydrological submodel, which was developed by Lundquist (1977), consists simply of two soil reservoirs and a snow reservoir when appropriate (Figure 5-7). Processes included in the model as applied to a tributary to Harp Lake, Ontario, are also shown in the figure (Rustad et al., 1986). Important trends in day-to-day and seasonal variations in water chemistry are explained by the model, but recent field studies indicate that modifications may be necessary, especially in the aluminum submodel (Sullivan et al., 1986). The model has also tentatively been used to estimate effects of variations in sulfur depositions.

Model calculations indicated that a halving or a doubling of the SO_4^{2-} concentrations in stream water may lead to considerable changes in stream water pH, even if the concentration of exchangeable "base cations" of the soil is assumed not to vary. With one particular choice of model parameters, increases from about 0.4 pH units up to 0.9 units were obtained by halving the sulfate concentration for water with pH in the range 4.6 to 5.2 (Seip et al., 1986). Christophersen and

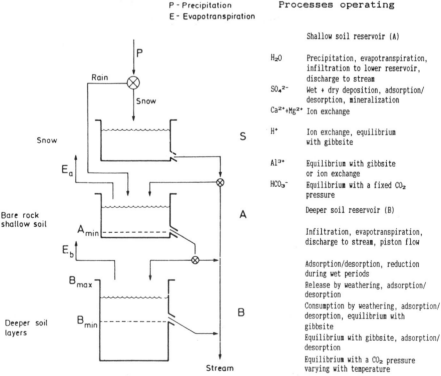

Figure 5-7. A simple hydrochemical model used for catchments sensitive to acidification (Rustad et al., 1986).

associates (1984) showed that the partial pressure of CO_2 assumed in soil and surface water in the model greatly affects the prediction of the pH shift corresponding to a given reduction in deposition.

MAGIC (model of acidification of groundwater in catchments), which is based on many of the same mechanisms as the Birkenes model, is mainly intended for long-term predictions (Cosby et al., 1985a, 1985b, 1985c). Changes in soil properties (e.g. base saturation) with time are therefore computed. Results of model calculations appear to agree fairly well with paleolimnological reconstructions of water acidification (Wright et al., 1986b).

4. Conclusions on Acidification Mechanisms

There is now consensus that the following factors are of crucial importance for water acidity: concentrations of mobile anions, cation exchange, weathering, and water pathways (Rosenqvist and Seip, 1986). Neutralization of acid precipitation does not only occur in the catchment but also in lakes (Hongve, 1978) and streams (Norton et al., 1987). Acidification causes leaching of aluminum from the catchment. It is clear that soil acidification caused by other factors than acid deposition, for example, afforestation, may affect water quality. The relative importance of the various factors in determining water acidity is still a matter of dispute. However, no systematic relations were found between changes in land use and regional lake acidification trends (or fish population decreases) in three Norwegian studies (Drabløs et al., 1980; Overrein et al., 1980). Although clearly reduced deposition of acidifying compounds will improve water quality in sensitive areas, the extent and rate of recovery cannot yet be satisfactorily predicted.

IV. Effects on Organisms and Biological Processes in the Soil

Effects of acid deposition on soil biota have been studied in the field experiments described previously and in laboratory experiments. Main groups of animals living in coniferous forest soils have been studied, as have mycorrhizal fungi and some fungi taking part in the decomposition of organic material. The processes studied are mainly decomposition of organic material but to some extent also nitrification and nitrogen fixation.

A. Soil Animals

Effects of soil acidity on collembola, mites, and enchytraeids were investigated by allowing the animals to colonize sterile soil cores pretreated with dilute sulfuric acid or lime (Hågvar and Abrahamsen, 1980). After 5 months the soil cores were extracted for animals. Some species preferred limed soil and others acidified soil, but the total number of enchytraeids and collembola was not significantly affected by any of the treatments. The mites were most abundant in the acidified cores. The number of species was little affected by the treatments (Table 5-3). The most significant effect was in the poor mull, where the number of Mesostigmata species in the acidified cores was only 50% of the number in the other cores.

The results from the laboratory experiment are supported by results from field experiments (Hågvar and Amundsen, 1981; Abrahamsen, 1983b; Hågvar, 1984). Limed plots generally had a lower abundance of collembola, mites, and enchytraeids. Plots supplied with "rain" with pH equal to or above 3 generally had the highest abundance. When the pH dropped to 2.5 and in particular 2, the abundance decreased. The most significant decrease was found for the enchytraeids, and the least decrease for the mites.

The abundance of protozoa, rotifera, and nematoda has also been studied in the field experiments (Stashurska-Hagen, 1980). In general, significant effects of the acidification were found only for the pH 2.5 and pH 2 treatments. No clear effect could be found for the acid treatment on the rotifera, but liming had a negative effect on the total number. In nematodes negative effects were only found for the pH 2 treatment and for liming.

The zoological studies in the field experiments indicate more severe effects on the soil fauna than the colonization experiment. In the latter, surplus acid and lime were leached from the soil before the animals invaded the cores. In the field experiments the animals were exposed directly to the artificial rain. The reason for the severe decrease in abundance of some species may therefore be direct contact with the most acidic water (Abrahamsen et al., 1980).

B. Fungi

The growth of fungi isolated from decomposing needles of lodgepole pine (*Pinus contorta*) was studied on nutrient agar with pH from 3 to 6 (Ishac and Hovland, 1976; Abrahamsen et al., 1980). The fungi modified the pH in the growth medium significantly, but there were significant differences among species in the acid tolerance.

Table 5-3. Number of species of enchytraeids, collembolas, and mites (*Oribatei, Mesostigmata, Astigmata*) that have colonized sterile soil cores pretrated with sulfuric acid (A), distilled water (W), or lime (L).

Groups of soil animal	Raw humus			Poor mull			Rich mull		
	A	W	L	A	W	L	A	W	L
Enchytraeidae	1	2	1	4	5	6	7	7	6
Collembola	12	12	13	11	12	14	10	10	9
Oribatei	18	17	16	18	16	18	15	13	12
Mesostigmata	10	8	10	7	14	14	8	6	8
Astigmata	1	1	1	2	2	2	2	2	2
Total no. of species	42	40	41	42	49	54	42	38	37

Hågvar and Abrahamsen, 1980.

Two years after terminating the acid application in one of the field experiments, mycorrhizal and other macromycetes were recorded (Høiland, 1986). Only two treatments, pH 5.6 and pH 2.5, were examined. The total production of fruit bodies increased in the acidified plots to about 160% of that in the control plots. The number of species, however, decreased from 67 in the control plots to 55 in the pH 2.5 plots. Thus the dominance of the acid-tolerant species increased significantly.

C. Decomposition of Organic Matter

The effect of acid deposition on the decomposition of organic material has been studied in the field and in laboratories. Bits of pure cellulose and sticks of aspen wood were placed in four of the five field experiments (Hovland and Abrahamsen, 1976). In one experiment the application of 50 mm month^{-1} of pH 2.5 water significantly reduced the decomposition rate, but in the other no effect was found of the acid application alone. In two of the three experiments including lime, liming increased the decomposition rate. On the limed plots, application of "rain" of pH 2.5 and 2 significantly decreased the decomposition rate.

In the laboratory, decomposition of lodgepole pine needles was studied by adding dilute sulfuric acid of pH 1.0, 1.8, 3.5, and 4.0 (Ishac and Hovland, 1976; Abrahamsen et al., 1980). The needles were inoculated with spores of *Trichoderma harzianum* and incubated at 25°C for 105 days. At pH 1 no decomposition occurred; at pH 1.8 the weight loss was 25% compared to 31% for the pH 3.5 and pH 4 treatments.

Decomposition of Norway spruce needles percolated with 100 or 200 mm month^{-1} of artificial rain of pH 5.6, 3, and 2 has been studied. Slightly reduced decomposition of lignin at the 200 mm "rain" was found when pH decreased, but otherwise the effect of pH was very small (Table 5-4). Other studies showed similar results (Hovland, 1981; Hågvar and Kjøndal, 1981). In conclusion, it is apparent that acid precipitation in Norway has little direct effect on the decompo-

Table 5-4. Decomposition of needles of Norway spruce exposed to artificial rain of different acidity. The dry matter and lignin are in percent of the starting material.

Incubation period	pH 5.6		pH 3		pH 2	
	100 mm	200 mm	100 mm	200 mm	100 mm	200 mm
			Dry matter			
16 weeks	76	76	73	78	73	77
38 weeks	63	61	62	65	63	66
			Lignin			
16 weeks	64	61	63	68	75	77
38 weeks	50	52	44	54	52	62

Hovland et al., 1980.

sition of organic material. However, continuing increased soil acidity is speculated to reduce decomposition rate.

D. Nitrogen Fixation and Nitrification

The effect of lime and acid deposition on the nonsymbiotic N fixation has been studied in soil from a field experiment (Hovland, 1978). The N fixation rate, measured by the acetylene reduction technique, was very low and often not detectable. The annual fixation was estimated at about 1.2 kg N ha^{-1}. No effect was found from the lime or the artificial rain.

A study on nitrification was carried out by perfusing $(NH_4)_2SO_4$ through soil from the field plots exposed to combinations of artificial rain and lime (Hovland and Ishac, 1975). Limed soil had a higher potential for nitrification, but no effect was found from the acid application.

V. Effects on Aquatic Organisms

The recent water acidification and the associated changes in water chemistry have had a profound effect on aquatic life on all trophic levels. Effects on freshwater fish are of most immediate concern to people living in the affected areas, and we will first discuss this topic.

A. Damage to Fish: Development in Space and Time

Early reports on losses of fish populations in lakes in southernmost Norway from around the turn of the century (e.g. Dahl, 1926, 1927) were presumably caused by water acidification. Catches of Atlantic salmon from several rivers in southernmost Norway also began to decline early in this century (Leivestad et al., 1976). Many of these rivers have now lost all their salmon. In the 1950s it was observed that brown trout populations in many lakes in southernmost Norway had either disappeared or been strongly reduced. Jensen and Snekvik (1972) were the ones to document regional declines and losses of salmon and trout stocks in southernmost Norway.

A survey of about 700 lakes in southernmost Norway was conducted in the early 1970s (Wright and Snekvik, 1978). Concentrations of major chemical components were determined in water samples, and information collected on fish population status (mainly brown trout) showed that fish had disappeared from about 40% of the lakes and that an additional 40% had sparse populations.

Another comprehensive survey showed that by the late 1970s all fish life was virtually extinct in lakes in an area of about 13,000 km^2 in Norway. In an additional area of about 20,000 km^2, many lakes had no fish or were only sparsely populated (Figure 5-8) (Sevaldrud and Muniz, 1980). Data on fish status were mainly based on interviews and questionnaires, but test fishing in some lakes supported the conclusions (Rosseland et al., 1980).

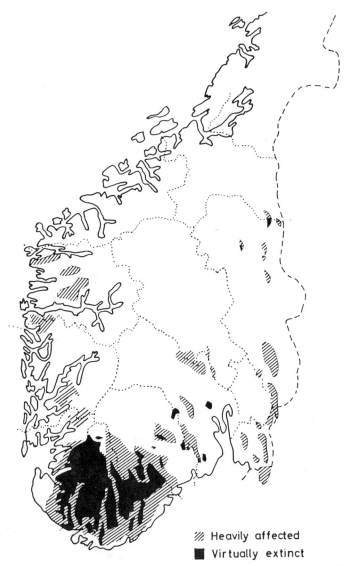

%. Heavily affected

█ Virtually extinct

Figure 5-8. Areas in south Norway where fish populations have been affected by acidification (Sevaldrud and Muniz, 1980).

Both surveys showed that the fish status was strongly related to pH. The frequency of lakes with no fish increased with decreasing pH. However, fish were found at lower pHs when the concentrations of "base cations" were also high. Sevaldrud and Muniz (1980) found that a dramatic increase in the number of lakes with lost fish populations had occurred in the late 1950s and 1960s (Figure 5-9). In addition, Muniz and Leivestad (1980) found that if the trend continues approxi-

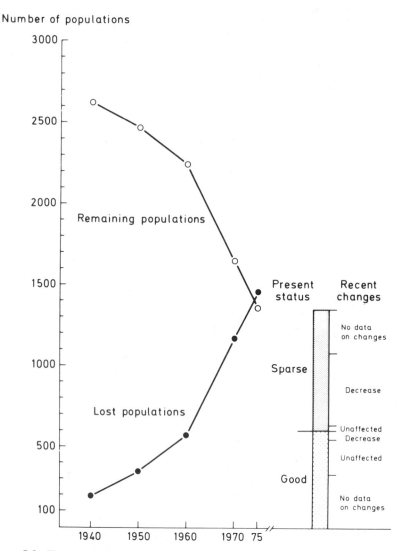

Figure 5-9. Time trend for population loss of brown trout in the affected areas in the four southernmost counties in Norway. *Recent changes* refer to changes during the period 1974–1979 (Sevaldrud and Muniz, 1980).

mately 80% of the trout populations in the affected areas of the four southernmost counties would disappear by 1990.

More recent studies have shown that the damage is still increasing, as predicted by Muniz and Leivestad (Rosseland et al., 1986a). During the period 1978 to 1983, 30% and 12% of the remaining populations of brown trout and perch (*Perca fluviatilis*), respectively, were lost in the two southernmost counties in Norway

(Sevaldrud and Skogheim, 1986). By 1983, 71% of the brown trout and 43% of the perch populations in these counties had been lost.

B. Other Aquatic Organisms

This section is partly based on a recent overview of the effect of acidification on benthic animals in lakes and streams (Økland and Økland, 1986). In general, acidification is accompanied by declining numbers of both planktonic and benthic invertebrates.

K. A. Økland (1980) found that the freshwater "shrimp," *Gammarus lacustris*, does not occur at pH below 6; at low elevations, the lower limit may be even higher (Figure 5-10). Similar observations were done by Raddum and Fjellheim (1984). Borgstrøm and associates (1976) did not find the tadpole shrimp (*Lepidurus arcticus*) in lakes with pH below 6.1 in Norway. For freshwater snails, J. Økland (1980) found a rapid drop in number of species around pH 6.0; he found no species in lakes with pH < 5.2. K. A. Økland (1980) found that no species of small mussels (*Sphaeriidae*) occurred below pH 4.7.

Many insect larvae are sensitive to low pH. For stone flies (*Plecoptera*) and caddis flies (*Trichoptera*), most field studies reveal a decreasing number of species with decreasing pH (Hendrey and Wright, 1976; Raddum and Fjellheim, 1984). Raddum (1980) found increased mean individual weight of mayflies (*Ephemer-*

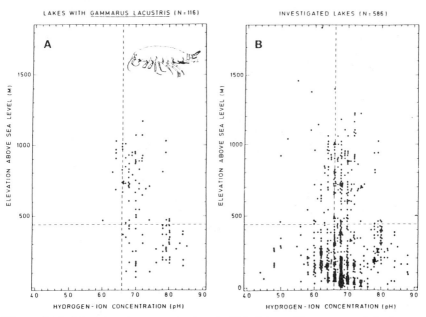

Figure 5-10. Elevation above sea level and pH in lakes with presence of freshwater "shrimp" (*Gamarus lacustris*) (A) and investigated lakes in southeastern Norway (B). Each dot represents one lake, showing summer pH in surface water (K. A. Økland, 1980).

optera), caddis flies (*Trichoptera*), and midges (*Chironomidae*) in acid lakes with low fish predation.

Hobæk and Raddum (1980) reported a reduced number of zooplankton species in acid lakes. In south Norway clearwater lakes with pH below 5.0 had an average of 7.1 species. Equally acid, humic lakes had a mean number of 11.7 species, and clearwater lakes with pH above 5.5 had on average 16.1 species. These numbers include crustaceans, rotifers and *Chaoborus* larvae.

Laake (1976) found indications of a shift from bacterial to fungal decomposition with acidification, and Traaen (1977) observed slow decomposition of wood and leaves in acidic clearwater lakes.

C. Mechanisms

The effects of acidification on aquatic biota are not a function of pH only (Leivestad, 1982; Muniz, 1984). As noted earlier, fish tolerate a lower pH if the concentration of "base cations," in particular the Ca concentration, is high. The same is true for many other species (Økland and Økland, 1986). In recent years it has also been established that aqueous aluminum plays a key role. The complicated interaction between species must also be taken into account when trying to explain changes in acidified communities (Nilssen et al., 1984; Økland and Økland, 1986).

Various fish species and life stages show different sensitivities (Grande et al., 1978; Rosseland and Skogheim, 1984; Rosseland et al., 1986b; Reite and Staurnes, 1987). High egg and fry mortality in acidic water leading to reduced younger age classes is regarded as a major cause of fish decline. However, high mortality among mature and old fish has also been observed in acidified lakes (Rosseland et al., 1980; Andersen et al., 1984). Loss of adult fish during acid spates, especially during snowmelt, is documented (Leivestad and Muniz, 1976; Rosseland et al., 1986a; Skogheim et al., 1984). Salmon seems to be particularly sensitive to acidification during the smolt period. Large numbers of dead Atlantic salmon and trout were found in the river Vikedal during spring in the years 1981 to 1984. Especially high mortality was found for 2-year-old salmon smolts (Hesthagen, 1986). During these periods pH fell to 5.1 to 5.2, and labile aluminum (see below) generally varied in the range 30 to 70 μg/L (SFT, 1986b).

Leivestad (1982), in his review of the physiological effects of acid stress on fish, concluded that gill tissue is the primary target organ. Failure in gill ion exchange resulting in decreasing blood levels of NaCl and related osmotic disturbances are dominating symptoms under moderate acid stress in softwater. Aluminum toxicity results in similar physiological symptoms.

Much of the recent work in Norway as well as in other countries has been directed towards clarifying the role of aluminum. The first Norwegian studies on the effects of aluminum on fish were reported by Muniz and Leivestad (1980). Aluminum may occur in several forms. The nonlabile fraction, consisting mainly of organically bound aluminum, seems to be virtually nontoxic to aquatic biota. Of the labile, mainly inorganic forms, the hydroxy monomers $Al(OH)^{2+}$ and

$Al(OH)_2^+$, are suspected to be most toxic to fish. Fivelstad and Leivestad (1984) reported that survival time for swim-up larvae and postlarvae of Atlantic salmon showed a particularly strong negative correlation with the concentration of $Al(OH)^{2+}$. Muniz (1981) found that Al levels as low as about 70 μg/L may be harmful to sensitive development stages of salmon. Rosseland et al. (1986b) reported 100% mortality of smolt exposed to acidic, aluminum-rich water (pH 5.1, total Al 225 μg/L, labile Al 135 μg/L). Brown trout and brook (*Salvelinus tontinalis*) trout showed no mortality under these conditions. Reite and Staurnes (1987) observed that low pH and increased Al concentration reduced the activity of two gill enzymes (NA-K-ATPase and carbonic anhydrase).

VI. Effects on Forest

A. Effects on Forest Trees

The work of Dahl and Skre (1971) gave rise to serious concern about the future development of the forest. The first attempt to test their hypothesis was based on tree-ring analyses (Tveite, 1975; Vestjordet, 1975; Abrahamsen et al., 1976b, 1977; Strand, 1980; Abrahamsen and Tveite, 1983). Tree-ring development was compared in areas exposed to different inputs of acid, and between sites speculated to differ in sensitivity to acidification. No indication of reduced growth was found. It is important to realize that interpretation of tree-ring series is very difficult. The main problem is in excluding alternative explanations to the effects observed. These explanations include possible differences in stand history, as well as different growth response to the weather in different altitudes and in consecutive years, and different age trends. Limited sample size of the groups to be compared may also represent a problem. For these reasons, further work on tree growth was based on experimental techniques.

The first experiment carried out addressed the effect of soil acidity on germination and establishment of seedlings (Teigen, 1975; Abrahamsen et al., 1976b). Seeds of Norway spruce and Scots pine were sowed in mineral soil previously leached with distilled water or dilute sulfuric acid of pH 4, 3, and 2. Germination and establishment of spruce (Figure 5-11) were significantly reduced by high soil acidity, but in pine no effect was found.

Rooted cuttings of clones of Norway spruce were exposed to lime and artificial rain in a greenhouse experiment (Ogner and Teigen, 1980). The "rain" was produced by mixing H_2SO_4 and neutral salts to distilled water to pH levels of 5.4, 4, 3, and 2.5. Seventeen mm of artificial rain were applied to the pots three times week^{-1} during a 29-month period. Plants that were supplied water of pH 2.5 and an amount of S equivalent to about 3,300 kg ha^{-1} had more yellowish needles than the other plants; otherwise, no significant effect was found. In the most acid treatments, the needles contained up to 0.13% Al.

Effects of artificial acid rain on tree growth were observed in the five field plot experiments described in the experimental studies section. In three of these

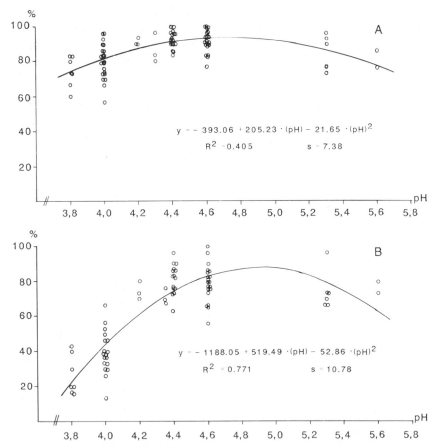

Figure 5-11. Seed germination (A) and seedling establishment (B) of Norway spruce related to soil pH(H$_2$O) (Teigen, 1975).

experiments, no effect was found (Tveite, 1980a), but in two experiments significant reduction in both height and diameter growth were observed (Abrahamsen et al., in press; Stuanes et al., 1988). In one experiment with Norway spruce in which irrigation went on from 1973 until 1978, little effect of the acid application could be seen during the first 6 years (Figure 5-12). In the following 4 years, however, growth declined in the most acidified plots and minimum growth was recorded in 1982. Since then the growth has improved. With Scots pine, greater effects were observed (Figure 5-13). When the irrigation stopped in 1981, growth of the trees in the two most acid treatments was significantly retarded. Thereafter, growth continued to decline until 1985 when minimum growth occurred. In 1986 there was indication of recovery.

It may be concluded that the reduced growth is due to decreased soil pH and base

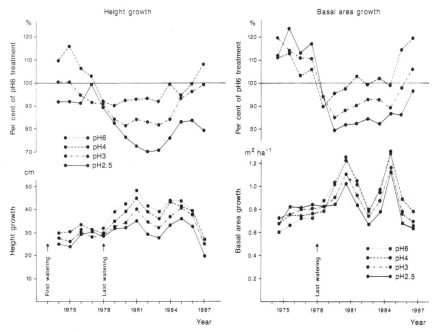

Figure 5-12. Effects of acid application on the height and basal area growth of Norway spruce. The two upper figures show the growth in percent of the control (pH 6); the two lower figures show the actual growth. More information on the pH of the treatments is given in Figure 5-6 (Stuanes et al., 1988).

saturation. The mechanism is not yet completely understood, but needle analyses indicate Mg deficiency. Other nutrients, except for N, appear to be in surplus. Another cause for the reduced growth in spruce could be high Al concentrations in soil solution. In lysimeters total Al concentrations up to 2 mM in the seepage water were found in the most acid treatments (Abrahamsen, 1983a). Such concentrations appear to be injurious to spruce but not to pine (Abrahamsen, 1984; Eldhuset, 1986).

B. Effects on Ground Cover Vegetation

Early field experiments showed that ground cover vegetation suffered from acid application. Mosses died when the pH of the applied water was 3 or below (Teigen et al., 1976; Abrahamsen, 1980b). Necrotic spots occurred only on birch, rose bay willow herb (*Chamaenerion augustifolium*), chickweed wintergreen (*Trientalis europea*), and may lily (*Maianthemum bifolium*) at pH 2.5 and 2. Recent studies have confirmed these results (Røsten, 1985; Nygaard, 1988). After the application of acid was ceased, significant recovery of the vegetation takes place.

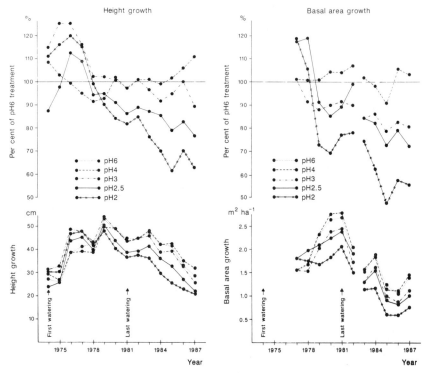

Figure 5-13. Effects of acid application on the height and basal area growth of Scots pine. The two upper figures show the growth in percent of the control (pH 6); the two lower figures show the actual growth. More information on the pH of the treatments is given in Figure 5-5 (Abrahamsen et al., in press).

C. Forest Decline in Norway

The main conclusion from the SNSF project, terminated in 1980, is that the main effect of acid deposition on forest was of a nutritional nature (Abrahamsen, 1980a). The deposition of N was considered beneficial due to its fertilizer value, whereas increased leaching of base cations, particularly Mg, might eventually result in nutrient deficiency. This deficiency would be promoted by increased growth caused by N deposition.

However, reports from West Germany on severe declines in fir (*Abies alba*), Norway spruce, and later other tree species renewed the interest in the 1980s in possible effects on Norwegian forest. Symptoms of forest decline are now reported from different places in Norway—even from areas where the deposition of air pollutants is minimal. A Norwegian survey of forest vigor à la German techniques was conducted in 1984 and 1985. The survey showed crown density decreasing with age, northerly latitude, and altitude (Horntvedt and Tveite, 1985). As pollution decreases from south to north, the symptoms of decline did not seem to be related to long-range transport of air pollutants. However, as acid deposition is likely to increase soil acidity, the forest will probably be affected negatively in the long run. It is also possible that interactions between pollutants such as the N deposition or ozone and natural factors such as drought or frost may affect the forest negatively in the future.

VII. Effects on Humans

Although high concentrations of many air pollutants may have harmful health effects, inhalation of air with pollutant levels typically caused by long-range transport to Norway is not likely to represent serious health problems. However, there is concern about mobilization or increased bioavailability of metals due to acidification. In Norway particular attention has been paid to possible effects of aluminum in drinking water.

Dementia has been observed in patients with chronic renal failure subject to dialysis if the water used has moderate or high concentrations of aluminum. It is also known that aluminum is elevated in specific parts of the brain in some people with various types of dementia. There is, however, disagreement on the importance of Al intake for healthy people. A statistical study by Vogt (1986) indicated a positive covariation between the frequency of dementia (including Alzheimer's disease) and the aluminum concentration in drinking water. She divided Norway into five zones with varying acid deposition and varying average aluminum concentrations in drinking water, primarily based on the regional lake surveys discussed previously. (Surface water is the main source of drinking water in Norway.) Frequencies of dementia were obtained from death certificates with dementia as an underlying or contributory cause of death. Her results are supported by a study by Flaten (1986, in press) who analyzed drinking water from nearly 400 Norwegian waterworks and used the same information on dementia as Vogt. Data were collected at the municipality level, but to obtain units of at least 10,000 inhabitants data for neighboring municipalities were pooled if necessary. Spearman's rank correlation coefficients for aluminum in drinking water and age-adjusted annual death rates with dementia were calculated for four periods: 1969–1973, 1974–1978, 1979–1983, and 1974–1983. For all periods, r was positive both for males (r interval 0.27 to 0.35) and females (r interval 0.43 to 0.59). For females r is significantly different from zero (the significant level, P, varied between 0.05 and 0.005), but for males $P < 0.1$ for two periods and $P > 0.1$ for the two other. Flaten points out a number of uncertainties in the method used, particularly related to the use of registered death rates with dementia as a measure for the frequency of these diseases.

Aluminum in drinking water normally gives only a minor contribution ($<10\%$) to the total aluminum intake (Alexander et al., 1986), the major sources being food and, for some people, drugs, especially antacids. The total intake may, however, be misleading in evaluating health effects because the bioavailability of various Al complexes may vary widely. Nordal and associates (1988a) gave a daily dose of

976 mg aluminum (as Al(OH)$_3$) to patients with renal failure. This dosage, which corresponds to approximately 100 times the average daily dietary intake, caused only a minor increase in serum aluminum (sAl), but a significant increase in urinary Al excretion. However, intake of citric acid with the aluminum hydroxide led to a substantial increase in sAl, averaging 5.75 times the median baseline concentration. In another study, large seasonal variations in sAl were observed in humans. It was hypothesized that compounds in drinking water could affect uptake of aluminum (Nordal et al., 1988b).

VIII. Current and Future Research

Monitoring and research on long-range transported air pollutants and their environmental and biological effects are given high priority in Norway. Data on long-term trends in deposition, water quality, soil properties, and populations of aquatic biota are being obtained within the "Norwegian Monitoring Program for Long-Range Transported Air Pollutants." Measurements of concentrations of pollutants in the Arctic are to continue. A major effort to develop a comprehensive model for transport and deposition of nitrogen oxides, ammonia, and oxidizing reactants, in addition to sulfur oxides, is underway under the auspices of the "Tropospheric Ozone Research Programme."

Studies on mechanisms for acidification of soil and water are also given high priority. Ongoing activities include studies at field sites, laboratory experiments, artificial acidification of small catchments, and development of simulation models. One goal of this work is to improve the understanding of the extent and rate of change in soil and water acidity with changing deposition. Studies of effects of acid water with varying concentrations of aluminum species and calcium on fish and other organisms are continuing under the auspices of the British-Scandinavian "Surface Water Acidification Programme." Possible future harmful effects of nitrogen deposition give cause for concern and research in this field is increasing. A research program on the health effects of aluminum has just been launched.

The effects of forests are being investigated through the international research program "Forest and Environment—Growth and Vigour" and the national "Monitoring Programme for Forest Damage." Vitality characteristics like crown density and needle color are being observed together with analyses of growth, litterfall, soil solution, and soil physical and chemical properties. The main objective of the research program is to investigate the relation between forest vigor and growth on the one hand and climate, soil properties, pathogens, and air pollution on the other hand. Of special concern is the interaction between natural stress factors and air pollution.

Research has also been initiated in a catchment nearly unaffected by acid deposition (Høylandet, Nord Trøndelag county) to study catchment and lake processes under pristine conditions.

References

Abrahamsen, G. 1980a. Acid precipitation, plant nutrients and forest growth. *In* D. Drabløs and A. Tollan, eds. *Proc Int conf ecol impact acid precip*, 58–63, SNSF project, Oslo-Ås.

Abrahamsen, G. 1980b. Impact of atmospheric sulphur deposition on forest ecosystems. *In* D. S. Shriner, C. R. Richmond, and S. E. Lindberg, eds. *Atmospheric sulphur deposition. Environmental impact and health effects*, 397–415. Ann Arbor Science Publishers, Ann Arbor, Michigan.

Abrahamsen, G. 1983a. Sulphur pollution: Ca, Mg and Al in soil and soil water and possible effects on forest trees. *In* B. Ulrich and J. Pankrath, eds. *Effects of accumulation of air pollutants in forest ecosystems*, 207–218, D. Reidel Publishing Company, Dordrecht, Holland/Boston, U.S.A./London, England.

Abrahamsen, G. 1983b. Holarctic Ecology 6:247–254.

Abrahamsen, G. 1984. Phil Trans R Soc Lond B305:369–382.

Abrahamsen, G. 1985. Assessment of the long-term effect of acid deposition on leaching from forest soil. A methodical study. *In* R. Klimo and R. Saly, eds. *Air pollution and stability of coniferous forest ecosystems. International symposium*, 3–22. University of Agriculture, Brno, CSSR.

Abrahamsen, G., K. Bjor, R. Horntvedt, and B. Tveite. 1976b. Effects of acid precipitation on coniferous forests. *In* F. H. Bräkke, ed. *Impact of acid precipitation on forest and freshwater ecosystems in Norway*, 37–63, SNSF prosjektet FR 6/76, Oslo-Ås.

Abrahamsen, G., K. Bjor, and O. Teigen. 1976a. Field experiments with simulated acid rain in forest ecosystems. SNSF project, FR 4/76, Oslo-Ås, 15 p.

Abrahamsen, G., and G. J. Dollard. 1979. Effects of acidic precipitation on forest vegetation and soil. *In* M. J. Wood, ed. *Ecological effects of acid precipitation*. Report from a workshop held at Cally Hotel, Gatehouse-of-Fleet, Galloway, the UK, 4–7 September, 1978, 17 p.

Abrahamsen, G., R. Horntvedt, and B. Tveite. 1977. Water, Air, and Soil Pollution 8:57–73.

Abrahamsen, G., J. Hovland, and S. Hågvar. 1980. Effects of artificial acid rain and liming on soil organisms and the decomposition of organic matter. *In* T. C. Hutchinson and M. Havas, eds. *Effects of acid precipitation on terrestrial ecosystems*, 341–362, Plenum Press, New York.

Abrahamsen, G., and A. O. Stuanes. 1980. Effects of simulated rain on the effluent from lysimeters with acid, shallow soil, rich in organic matter. *In* D. Drabløs and A. Tollan, eds. *Proc Int conf ecol impact acid precip*, 152–153, SNSF project, Oslo-Ås.

Abrahamsen, G., and A. O. Stuanes. 1986. Water, Air and Soil Pollution 31:865–878.

Abrahamsen, G., A. Stuanes, and K. Bjor. 1979. Water, Air, and Soil Pollution 11:191–200.

Abrahamsen, G., and B. Tveite. 1983. Effects of air pollutants on forest and forest growth. *In Ecological Effects of Acid Deposition*, 199–219. National Swedish Environment Protection Board, Report PM 1636.

Abrahamsen, G., B. Tveite, and A. O. Stuanes (in press). Wet acid deposition effects on soil properties in relation to forest growth. Experimental results. Paper given at the IUFRO Conference: Woody Plant Growth in a Changing Physical and Chemical Environment, Vancouver July 27–31, 1987.

Alexander, J., D. Hongve, and M. Lægreid. 1986. Helsemessig betydning av aluminum i drikkevann (Health effects of aluminum in drinking water). SIFF-VANN report 54, Statens Institutt for Folkehelse, Oslo.

Andersen, R., I, P. Muniz, and J. Skurdal. 1984. Effects of acidification on age class composition in Arctic char (*Salvelinus alpinus* L.) and brown trout (*Salmo trutta* L.) in a coastal area, SW Norway. Report Inst. Freshwater Res., Drottningholm, 61:5–15.

Barrie, L. 1986. Atmos Environ 20:643–663.

Bjor, K., and O. Teigen. 1980. Effects of acid precipitation on soil and forest. 6. Lysimeter experiment in greenhouse. *In* D. Drabløs and A. Tollan, eds. *Proc Int conf ecol impact acid precip,* 200–201. SNSF Project, Oslo-Ås.

Blakar, I. A., and I. Digernes. 1984. Verh Internat Verein Limnol 22:679–685.

Borgstrøm, R., J. Brittain, and A. Lillehammer. 1976. Everterbrater og surt vann. Oversikt over innsamlingslokaliteter (Invertebrates and acid water. Review of sampling sites). SNSF Project, IR 21/76, Oslo-Ås, 33 p.

Br. 1881. Naturen 5:47.

Brantseg, A., A. Brekka, and H. Braastad. 1970. Medd Nor inst skogforsk 27:537–607.

Christophersen, N., S. Kjærnsrød, and A. Rodhe. 1985. Preliminary evaluation of flow patterns in the Birkenes catchment using ^{18}O as a tracer. *In* I. Johansson, ed. *Hydrological and hydrogeochemical mechanisms and model approaches to the acidification of ecological systems*. Nordic Hydrol. Programme, Report No. 10.

Christophersen, N., S. Rustad, and H. M. Seip. 1984. Phil Trans R Soc Lond B 305:427–439.

Christophersen, N., H. M. Seip, and R. F. Wright. 1982b. Water Resour Res 18:977–996.

Christophersen, N., and R. F. Wright. 1981. Water Resour Res 17:377–389.

Cosby, B. J., G. M. Hornberger, J. N. Galloway, and R. F. Wright. 1985a. Water Resour Res 21:51–63.

Cosby, B. J., G. M. Hornberger, J. N. Galloway, and R. F. Wright. 1985c. Environ Sci Technol 19:1144–1149.

Cosby, B. J., G. M. Hornberger, R. F. Wright, and J. N. Galloway. 1985b. Water Resour Res 22:1283–1291.

Dahl, E. 1987a. Acidification of soils in the Rondane Mountains, South Norway, due to acid precipitation. Expected to be published as a report from the Ministry of Environment, 15 p.

Dahl, E. 1987b. The Høylandet project, Surface Water Acidification Programme, Midterm review conference, 22–26 June, Bergen, Norway, 79–80, The Royal Society.

Dahl, J. B., C. Qvenild, H. M. Seip, and O. Tollan. 1979. Omsetting av kalsium og sulfat i smeltevann og i regnvann på små felter undersøkt ved hjelp av radioaktive tracere (Movement of calcium and sulphate from rain and meltwater through minicatchments, studied by use of radioactive tracers). SNSF Project IR 49/79, Oslo-Ås. 65 p.

Dahl, E., and O. Skre. 1971. En undersøkelse over virkningen av sur nedbør på produktiviteten i landbruket (An investigation of the effect of acid precipitation on land productivity). *In Konferens om avsvavling*. Stockholm, 11 Nov., 1969. Nordforsk, Miljøvårdssekretariatet, Publikation 1971 (1):27–40.

Dahl, K. 1926. Tidskrift norske Landbr 232–242.

Dahl, K., 1927. Salm Trout Mag 46:35–43.

Dale, T., A. Henriksen, E. Joranger, and S. Krog. 1974. Vann- og nedbør-kjemiske studier i Birkenesfeltet for perioden 20. juli 1972 til 31. april 1973 (Chemical investigations of precipitation and runoff in the Birkenes basin, 20 July, 1972–31 April, 1973). SNSF Project, TN 1/74, Oslo-Ås. 45 p.

Dannevig, A. 1959. Jeger Fisker 3:116–118.

Davis, R. B., D. S. Andersson, and F. Berge. 1985. Nature 316:436–438.

Davis, R. B., and F. Berge. 1980. Atmospheric deposition in Norway during the last 300 years as recorded in SNSF lake sediments. II. Diatom stratigraphy and inferred pH. *In* D. Drabløs and A. Tollan, eds. *Proc Int conf ecol impact acid precip*, 270–271. SNSF Project, Oslo-Ås.

Dollard, G. F., and V. Vitols. 1980. Wind tunnel studies of dry deposition of SO_2 and H_2SO_4 aerosols. *In* D. Drabløs and A. Tollan, eds. *Proc Int conf ecol impact acid precip*, 108–109. SNSF Project, Oslo-Ås.

Dovland, H., and A. Eliassen. 1976. Atm Environ 10:783–785.

Dovland, H., E. Joranger, and A. Semb. 1976. Deposition of air pollutants in Norway. *In* F. H. Brække, ed. *Impact of acid precipitation on forest and freshwater ecosystems in Norway*, 15–35, SNSF Project FR 6/76, Oslo-Ås.

Dovland, H., and A. Semb. 1978. Deposition and runoff of sulphate in the Tovdal river. A study of the mass balance for September 1974–August 1976. SNSF project IR 38/78, Oslo-Ås, 22 p.

Drabløs, D., I. Sevaldrud, and J. A. Timberlid. 1980. Historical land-use changes related to fish status development in different areas in southern Norway. *In* D. Drabløs and A. Tollan, eds. *Proc Int conf ecol impact acid precip*, 367–368, SNSF Project, Oslo-Ås.

Eldhuset, T. D. 1986. Effects of aluminum on growth and nutrient uptake of *Betula pendula, Pinus sylvestris* and *Picea abies* seedlings. Nor. inst. skogforsk., 1432 Ås-NLH, 82 p.

Elgmork, K., A. Hagen, and A. Langeland. 1973. Environ Pollut 4:41–52.

Eliassen, A. 1978. Atmos Environ 12:479–487.

Eliassen, A., Ø. Hov, I. S. A. Isaksen, J. Saltbones, and F. Stordal. 1982. J Appl Met 21:1645–1661.

Eliassen, A., and J. Saltbones. 1975. Atmos Environ 9:425–429.

Eliassen, A., and J. Saltbones. 1976. Concentration of sulphate in precipitation and computed concentrations of sulphur dioxide. *In* M. Benarie, ed. *Atmospheric Pollution*, 123–133. Elsevier, Amsterdam.

Eliassen, A., and J. Saltbones. 1983. Atmos Environ 17:1457–1473.

Fivelstad, S., and H. Leivestad. 1984. Aluminum toxicity to Atlantic salmon (*Salmo salar* L.) and Brown trout (*Salmo trutta* L.): Mortality and physiological response. Report Inst. Freshwater Res., Drottningholm, 61:69–77.

Flaten, T. P. 1986. An investigation of the chemical composition of Norwegian drinking water and its possible relationships with the epidemiology of some diseases. Thesis, University of Trondheim, Norway.

Flaten, T. P. (in press) J Environ Geochem Health.

Førland, E. J. 1973. Tellus 25:291–299.

Førland, E. J., and E. Joranger. 1980. Målefeil på NILU's nedbørsmålere for snø og regn. Sammenligning mellom målertyper (Measurement errors of precipitation gauges used by the Norwegian Institute for Air Research. Comparisons of gauges). SNSF project TN 54/80, Oslo-Ås. 48 p.

Gjessing, E. T., A. Henriksen, M. Johannessen, and R. F. Wright. 1976. Effects of acid precipitation on freshwater chemistry. *In* F. H. Brække, ed. *Impact of acid precipitation on forest and freshwater ecosystems in Norway*, 65–85. SNSF Project FR 6/76, Oslo-Ås.

Glømme, H. 1928. Meddr norske SkogforsVes. III (10):1–216.

Grande, M., I. P. Muniz, and S. Andersen. 1978. Verh int Verein Limnol 20:2076–2084.

Grennfelt, P., A. Eliassen, Ø. Hov, R. Berkowicz, and G. Nordlund. 1986. Atmospheric chemistry, transport and deposition of nitrogen oxides. Nordisk Ministerråd, Miljørapport 1987 (1), 89 p.

Grennfelt, P., and J. Schjoldager. 1984. Ambio 13:61–67.

Hågvar, S. 1984. Pedobiologia 27:341–354.

Hågvar, S., and G. Abrahamsen. 1980. Oikos 34:245–258.

Hågvar, S., and T. Amundsen. 1981. Oikos 37:7–20.

Hågvar, S., and B. R. Kjøndal. 1981. Pedobiologia 22:232–245.

Hanssen, J. E., J. P. Rambæk, A. Semb, and A. Steinnes. 1980. Atmospheric deposition of trace elements in Norway. In D. Drabløs and A. Tollan, eds. Proc Int conf ecol impact acid precip, 116–117. SNSF Project, Oslo-Ås.

Hendrey, G. R., and R. F. Wright. 1976. J Great Lakes Res, 2, Suppl. 1:192–207.

Henriksen, A. 1979. Nature 278:542–545.

Henriksen, A. 1980. Acidification of freshwaters—a large scale titration. In D. Drabløs and A. Tollan, eds. Proc Int conf ecol impact acid precip, 68–74, SNSF Project, Oslo-Ås.

Hesthagen, T. 1986. Water, Air, and Soil Pollution 30:619–628.

Hobæk, A., and G. G. Raddum. 1980. Zooplankton communities in acidified lakes in South Norway. SNSF Project, IR 75/80, Oslo-Ås, 132 p.

Høeg, H. I. 1987. The pollenanalytic research in Tveitå Fiskeløs, Bygland. Surface Water Acidification Programme, Mid-term Review Conference, Bergen, Norway, 386–390, The Royal Society, London.

Høiland, K. 1986. Storsoppens reaksjon overfor forsuring, med spesiell vekt på mykorrhizasoppene. Undersøkelse foretatt i Norsk institutt for skogforsknings forskningsfelter i Åmli, Aust-Agder. (Effects of acidification on the macrofungi in a field experiment with Scots pine). Miljøver departementet, Avdeling for naturvern og friluftsliv, Rapport T-671, 62 p.

Hongve, D. 1978. Verh Int Verein Limnol 20:743–748.

Horntvedt, R., G. J. Dollard, and E. Joranger. 1980. Effects of acid precipitation on soil and forest. 2. Atmosphere-vegetation interactions. In D. Drabløs and A. Tollan, eds. Proc Int conf ecol impact acid precip, 192–193, SNSF Project, Oslo-Ås.

Horntvedt, R., and B. Tveite. 1985. Overvåking av skogskader (Monitoring of forest damage). Norsk institutt for skogforskning, Årbok 1985:38–44.

Hov, Ø., F. Stordal, and A. Eliassen. 1985. Photochemical oxidant control strategies in Europe: a 19 days' case study using a Lagrangian trajectory model. Norwegian Institute for Air Research, Lillestrøm, TR 5/85.

Hovland, J. 1978. Nitrogenfiksering i noen jordtyper fra norsk barskog (Nitrogen fixation in some Norwegian coniferous soils). SNSF Project, IR 35/78, Oslo-Ås, 16 p.

Hovland, J. 1981. Soil Biol Biochem 13:23–26.

Hovland, J., and G. Abrahamsen. 1976. Eksperimentelle forsuringsforsøk i skog. 1. Nedbrytning av cellulose og ved (Acidification experiments in conifer forest. 1. Decomposition of cellulose and wood material). SNSF Project, IR 27/76, Oslo-Ås, 16 p.

Hovland, J., G. Abrahamsen, and G. Ogner. 1980. Plant and Soil 56:365–378.

Hovland, J., and Y. Z. Ishac. 1975. Effects of simulated acid precipitation and liming on nitrification in forest soil. SNSF Project, IR 14/75, Oslo-Ås. 15 p.

Ishac, Y. Z., and J. Hovland. 1976. Effects of simulated acid precipitation and liming on pine litter decomposition. SNSF Project, IR 24/76, Oslo-Ås, 20 p.

Jensen, K. W., and E. Snekvik. 1972. Ambio 1:223–225.

Johannessen, M. 1974. Noen hydrokjemiske resultater fra Fyresdal forsøksområde (Some hydrochemical results from the Fyresdal research basins). SNSF Project, IR 3/74, 5–13, Oslo-Ås.

Johannessen, M., and A. Henriksen. 1978. Water Res 14:615–619.

Joranger, E., J. Schaug, and A. Semb. 1980. Deposition of air pollutants in Norway. *In* D. Drabløs and A. Tollan, eds. *Proc Int conf ecol impact acid precip*, 120–121. SNSF Project, Oslo-Ås.

Laake, M. 1976. Effekter av lav pH på produksjon, nedbryting og stoffkretsløp littoral-sonen. Resultater av feltforsøk i Tovdal 1974–75 (Effects of low pH on primary production, decomposition and nutrient cycling in the littoral zone of softwater lakes, Tovdal, south Norway). SNSF project IR 29/76, Oslo-Ås, 75 p.

Låg, J. 1963. Forsk Fors Landbr 14:553–563.

Larssen, S., and J. E. Hanssen. 1980. Annual variations and origin of aerosol components in the Norwegian Arctic and Sub-Arctic. WHO Tech. Conf. on Regional and Global Observation of Atmospheric Pollution Relative to Climate. Boulder, Colo. (WHO No. 549), 251–258.

Leivestad, H. 1982. *In* T. A. Haines and R. E. Johnson, eds. *Acid rain/fisheries*, 157–164. Amer Fisheries Soc., Bethesda, Maryland.

Leivestad, H., G. Hendrey, I. P. Muniz, and E. Snekvik. 1976. Effects of acid precipitation on freshwater organisms. *In* F. H. Bräkke, ed. *Impact of acid precipitation on forest and freshwater ecosystems in Norway,* 87–111. SNSF Project FR 6/76, Oslo-Ås.

Leivestad, H., and I. P. Muniz. 1976. Nature 259:391–392.

Lundquist, D. 1977. Modellering av hydrokjemi i nedbørfelter (Hydrochemical modeling of drainage basins). SNSF Project IR 31/77, Oslo-Ås, 27 p.

Muniz, I. P. 1981. Acidification and the Norwegian salmon. *In Acid Rain and the Atlantic Salmon*. Int. Atlantic Salmon Found. Spec. Publ. 10, 65–72.

Muniz, I. P. 1984. Phil Trans R Soc Lond B305:517–528.

Muniz, I. P., and H. Leivestad. 1980. Acidification—effects on freshwater fish. *In* D. Drabløs and A. Tollan, eds. *Proc Int conf ecol impact acid precip*, 93–98, SNSF Project, Oslo-Ås.

Nilssen, J. P., T. Østdahl, and W. T. W. Patts. 1984. Species replacement in acidified lakes: physiology, predation or competition. Report Inst. Freshwater Res., Drottning-holm, 61:148–153.

Nilssen, J. P., and S. Sandøy. 1986. Hydrobiol 143:349–354.

Nordal, K. P., E. Dahl, K. Sørhus, K. J. Berg, Y. Thomassen, I. Kofstad, and J. Halse. 1988a. Pharmacology & Toxicology, in press.

Nordal, K. P., E. Dahl, Y. Thomassen, E. K. Brodwall, and J. Halse. 1988b. Pharmacology & Toxicology 62:80–83.

Nordø, J. 1974. Ann Met(NF) 9:71–77.

Nordø, J. 1977. En statistisk undersøkelse av surheten i en bekk nær Birkenes i Aust-Agder. *In* I. Th. Rosenqvist, ed. *Sur jord—surt vann (Acid soil—acid water),* 106–110. Ingeniørforlaget A/S, Oslo.

Nordø, J., A. Eliassen, and J. Saltbones. 1973. Adv Geophys 18B:137–150.

Nordø, J., and K. Hjortnæs. 1967. Geofys Publ 26(12):1–46.

Norton, S. A., A. Henriksen, B. M. Wathne, and A. Veidel. 1987. Aluminum dynamics in response to experimental additions of acid to a small Norwegian stream. *In Acidification and water pathways*, 249–258. Bolkesjø, Norwegian National Com. for Hydrology.

Nygaard, P. H. 1988. Virkning av kunstig forsuring på feltvegetasjon, bunnvegetasjon og jord i en furuskog (Vaccinio-Pinetum) i Aust-Agder (Effect of artificial acidification on the ground cover vegetation and soil in a Scots pine forest in Åmli, Aust-Agder). Thesis, University of Oslo, 85 p.

Oden, S. 1968. Ecological Bull (Stockholm) 1:68.

OECD. 1977. The OECD programme on long-range transport of air pollutants. Measurements and findings. Paris, Organisation for Economic Co-operation and Development.

Ogner, G., and O. Teigen. 1980. Plant and Soil 57:305–321.

Økland, J. 1980. Environment and snails (Gastropoda): Studies of 1000 lakes in Norway. *In* D. Drabløs and A. Tollan, eds. *Proc Int conf ecol impact acid precip*, 322–323, SNSF Project, Oslo-Ås.

Økland, J., and K. A. Økland. 1986. Experientia 42:471–486.

Økland, K. A. 1980. Mussels and crustaceans: Studies of 1000 lakes in Norway. *In* D. Drabløs and A. Tollan, eds. *Proc Int conf ecol impact acid precip*, 324–325. SNSF Project, Oslo-Ås.

Ottar, B. 1978a. Årsakene til nedbørens forsuring (Origin of precipitation acidification). Nordforsk, Miljøvårdssekretariatet, Publ. 1975:10, 110 p.

Ottar, B. 1978b. Atm Environ 12:445–454.

Ottar, B., Y. Gotaas, Ø. Hov, T. Iversen, E. Joranger, M. Oehme, J. Pacyna, A. Semb, W. Thomas, and V. Vitols. 1986. Air pollution in the Arctic. Final report of a research program conducted on behalf of British Petroleum, Ltd., Lillestrøm (NILU Or 30/86). 80 p.

Overrein, L. N., H. M. Seip, and A. Tollan. 1980. Acid precipitation—Effects on forest and fish. SNSF Project FR 19/80, 175 p.

Pacyna, J. M. 1984. Atmos Environ 18:41–50.

Pacyna, J. M., A. Semb, and J. E. Hanssen. 1984. Tellus 36B:163–178.

Raddum, G. G. 1980. Comparison of benthic invertebrates in lakes with different acidity. *In* D. Drabløs and A. Tollan, eds. *Proc Int conf ecol impact acid precip*, 330–331, Oslo-Ås.

Raddum, G. G., and A. Fjellheim. 1984. Verh int Verein Limnol 22:1973–1980.

Rahn, K. A., E. Joranger, A. Semb, and T. J. Conway. 1980. Nature 287:824–825.

Reite, O. B., and M. Staurnes. 1987. Acidified water: effects on physiological mechanisms in the gills of salmonids. Surface Water Acidification Programme, Mid-term Review Conference, Bergen, Norway, 298–304, The Royal Society, London.

Rosenqvist, I. Th. 1977. Sur jord—surt vann (Acid soil—acid water). Ingeniørforlaget A/S, Oslo, 123 p.

Rosenqvist, I. Th. 1978. Sci Total Environ 10:39–49.

Rosenqvist, I. Th. 1985. Land Use Policy 2:70–73.

Rosenqvist, I. Th. 1987. Acidity of surface water in Norway. *In Acidification and water pathways*, 223–235, Bolkesjø, Norway, May 1987, Norwegian National Commitee for Hydrology.

Rosenqvist, I. Th., and H. M. Seip. 1986. Forsurning av vassdrag—hvor stor er uenigheten? Kjemi 46(3):13–18. (CEGB Translation, T 16668(R): Acidification of watercourses—how great is the disagreement? Central Electricity Res. Lab., Kelvin Av., Leatherhead, Surrey KT22 7SE, England).

Rosseland, B. O., I. Sevaldrud, D. Svalastog, and I. P. Muniz. 1980. Studies of freshwater fish populations—effects of acidification on reproduction, population structures, growth and food selection. *In* D. Drabløs and A. Tollan, eds. *Proc Int conf ecol impact acid precip*, 336–337. SNSF Project, Oslo-Ås.

Rosseland, B. O., and O. K. Skogheim. 1984. A comparative study on salmonid fish species in acid aluminum-rich water. II. Physiological stress and mortality of one- and two-year-old fish. Report Inst Freshwater Res., Drottningholm 61:186–194.

Rosseland, B. O., O. K. Skogheim, F. Kroglund, and E. Hoell. 1986b. Water, Air, and Soil Pollution 30:751–756.

Rosseland, B. O., O. K. Skogheim, and I. H. Sevaldrud. 1986a. Water, Air, and Soil Pollution 30:65–74.

Røsten, E. 1985. Kunstig forsuring og kalking i skog. Virkninger på markvegetasjonen i en tidlig suksesjonsfase på et plantefelt med bjørk, furu og gran (Effects of artificial acidification and liming on the ground flora in young plantations of Norway spruce, Scots pine and silver birch). Thesis, Agricultural University of Norway, 52 p.

Royal Ministry for Foreign Affairs. 1971. Air pollution across national boundaries. The impact on the environment of sulphur in air and precipitation. Sweden's case study for the United Nations conference on the human environment. Royal Ministry for Foreign Affairs, Royal Ministry of Agriculture, Stockholm. 93 p.

Rustad, S., N. Christophersen, H. M. Seip, and P. J. Dillon. 1986. Can J Fish Aquat Sci 43:625–633.

Schjoldager, J., H. Dovland, P. Grennfelt, and J. Saltbones. 1984. Photochemical oxidants in the northwestern Europe 1976–1979. A pilot study. Proc. EPA-OECD, Int. Conf. Long-Range Transport Models for Photochemical Oxidants and their Precursors, 439–473. Research Triangle Park, April 12–14, 1983.

Seip, H. M. 1980. Acidification of freshwater: sources and mechanisms. In D. Drabløs and A. Tollan, eds. Proc Int conf ecol impact acid precip, 358–365. SNSF Project, Oslo-Ås.

Seip, H. M. 1987. Norwegian SWAP sites: Description and review of field studies. Surface Water Acidification Programme, Mid-term review conference, 22–26 June, Bergen, Norway, 59–68. The Royal Society, London.

Seip, H. M., G. Abrahamsen, N. Christophersen, E. T. Gjessing, and A. O. Stuanes. 1980b. Snowmelt and meltwater chemistry in minicatchments. SNSF Project IR 53/80, Oslo-Ås, 51 p.

Seip, H. M., G. Abrahamsen, E. T. Gjessing, and A. Stuanes. 1978. Studies of soil, precipitation and runoff chemistry in six small natural plots ("minicatchments"). SNSF Project IR 46/78, Oslo-Ås, 62 p.

Seip, H. M., S. Andersen, and B. Halsvik. 1980a. Snowmelt studied in a minicatchment with neutralized snow. SNSF Project IR 65/80, Oslo-Ås, 20 p.

Seip, H. M., N. Christophersen, and S. Rustad. 1986. Water, Air, and Soil Pollution 31:239–246.

Semb, A. 1978a. Atmos Environ 12:455–460.

Semb, A. 1978b. Deposition of trace elements from the atmosphere. SNSF Project, FR 13/78, Oslo-Ås, 28 p.

Semb, A., R. Brekkan, and E. Joranger. 1984. Geophys Res Lett 11:445–448.

Semb, A., and H. Dovland. 1986. Water, Air, and Soil Pollution 30:5–16.

Sevaldrud, I. H., and I. P. Muniz. 1980. Sure vatn og innlandsfisket i Norge. Resultater av intervjuundersøkelsene 1974–1979 (Acid lakes and freshwater fishery in Norway. Results of interviews 1974–1979). SNSF Project IR 77/80, Oslo-Ås, 95 p.

Sevaldrud, I. H., and O. K. Skogheim. 1986. Water, Air, and Soil Pollution 30:381–386.

SFT. 1985. Overvåking av langtransportert forurenset luft og nedbør. Årsrapport 1984, Rapport 201/85, 189 pp. (The Norwegian monitoring program for long-range transported air pollutants. Annual report 1984, Report 201/85, 189 pp.). The Norwegian State Pollution Control Authority (SFT).

SFT. 1986a. Overvåking av langtransportert forurenset luft og nedbør. Årsrapport 1985, Rapport 256/86, 199 pp. (The Norwegian monitoring program for long-range transported air pollutants. Annual report 1985, Report 256/86, 199 pp.). The Norwegian State Pollution Control Authority (SFT).

SFT. 1986b. The Norwegian monitoring programme for long-range transported air pollutants. Results 1980–1984. The Norwegian State Pollution Control Authority (SFT), TA-number: TA-606 Report 230/86, 93 pp.

SFT. 1987. 1000 lakes survey 1986 Norway. The Norwegian State Pollution Control Authority (SFT), Report 283/87, 33 p.

Singh, B. R. 1980. Effects of acid precipitation on soil and forest. 3. Sulfate sorption by acid forest soils. *In* D. Drabløs and A. Tollan, eds. *Proc Int conf ecol impact acid precip*, SNSF Project, 194–195, Oslo-Ås.

Singh, B. R. 1984a. Soil Science 138:189–197.

Singh, B. R. 1984b. Soil Science 138:294–297.

Singh, B. R., G. Abrahamsen, and A. O. Stuanes. 1980. Soil Sci Soc Am J 44:75–80.

Skartveit, A. 1981. Nord Hydrol 12:65–80.

Skartveit, A. 1982. Atmos Environ 16:2715–2724.

Skogheim, O. K., B. O. Rosseland, and I. H. Sevaldrud. 1984. Deaths of spawners of Atlantic salmon (*Salmo salar* L.) in river Ogna, SW Norway, caused by acidified aluminum-rich water. Report Inst Freshwater Res, Drottnigholm, 61:195–202.

Snekvik, E. 1972. Vann 7:59–67.

Stabel, B. 1987. Development of the diatom flora in Tveitå Fiskeløstjern, Agder, southern Norway. Surface Water Acidification Programme, Mid-term Review Conference, 381–385, Bergen, Norway, The Royal Society, London.

Stachurska-Hagen, T. 1980. Acidification experiments in conifer forest. 8. Effects of acidification and liming on some soil animals: Protozoa, Rotifera and Nematoda. SNSF Project, IR 74/80, Oslo-Ås, 23 p.

Steinnes, E., and J. P. Rambæk. 1980. Kartlegging av tungmetaller i Norge ved analyse av mose (Mapping of heavy metals in Norway by analyses of moss samples). Kjeller, Institutt for atomenergi.

Strand, L. 1980. The effects of acid precipitation on tree growth. *In* D. Drabløs and A. Tollan, eds. *Proc Int conf ecol impact acid precip*, 64–67. SNSF Project, Oslo-Ås.

Stuanes, A. O., G. Abrahamsen, and A. Stormoen. 1987. Medd Nor inst skogforsk 39(13):233–249.

Stuanes, A. O., G. Abrahamsen, and B. Tveite (in press). Effect of artificial rain on soil chemical properties. *In* P. Mathy, ed. *Air Pollution and Ecosystems*, D. Reidel Publishing Company, Dordrecht, Boston.

Stuanes, A. O., and T. E. Sveistrup. 1979. Field experiments with simulated acid rain in forest ecosystems. 2. Description and classification of the soil used in field, lysimeter and laboratory experiments. SNSF Project FR 15/79, Oslo-Ås, 35 p.

Sture, S. 1984. Gjødslingsforsøk i gran- og furuskog (Fertilizer trials in forests of Norway spruce and Scots pine). Norsk institutt for skogforskning, Report, 528 p.

Sullivan, T. J., N. Christophersen, R. P. Hooper, H. M. Seip, I. P. Muniz, P. D. Sullivan, and R. D. Vogt. 1987. Episodic variation in streamwater chemistry at Birkenes, southernmost Norway: Evidence for the importance of water flow paths. *In Acidification and Water Pathways*, 269–279. Proc Int Symp, Bolkesjø, Norway.

Sullivan, T. J., N. Christophersen, I. P. Muniz, H. M. Seip, and P. D. Sullivan. 1986. Nature 323:324–327.

Teigen, O. 1975. Spire og etableringsforsøk med gran og furu i kunstig forsuret mineraljord (Germination and establishment of spruce and pine in artificially acidified mineral soil). SNSF Project, IR 10/75, Oslo-Ås, 36 p.

Teigen, O., G. Abrahamsen, and O. Haugbotn. 1976. Eksperimentelle forsuringsforsøk i skog. 2. Lysimeterundersøkelser (Acidification experiments in conifer forest. 2. Lysimeter investigations). SNSF Project IR 26/76, Oslo-Ås, 45 p.

Tollan, A. 1981. Annotated bibliography 1974–1980. SNSF Project. Acid precipitation—effect on forest and fish. SNSF Project, Oslo-Ås, 42 p.

Torgersen, H. 1934. Forsøk med oparbeidelse av sure, fisketomme vann (Stocking trials in previously fishfree lakes). Stangfiskeren 38–46.

Traaen, T. 1977. Nedbryting av organisk materiale (Decomposition of organic matter). SNSF Project TN 35/77, Oslo-Ås, 12 p.

Tveite, B. 1975. Sur nedbør—skogproduksjon. Regionale årring-undersøkelser (Acid precipitation—tree growth. Regional tree-ring investigations). SNSF Project, TN 11/75, Oslo-Ås, 49 p.

Tveite, B. 1980a. Effects of acid precipitation on soil and forest. 9. Tree growth in field experiments. In D. Drabløs and A. Tollan, eds. Proc Int conf ecol impact acid precip, 206–207, SNSF Project, Oslo-Ås.

Vestjordet, E. 1975. Sur nedbør—skogproduksjon. Utvikling av årringbredden i furu—og granbestand på Sørlandet for tidsrommet 1931–1971 (Development of annual ring width in stands of Scots pine and Norway spruce in south Norway, 1931–1971). SNSF Project, IR 12/75, Oslo-Ås, 35 p.

Vogt, T. 1986. Vannkvalitet og helse. Analyse av en mulig sammenheng mellom aluminium i drikkevann og alders demens (Water quality and health. Study of a possible relation between aluminum in drinking water and dementia). Sosiale og økonomiske studier 61, Statistisk Sentralbyrå, Oslo.

Wright, R. F. 1977. Historical changes in the pH of 128 lakes in southern Norway and 130 lakes in southern Sweden over the period 1923–1976. SNSF Project, TN 34/77, Oslo-Ås, 71 p.

Wright, R. F. 1986. RAIN Project, Appendix to annual report for 1986. Norwegian Inst. Water Research, Oslo.

Wright, R. F., B. J. Cosby, G. M. Hornberger, and J. N. Galloway. 1986b. Water, Air, and Soil Pollution 30:367–380.

Wright, R. F., T. Dale, A. Henriksen, G. H. Hendrey, E. T. Gjessing, M. Johannessen, C. Lysholm, and E. Støren. 1977. Regional surveys of small Norwegian lakes. SNSF Project IR 33/77, Oslo-Ås, 135 p.

Wright, R. F., E. Gjessing, N. Christophersen, E. Lotse, H. M. Seip, A. Semb, B. Sletaune, R. Storhaug, and K. Wedum. 1986a. Water, Air, and Soil Pollution 30:47–63.

Wright, R. F., and A. Henriksen. 1978. Limnol Oceanogr 23:487–498.

Wright, R. F., and A. Henriksen. 1983. Nature 305:422–424.

Wright, R. F., and E. Snekvik. 1978. Verh int Verein Limnol 20:765–775.

Chemistry of Rocky Mountain Lakes

J.T. Turk* and N.E. Spahr*

Abstract

Lakes of the Rocky Mountain region are not chronically acidic. Data generally are unavailable to determine whether episodic acidity is a problem and whether such acidity is anthropogenic. Researchers have hypothesized that decreases in pH of about 0.2 unit have occurred in lakes along the Colorado Front Range between the late 1930s and 1979. However, confirmation of this hypothesis is impossible with present (1987) data.

Many lakes in the Rocky Mountains have little alkalinity to buffer any future increase in acid deposition. Loss of alkalinity, caused by acidic deposition, also has been hypothesized for lakes along the Colorado Front Range and in west-central Colorado. Similarly, confirmation of this hypothesis is impossible with present data. Maximum loss of alkalinity for lakes in Rocky Mountain National Park and the Mt. Zirkel Wilderness Area, Colorado is estimated to be 10 μeq/L or less. Maximum acidification of lakes is about 10 μeq/L for the Colorado Front Range, 16 μeq/L for the Wind River Range, and about 9 μeq/L for the Bitterroot Range.

Wetfall is the primary control on chloride and sulfate concentrations in lakes in the Rocky Mountains. Comparison of chloride concentrations in lakes and in wetfall indicates that evapotranspiration is small in most high-elevation lakes having alkalinity less than 200 μeq/L. In addition, comparison of sulfate concentrations in lakes and in wetfall indicates that processes that add sulfate to or remove it from the sulfate concentrations supplied by wetfall rarely change lake sulfate concentrations by more than 50% from wetfall-sulfate concentrations. In particular, dryfall contributions of sulfate are small compared to wetfall contributions for high-elevation lakes that have alkalinity less than 200 μeq/L.

Data generally are inadequate to calibrate and verify models that might be used to predict the effects of changes in atmospheric deposition of acidity and

*U.S. Geological Survey, Bldg. 53, MS 415, Denver Federal Center, Lakewood, CO 80225 USA.

associated constituents on lake chemistry. The few watersheds for which sufficient data exist still are inadequate to represent the range of conditions that occur in the Rocky Mountains.

Physical- and legal-access restrictions hamper the collection of data pertaining to important watershed processes, lake chemistry during periods of ice cover and snowmelt, and quantity and chemistry of atmospheric deposition. Alternative instrumentation and approaches are necessary to measure the relative and absolute magnitudes of sources, sinks, and processes important to watersheds in the Rocky Mountains, especially for remote watersheds.

I. Introduction

The use of coal resources for power generation, retorting of oil shale, and smelting of sulfide ores, along with population growth, likely will result in an increase of atmospheric emissions in the Rocky Mountain region. These emissions, if not sufficiently dispersed or neutralized, could acidify atmospheric deposition to watersheds. The Rocky Mountain region contains many lake watersheds that may not be capable of neutralizing all added acid. Thus, a loss of alkalinity and a decrease in pH to values that are harmful to aquatic and terrestrial life may occur in these lakes and in watershed soils.

II. Purpose

This chapter summarizes the results of previous work on the acidification of lakes in the Rocky Mountains and interprets data from the 1985 Western Lake Survey to determine controls on the chemistry of these lakes.

III. Discussion

A. Sensitivity of Lakes to Acidification

If increased atmospheric emissions acidify atmospheric deposition in the Rocky Mountain region, a large range of response is likely to occur in the lakes of the region. Early studies (Hendrey et al., 1980) indicate that, even in watersheds receiving the most acidic atmospheric deposition, not all surface water becomes acidified. The most obvious difference between watersheds in which acidification has occurred and watersheds in which acidification has not occurred is bedrock geology. Lakes in watersheds underlain by limestone or other rapidly weathering rock types indicate little evidence of acidification, even in response to wetfall as acidic as 4 pH units. However, watersheds underlain by granite may contain large numbers of acidified lakes if exposed to acidic wetfall (Galloway and Cowling, 1978; Hendrey et al., 1980).

Models of lake sensitivity as a function of bedrock geology predict large variations in sensitivity in the Rocky Mountain region (Galloway and Cowling, 1978; National Atmospheric Deposition Program, 1982). These models have been modified or replaced by other predictive techniques because of the large variations in lake sensitivity within areas of homogeneous bedrock geology. Some models, such as the Integrated Lake Watershed Acidification Study (ILWAS) model (Gherini et al., 1985), require too much data to be applied widely in the wilderness areas common to the Rocky Mountain region. Examples of empirical maps and models that use alkalinity as a measure of sensitivity to acidification are:

1. Alkalinity maps based on existing alkalinity data and empirical associations of alkalinity with watershed characteristics (Omernik and Griffith, 1986).
2. Alkalinity as a function of lake elevation for lakes on basalt mesas (Turk and Adams, 1983; Turk and Campbell, 1984).
3. Alkalinity as a function of lake elevation for lakes on granite (Turk and Campbell, 1987).
4. Alkalinity as a function of computed flow path of water on basalt mesas (Turk and Campbell, 1984).
5. Alkalinity as a function of dominant, mapped-geologic unit for lakes on granite (Turk and Campbell, 1987).

These empirical maps and models of lake alkalinity indicate that alkalinity is a function of bedrock geology and of hydrology. Bedrock geology can affect lake alkalinity through control of weathering rates per unit area and volume of watershed minerals. To have the least alkalinity and to be the most sensitive to acidification, a lake needs to be on any one of the slow-weathering bedrock types common in the Rocky Mountain region, for example, quartzite, quartz monzonite, granite, or basalt. These slow-weathering bedrock types result in minimal lake alkalinity because of slow rates of weathering per unit area of mineral. Within each of these bedrock types, the most sensitive lakes have minimal amounts of material that can form the matrix of an aquifer, such as glacial till, glacial gravel, or alluvium. These materials not only provide a large surface area per unit volume for mineral dissolution and alkalinity production but also provide a continuous flow of alkalinity into a lake or stream.

Hydrology also can affect lake alkalinity. The most sensitive lakes likely occur at the beginning of hydrologic flow paths, rather than farther along these flow paths. Longer flow paths provide additional time for reaction between groundwater and the minerals that dissolve to produce alkalinity. Thus, sensitive lakes likely are on or near topographic highs, such as saddles, cirques, and mesa tops, that are groundwater recharge areas, rather than in groundwater discharge areas, such as stream valleys. Temporal variations in hydrologic processes, such as snowmelt, may cause changes in flow paths that can decrease alkalinity by minimizing the mineral surface area exposed to reaction and contact time of water and mineral surface. For example, saturation of soils during snowmelt may cause a large fraction of the flow to a lake to occur as overland flow, whereas, as the zone of saturation thins during summer, most of the flow might be groundwater.

Because of the many factors that affect alkalinity, the distribution of sensitive lakes is very site-specific. The design criteria of any lake-sampling program affect the proportion of sensitive to nonsensitive lakes sampled. Thus, if the results of one program for one area are compared to the results of another program for another area, a sampling bias may determine which area seems to be the most sensitive to acidification. In this chapter, comparisons of sensitivity among areas use a single data base, the Western Lake Survey, conducted by the U.S. Environmental Protection Agency during the fall of 1985. The Western Lake Survey included the Rocky Mountain region. Additional information for particular regions is used to address questions specific to the individual regions.

B. Data Indicating Possible Acidification of Lakes

Precipitation chemistry of the Rocky Mountain region is documented at a few sites but only for the period since about 1979. Similarly, quality-assured data about the chemistry of lakes only recently have become available on a regional basis. Few lakes or streams have been sampled sufficiently to enable determination of the presence or absence of trends related to acidification.

1. Precipitation-Quality Data

The first documentation of trends in the chemistry of precipitation for the region is for the Como Creek watershed, 6 km east of the Continental Divide in Boulder County, Colorado (Lewis and Grant, 1980). During 150 weeks beginning in 1975, analysis of weekly bulk-deposition samples indicated a linear decrease in pH of about 0.8 unit. The magnitude of the increase in hydrogen-ion loading was balanced by an equivalent increase in the nitrate loading. Sulfate concentration did not have a statistically significant trend with time; sulfate was only slightly larger in loading rate than nitrate was: 196 (sulfate) compared to 129 (nitrate) (eq/ha)/yr (Grant and Lewis, 1982).

The Como Creek precipitation data are considered affected by the densely urbanized Denver Front Range metropolitan area. From January to July, when easterly winds are the most prevalent, a 100-fold increase in concentrations of pollutants in air masses from the metropolitan area is the main factor that affects average pollutant concentration during collection periods of about a week, such as used in the work of Lewis and Grant (1980) on Como Creek (Kelly and Stedman, 1980). The large effect of the metropolitan area on precipitation chemistry along the Front Range was documented further by Huebert and associates (1983), who concluded that more than 60% of the total concentration of NO_x and HNO_3 was attributable to the nearby metropolitan area.

Bulk-deposition samples were collected from a statewide network of 42 sites in Colorado between May 1982 and May 1983 (Lewis et al., 1984). The calculated deposition rate of oxides of nitrogen and sulfur was about equal to emissions from power plants in and near Colorado. A negative correlation of pH, as a function of elevation, was attributed to the preferential removal of alkaline crustal materials relative to acidic particles from air masses by orographic effects. Five stations had an average pH less than 5.0; the minimum average pH was 4.7. Measurements of wetfall at a site in west-central Colorado indicate an average pH of 4.9 for 1981 to 1984 (Harte et al., 1985).

Atmospheric deposition of sulfate in wetfall has been correlated with SO_2 emissions from nonferrous smelters. The annual, volume-weighted concentration of sulfate at eight wetfall sites in Arizona, Colorado, Idaho, and Montana correlated with annual SO_2 emissions from smelters in Arizona, New Mexico, Nevada, and Utah for 1980 to 1983 (Oppenheimer et al., 1985). This correlation approach was modified to examine correlation between monthly volume-weighted concentrations of sulfate at five wetfall sites in Colorado and monthly SO_2 emissions from smelters during 1980 to 1984 (Epstein and Oppenheimer, 1987). Little correlation occurred between monthly smelter emissions of SO_2 and monthly volume-weighted concentration of sulfate. Modification of the data set to eliminate months with incomplete sample collection and to deseasonalize the data resulted in explanation of 41% of the variance in wetfall-sulfate concentration by the smelter emissions ($P<0.001$).

Wetfall chemistry data are available for the Rocky Mountain region from the National Atmospheric Deposition Program (Figures 6-1 through 6-3). These data provide an internally consistent regional assessment of the acidity status of wetfall. Although most of the stations are located at low altitudes, a few stations are located at altitudes typical of many sensitive aquatic systems. The sparsity of high-altitude stations makes regional contour maps ambiguous.

2. Surface-Water Data

At the Como Creek site discussed by Lewis and Grant (1980), the increase in hydrogen-ion loading to the watershed was matched by an approximately equivalent decrease in stream transport of bicarbonate. The stream loading of sulfate, nitrate, ammonia, and dissolved organic matter also indicated significant increases. The decrease in total cation loading in the stream accounted for only about one-third of the measured decrease in alkalinity, indicating that much of the decrease in alkalinity was attributable to neutralization by hydrogen ions (Lewis and Grant, 1979).

In a 1979 reconnaissance of Colorado lakes initially sampled during 1938 to 1942 and 1949 to 1952, a 17% decrease in alkalinity and a 0.2-unit decrease in pH occurred. Both decreases were attributed to acidification by precipitation (Lewis, 1982). Differences in methods and a lack of data about major ion chemistry have prevented determination of whether this apparent acidification is plausible.

The alkalinity of six ponds in central-western Colorado, initially measured in 1972, was remeasured in 1981 to 1983; a trend toward decreased alkalinity in the more recent data was evident. Seasonal variance of alkalinity in these systems and differences in methods hamper determination of whether a significant trend exists (Schneider, 1984; Harte et al., 1985).

A reconnaissance of lakes in Rocky Mountain National Park, Colorado, and

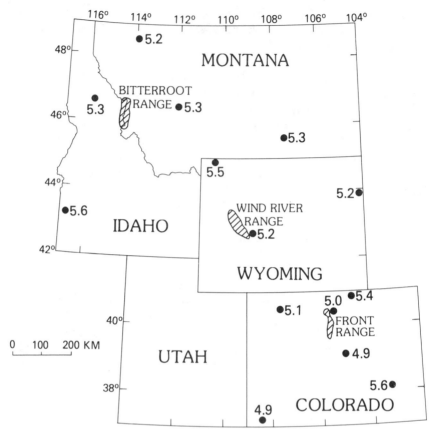

Figure 6-1. Precipitation-weighted mean hydrogen-ion concentrations as pH for 1985.

Yellowstone National Park, Wyoming, indicated no evidence of chronic acidification in either area (Gibson et al., 1983). Maximum possible acidification,
estimated from the concentrations of nitrate and sulfate, was less than 10 μeq/L in
seven of eight watersheds studied. One watershed was calculated to have lost a
maximum of 18 μeq/L.

A 1983 reconnaissance of 92 lakes in the Wind River Range of Wyoming
determined that, although the area contained lakes with little alkalinity, the anion
chemistry was dominated by bicarbonate. Four lakes routinely were monitored
during 1983; they did not show evidence of acidification (Stuart, 1984).

Monitoring of 13 lakes in Glacier National Park, Montana, began in 1984 (Ellis
et al., 1986). Although the study is not designed specifically to detect acidification, the data may be used for that purpose. During 1984 and 1985, the minimum
alkalinity was 80 μeq/L and the minimum pH was 6.5. In the lakes that have the
smallest alkalinity, sulfate concentrations were only about 10 μeq/L, and nitrate
concentrations were only a few μeq/L. Thus, little alkalinity could have been lost

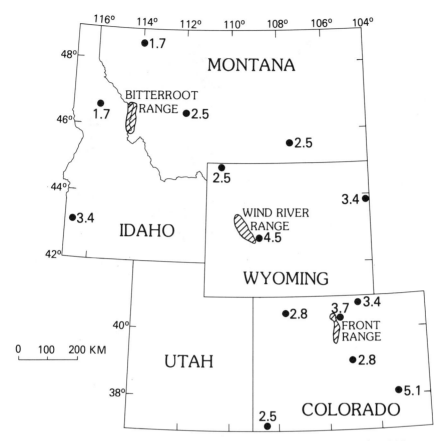

Figure 6-2. Precipitation-weighted mean chloride-ion concentration for 1985.

by neutralization of acidity, which possibly was associated with the sulfate and nitrate.

A 1983 reconnaissance of 70 lakes in the Mt. Zirkel Wilderness Area of Colorado indicated that, although the lakes had small alkalinity concentrations (about one-half the lakes had alkalinity less than 100 µeq/L), there was no evidence of acidification. Maximum possible acidification, estimated from the concentrations of nitrate, sulfate, chloride, and hydrogen ion in wetfall and the concentrations of sulfate and chloride in lakes, was no more than 9 µeq/L (Turk and Campbell, 1987).

A statewide reconnaissance of 175 Colorado lakes in 1984 was used to develop a classification system for all Colorado lakes (Chappell et al., 1985; Chappell et al., 1986). Although the study does not address directly the acidification status of lakes, the data can be used for this purpose. The minimum alkalinity for samples collected during the spring was 22 µeq/L, whereas that for samples collected during the fall was 130 µeq/L. Minimum pH during the spring was 6.4; minimum

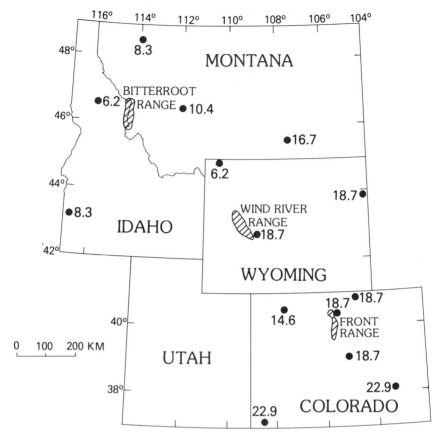

Figure 6-3. Precipitation-weighted mean sulfate-ion concentration for 1985.

pH during the fall was 6.9. Factor analysis explained the differences in lake-water chemistry with factors interpreted to be related to: (1) bedrock geology, (2) vegetation, (3) hydrologic setting, and (4) human use. No factors associated with atmospheric deposition were determined.

The presence of long-term trends in acidification was investigated by diatom and metal stratigraphy for four lakes in Rocky Mountain National Park, Colorado (Baron et al., 1986). The stratigraphic data indicated no historical effect on pH attributable to atmospheric deposition; however, there was evidence of enhanced lead deposition between 1855 and 1985.

Stream chemistry data are available in the Rocky Mountain region for several headwater streams in the national Hydrologic Benchmark Network (Smith and Alexander, 1983). Of the nine streams in the region, only four in Colorado and northern New Mexico have alkalinity values small enough to be likely to respond to acidification. Data from three of the four streams indicate an increasing trend in sulfate; data from all four indicate a decreasing trend in alkalinity between the mid-1960s and 1981.

C. Western Lake Survey

Recent data collected by the U.S. Environmental Protection Agency during the Western Lake Survey in 1985 (Landers et al., 1986; Eilers et al., 1986) provide an internally consistent regional assessment of the acidification status of lakes. The data were collected using a stratified random-sampling design to enable estimation of characteristics of all lakes with surface areas larger than about 1 ha. In this discussion, reference is made to actual lake analyses rather than to estimated lake characteristics.

The distributions of measured pH, alkalinity, chloride, and sulfate for lakes in the Rocky Mountain region are shown in Figure 6-4. For comparison, the regional distribution of pH, chloride, and sulfate in wetfall is shown in Figures 6-1 through 6-3. Of the 424 lakes sampled in this region, only Fern Lake, Wyoming, had a pH less than 6. The sulfate concentration of Fern Lake was 818 μeq/L, indicating that watershed sources of acidity such as mineral weathering, rather than atmospheric deposition, caused the low pH of 4.7. The regional distribution of pH indicates that no chronic acidity problem, caused by atmospheric deposition, presently (1985) exists in the Rocky Mountain region.

The frequency distributions of lake-chloride and lake-sulfate concentrations indicate skewed distributions, with frequent occurrences of chloride concentrations of about 2 μeq/L (Figure 6-4C) and sulfate concentrations of about 20 μeq/L (Figure 6-4D). These concentrations are similar to the concentrations in wetfall (Figures 6-2, 6-3); much of the chloride and sulfate in many of the lakes of this region is possibly controlled by wetfall.

Lakes likely to be sensitive to changes in atmospheric deposition of acidity can be isolated from the data set on the basis of alkalinity; those with alkalinity less than 200 μeq/L are discussed here (Hendry et al., 1980; Turk and Adams, 1983). Most of the lakes sampled have alkalinity less than 200 μeq/L (Figure 6-4B). The most common alkalinity class for lakes that have alkalinity less than 200 μeq/L is 37.5 to 62.5 μeq/L; thus, lakes sensitive to acidification are well represented in this data set. Based on alkalinity, most of the lakes sampled are sensitive to acidification; based on pH, most of the lakes are not acidic during the autumn, which was when the lakes were sampled during the Western Lake Survey (Figure 6-4A). The sampled lakes may be more acidic during periods of snowmelt or intense precipitation; however, data generally are lacking for such episodes.

To understand how a lake might respond to changes in the acidity of atmospheric deposition, it is necessary to know which processes affect the chemistry of the lake. The subset of lakes that have alkalinity less than 200 μeq/L is used here to define lakes sensitive to changes in the acidity of atmospheric deposition. If a watershed has no significant production or uptake of an element, the concentration of that element within the lake is determined by wetfall and evapotranspiration. For a group of lakes in such watersheds, a frequency distribution of the concentration of that element is likely to peak at about the concentration of the element in wetfall, after correction for the amount of evapotranspiration. Figure 6-5 indicates how several processes might alter the frequency distribution of an element within the lakes of a region. Processes that tend to increase the lake

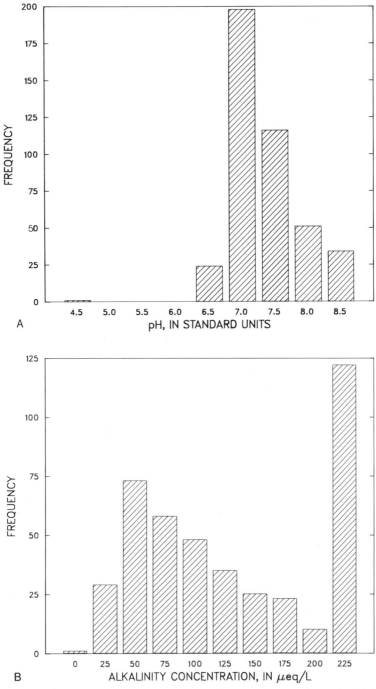

Figure 6-4. (A) Frequency distribution of pH for lake samples collected in the Rocky Mountains during the Western Lake Survey, 1985; (B) frequency distribution of alkalinity for lake samples collected in the Rocky Mountains during the Western Lake Survey, 1985;

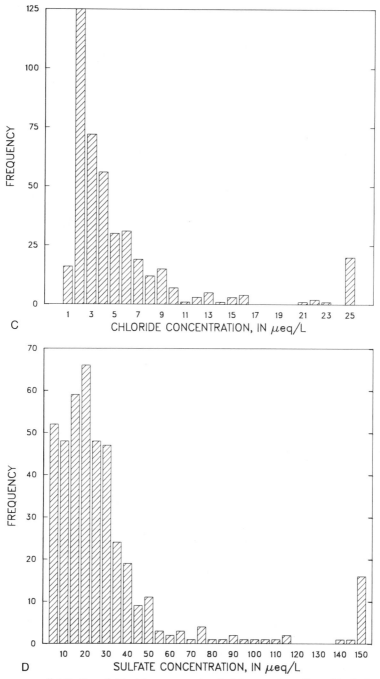

(C) frequency distribution of chloride concentration for lake samples collected in the Rocky Mountains during the Western Lake Survey, 1985; (D) frequency distribution of sulfate concentration for lake samples collected in the Rocky Mountains during the Western Lake Survey, 1985.

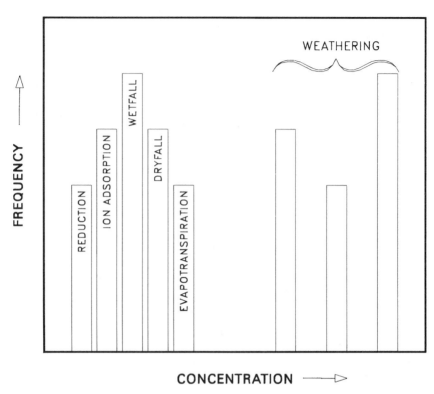

Figure 6-5. Hypothetical frequency distribution of an element controlled by several processes.

concentrations of the element relative to the wetfall concentration, such as evapotranspiration and deposition of dust and gases as dryfall, skew the initial distribution to larger concentrations. Processes that tend to decrease the minimum initial concentration, such as reduction and ion adsorption, skew the initial distribution to smaller concentrations. Because several processes may be occurring in either the same, or different, directions, determining which processes are most effective by simple examination of a frequency distribution is not always possible. Note that in Figure 6-5, contributions from weathering of the element are portrayed as producing lake concentrations considerably larger than concentrations that occur in watersheds with no weathering of the element. The assumption is that if the element is present and available for weathering, this process likely will dominate most other processes. This assumption is likely to be true in those lake watersheds that have readily weatherable deposits of an element, such as sulfate contributed by weathering of ore deposits or sulfate-rich sedimentary rock. This assumption is less likely to be true if trace quantities of sulfate are disseminated throughout slowly weathering minerals.

Historical data generally are not available for lakes sampled during the Western

Lake Survey. Thus, to determine whether regional trends consistent with acidification occur is not possible. Although the pH and alkalinity data indicate only one lake that has a pH value commonly associated with biological damage, some loss of alkalinity compared to historical concentrations is possible. In the absence of historical data, present lake chemistry was used to estimate the possible acidification of lakes in the Mt. Zirkel Wilderness Area, Colorado (Turk and Campbell, 1987). The approach used to interpret the Mt. Zirkel Wilderness Area data compares lake concentrations of chloride and sulfate to wetfall concentrations of the same constituents to evaluate the importance of watershed processes that might affect acidification.

In the Mt. Zirkel Wilderness Area, most lake-chloride concentrations were less than or equal to the wetfall-chloride concentrations; thus, evapotranspiration was known to be an insignificant process affecting most lakes. Similarly, most lake-sulfate concentrations were less than or equal to the wetfall-sulfate concentrations; thus, processes affecting sulfate usually were limited to processes removing sulfate from the water, such as sulfate reduction or adsorption. Because the most frequently measured lake-sulfate concentration was equal to wetfall-sulfate concentration, either sulfate was conservative, and wetfall was the only significant source, or a fortuitous balance of additional processes served to add to, and remove from, the wetfall sulfate. Because no evidence occurred for sources or processes that might add sulfate and associated acidity, the maximum acidification attributable to past increases in deposition of sulfate and acidity could be estimated from the present concentration of sulfate in wetfall.

The method used to determine which processes were significant in affecting lake-chloride and lake-sulfate concentrations in the Mt. Zirkel Wilderness Area can be applied to lakes sampled during the Western Lake Survey. To address processes relevant to lakes that are potentially sensitive to acidification, only lakes with alkalinity less than or equal to 200 μeq/L are used. Frequency distributions of pH, alkalinity, chloride, and sulfate for lakes along the Colorado Front Range that have alkalinity less than or equal to 200 μeq/L are shown in Figure 6-6. The pH distribution indicates that no problem with chronic acidity occurs; however, the alkalinity distribution indicates that many lakes have little alkalinity to buffer additions of acid. The chloride concentration of most lakes is about 2 μeq/L; the chloride concentration in wetfall for this area was about 3 μeq/L during 1985 (Figure 6-2). This similarity in chloride concentrations for lakes and wetfall indicates that the effect of evapotranspiration can be disregarded for most of these lakes. The sulfate concentration of most lakes is about 20 μeq/L; the sulfate concentration in wetfall for this area was about 19 μeq/L during 1985 (Figure 6-3). This similarity in sulfate concentrations for lakes and wetfall indicates that either no processes affecting the wetfall contribution of sulfate are significant in these watersheds or a fortuitous balance exists between processes adding and removing sulfate. If lake-sulfate concentrations are adjusted for the effect of evapotranspiration, calculated as the ratio of lake-chloride concentration to wetfall-chloride concentration for lakes with chloride concentration larger than wetfall concentration, the distribution shown in Figure 6-7 is obtained. The distribution in Figure

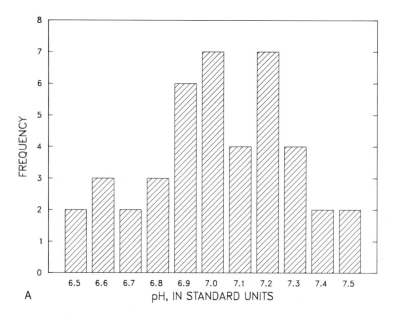

A

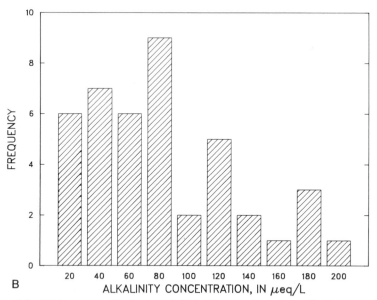

B

Figure 6-6. (A) Frequency distribution of pH for lake samples collected in the Front Range of Colorado during the Western Lake Survey, 1985; (B) frequency distribution of alkalinity for lake samples collected in the Front Range of Colorado during the Western Lake Survey, 1985; (C) frequency distribution of chloride concentration for lake samples collected in the

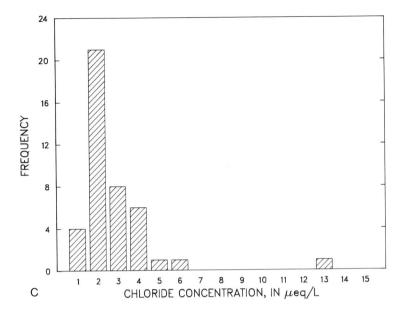

C

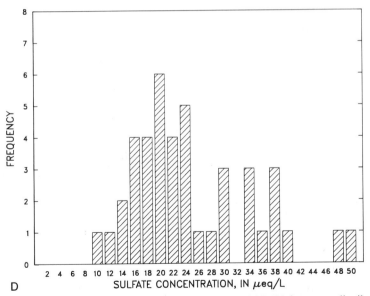

D

Front Range of Colorado during the Western Lake Survey, 1985; (D) frequency distribution of sulfate concentration for lake samples collected in the Front Range of Colorado during the Western Lake Survey, 1985.

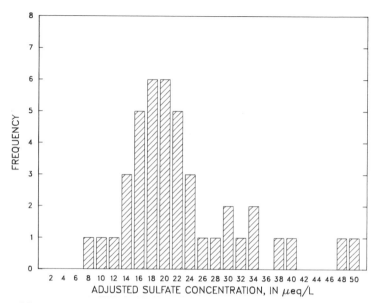

Figure 6-7. Frequency distribution of sulfate concentration, adjusted for evapotranspiration, for lake samples collected in the Front Range of Colorado during the Western Lake Survey, 1985.

6-7 indicates that the wetfall concentration of sulfate is the dominant control on lake-sulfate concentration. Processes that add sulfate, such as mineral weathering and dryfall, rarely contribute more than 25% of the sulfate contributed by wetfall in the sampled lakes. Some lakes that have sulfate concentrations less than those in wetfall also probably have significant sulfate-removal processes, such as sulfate reduction in the lake or the lake watershed. However, such systems are less common than systems controlled only by wetfall. This conclusion is especially significant because of its implications to dryfall contributions of sulfate. Dryfall is difficult to measure directly, even under the most ideal conditions. The considerably varying winds and surfaces of exposed vegetation, soils, rock, and snow that collect dryfall in these watersheds would make direct measurement almost impossible. However, to predict the response of these watersheds to changes in atmospheric deposition of acidity, it is necessary to know how important dryfall is compared to wetfall, which is measured more readily. Thus, the knowledge that dryfall is a minor source of sulfate compared to wetfall considerably simplifies predicting lake-watershed response to changes in acid deposition.

The lake-sulfate concentration of the Colorado Front Range is controlled by wetfall-sulfate concentration. Because other processes that affect lake-sulfate concentration are small compared to the deposition of sulfate by wetfall, these other processes cannot have contributed much acidity compared to that contributed by wetfall. The pH of Colorado Front Range wetfall is about 5.0, or about 10 μeq/L of acidity. Without addition of sulfate and acidity by dryfall, weathering of sulfur minerals, and the concentrating effect of evapotranspiration, the maximum amount of acidification of Colorado Front Range lakes is also about 10 μeq/L. Because some of the present acidity within wetfall probably is not anthropogenic, actual acidification is probably less than 10 μeq/L.

Comparison of frequency distributions of lake pH, alkalinity, chloride, and sulfate among areas within the Rocky Mountain region may provide information about which processes are regionally important and which are specific to a particular area. Because the Western Lake Survey data were collected within a few weeks, using equivalent sampling and analysis protocol, they enable an internally consistent comparison of areas within the Rocky Mountain region. Data for the Colorado Front Range indicate that lakes within the specified alkalinity range are similar with respect to which processes control sulfate concentrations in these lakes. By selecting areas remote from the Colorado Front Range, a test of differences in controlling processes, as a function of location within the region, can be made. Such differences in location may make it possible to test the effects of variations in atmospheric deposition on lake chemistry, because atmospheric deposition may vary with location (Figures 6-1 through 6-3).

The Wind River Range of Wyoming contains about the same number of sampled lakes, which have alkalinity less than or equal to 200 μeq/L, as does the Colorado Front Range. Frequency distributions of lake pH, alkalinity, chloride, and sulfate are shown in Figure 6-8. The range in pH is similar to that for lakes in the Colorado Front Range (Figures 6-6A and 6-8A); however, a greater frequency of lakes in the more alkaline part of the distribution occurs for the Wind River Range. The alkalinity distribution for the Wind River Range (Figure 6-8B) is more normal than that for the Colorado Front Range. The chloride distribution is shifted to larger concentrations for the Wind River Range (Figure 6-8C), the most commonly measured concentrations being about 6 μeq/L. The chloride concentration in wetfall for this area in 1985 was about 5 μeq/L (Figure 6-2); thus, evapotranspiration can be disregarded for most of these lakes. The sulfate distribution for the Wind River Range (Figure 6-8D) is similar to that for the Colorado Front Range (Figure 6-6D); the most commonly measured concentration was about 22 μeq/L. The sulfate concentration in wetfall for this area in 1985 was about 19 μeq/L (Figure 6-3); thus, wetfall is likely the most important control on lake-sulfate concentration. After adjusting lake-sulfate concentrations for the calculated effect of evapotranspiration, about one-third of the sampled lakes in the Wind River Range have sulfate concentrations slightly larger than sulfate concentrations in wetfall (Figures 6-3 and 6-9); thus, dryfall or minor amounts of mineral weathering may affect some of these watersheds. Wetfall remains the likely dominant control on lake-sulfate concentration in these watersheds.

The lake-sulfate concentration of the Wind River Range is controlled by wetfall-sulfate concentration. All other processes increase lake-sulfate concentration by no more than 50% of wetfall-sulfate concentration. The pH of Wind River Range wetfall is about 5.2, or about 6 μeq/L of acidity. If processes that increase lake-sulfate concentration compared to wetfall-sulfate concentration affect acidity in proportion to sulfate, then total acidity supplied to lakes is no more than about 9

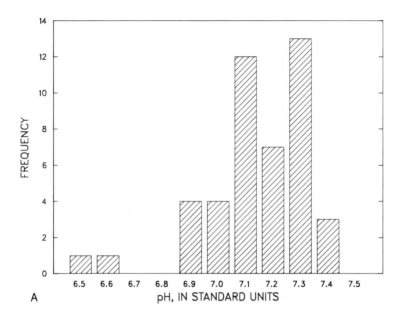

A

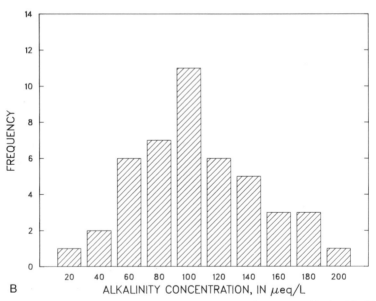

B

Figure 6-8. (A) Frequency distribution of pH for lake samples collected in the Wind River Range of Wyoming during the Western Lake Survey, 1985; (B) frequency distribution of alkalinity for lake samples collected in the Wind River Range of Wyoming during the Western Lake Survey, 1985; (C) frequency distribution of chloride concentration for lake

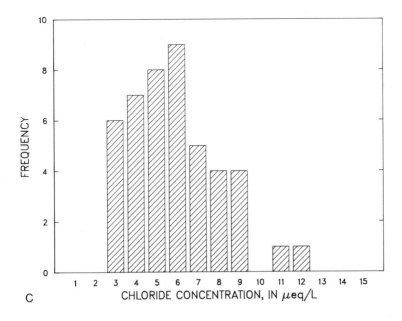

C   CHLORIDE CONCENTRATION, IN μeq/L

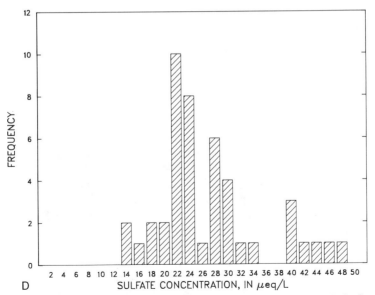

D   SULFATE CONCENTRATION, IN μeq/L

samples collected in the Wind River Range of Wyoming during the Western Lake Survey, 1985; (D) frequency distribution of sulfate concentration for lake samples collected in the Wind River Range of Wyoming during the Western Lake Survey, 1985.

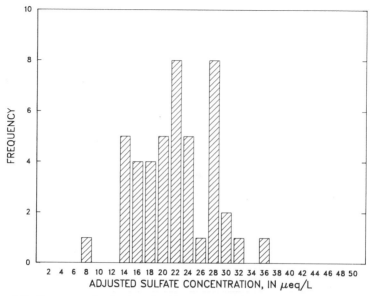

Figure 6-9. Frequency distribution of sulfate concentration, adjusted for evapotranspiration, for lake samples collected in the Wind River Range of Wyoming during the Western Lake Survey, 1985.

μeq/L (6 μeq/L from wetfall plus 3 μeq/L from all other processes). If these same processes increase acidity equivalent to the increase in lake-sulfate concentration compared to wetfall-sulfate concentration, then total acidity supplied to lakes is about 6 μeq/L from wetfall and about 10 μeq/L (0.5 × wetfall-sulfate concentration) from all other processes. Because some of the present acidity within wetfall probably is not anthropogenic, actual acidification is probably less than 6 to 16 μeq/L.

The Bitterroot Mountains of Idaho and Montana have a slightly smaller number of sampled lakes within the specified alkalinity range than the Colorado Front Range or the Wind River Range. The distributions of pH, alkalinity, chloride, and sulfate are shown in Figure 6-10. The pH distribution (Figure 6-10A) is normal and tends to be slightly more acidic than those for the other areas. The alkalinity distribution (Figure 6-10B) tends to be shifted toward smaller concentrations than those for the other areas. The chloride distribution (Figure 6-10C) is almost identical to that for the Colorado Front Range (Figure 6-6C); the most commonly measured concentration was 2 μeq/L. The chloride concentration in wetfall for this area in 1985 was about 2 μeq/L (Figure 6-2); thus, evapotranspiration can be disregarded for most of these lakes. The sulfate distribution (Figure 6-10D) is shifted to smaller concentrations than those for the other areas; the most commonly measured concentration was about 6 μeq/L. The sulfate concentration in wetfall for this area in 1985 was about 6 to 8 μeq/L (Figure 6-3). After adjusting for the effects of evapotranspiration (Figure 6-11), about one-fourth of the lakes may have

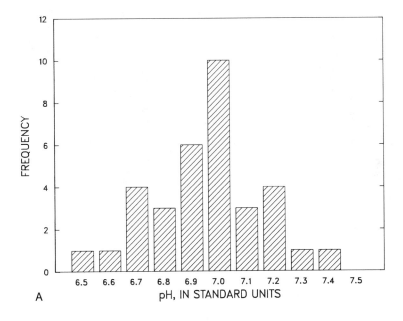

A

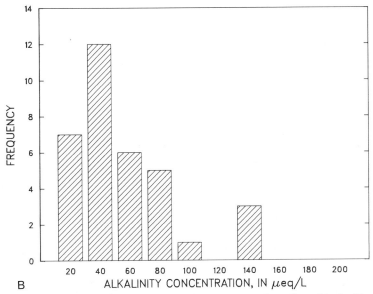

B

Figure 6-10. (A) Frequency distribution of pH for lake samples collected in the Bitterroot Range of Idaho and Montana during the Western Lake Survey, 1985; (B) frequency distribution of alkalinity for lake samples collected in the Bitterroot Range of Idaho and Montana during the Western Lake Survey, 1985.

(Continued)

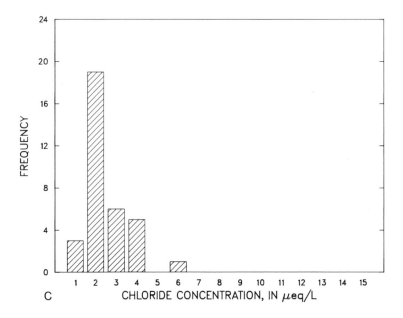

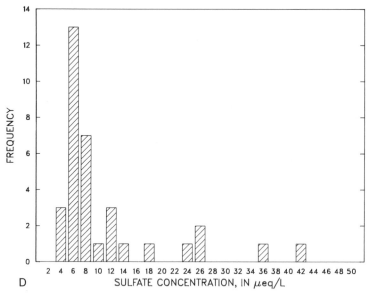

Figure 6-10 (continued). (C) Frequency distribution of chloride concentration for lake samples collected in the Bitterroot Range of Idaho and Montana during the Western Lake Survey, 1985; (D) frequency distribution of sulfate concentration for lake samples collected in the Bitterroot Range of Idaho and Montana during the Western Lake Survey, 1985.

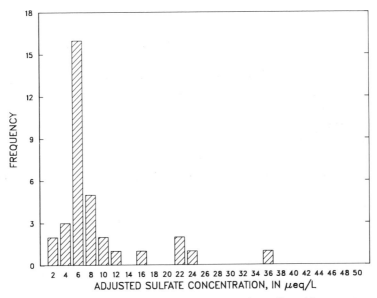

Figure 6-11. Frequency distribution of sulfate concentration, adjusted for evapotranspiration, for lake samples collected in the Bitterroot Range of Idaho and Montana during the Western Lake Survey, 1985.

sources such as dryfall or minor mineral weathering that contribute one-half as much sulfate as does wetfall. Sulfate-removal processes, such as reduction, are unlikely to be significant unless they are balanced by processes that apply sulfate in addition to the wetfall source.

The lake-sulfate concentration of the Bitterroot Range is controlled by the wetfall-sulfate concentration. All other processes increase lake-sulfate concentration by no more than 50% of the wetfall-sulfate concentration. The pH of Bitterroot Range wetfall is about 5.3, or about 5 μeq/L of acidity. If processes that increase lake-sulfate concentration compared to wetfall-sulfate concentration affect acidity in proportion to sulfate, then total acidity supplied to lakes is no more than about 7 μeq/L (5 μeq/L from wetfall plus 2 μeq/L from all other processes). If these same processes increase acidity equivalent to the increase in lake-sulfate concentration compared to wetfall-sulfate concentration, then total acidity supplied to lakes is about 5 μeq/L from wetfall and about 4 μeq/L (0.5 × wetfall-sulfate concentration) from all other processes. Because some of the present acidity within wetfall probably is not anthropogenic, actual acidification probably is less than 7 to 9 μeq/L.

In summary, the Western Lake Survey data indicate that no present problem of chronic acidity occurs in lakes in the Rocky Mountain region, even in lakes close to major urban or industrial sources of atmospheric emissions. Many of the sampled lakes in this region have only small concentrations of alkalinity; they may become acidified if atmospheric deposition becomes more acidic. Lake-chloride concentration is controlled primarily by the wetfall-chloride concentration. Variations of wetfall-chloride concentrations are about threefold; they are matched by variations in the lake-chloride concentration. Evapotranspiration has only a minor effect on most lakes that have alkalinity less than 200 μeq/L, although presumably evapotranspiration would have more effect on lower-elevation lakes, which probably would be more alkaline. Lake-sulfate concentration is controlled primarily by the wetfall-sulfate concentration. Variations of wetfall-sulfate concentrations are similar in magnitude to variations of wetfall-chloride concentrations; however, maximum concentrations of each do not coincide geographically. Nonetheless, lake-sulfate concentrations match the variations of wetfall-sulfate concentrations. Watershed or in-lake processes, such as chemical reduction, ion adsorption, mineral weathering, and dryfall probably are significant controls on lake-sulfate concentrations in some watersheds. Even in the watersheds in which these processes are significant, the wetfall-sulfate concentrations still tend to be the primary control on lake-sulfate concentrations.

IV. Future Research Needs

Lakes and streams in the Rocky Mountain region indicate little evidence of acidification; however, many are likely to respond rapidly to changes in the atmospheric deposition of acidity. Considering the number of lakes and streams in the region, and the number of important watershed characteristics that may affect watershed response to acidity, few systems are monitored routinely. Thus, the preparation of any watershed model appropriate for the evaluation of regional acidification is hampered. Such models would need to be calibrated as a function of the variability of geology, soil, vegetation, climate, atmospheric deposition, and hydrologic characteristics common to the region. As of 1987, data do not exist to calibrate and verify such models, except for a few individual watersheds.

To determine if regional changes in lake and stream chemistry occur in response to changes in atmospheric deposition, an appropriate measure of atmospheric deposition needs to be available. The present monitoring networks have few sites at elevations comparable to the elevations of watersheds likely to be affected by atmospheric deposition. Data on the quantity and quality of wetfall at high elevations generally are unavailable. To determine adequately whether watersheds in a geomorphic unit are affected adversely by atmospheric deposition, some minimal number of sites at the elevation of sensitive watersheds is needed for each geomorphic unit of interest. Interpretations presented in this chapter indicate that data on wetfall deposition are more critical than data on dryfall deposition. Both physical and legal problems of access will hamper collecting such data, particularly in wilderness areas. Also, use of instruments suitable to lower elevations and more moderate climate is difficult or unreliable at elevations common to the Rocky Mountain region. Surrogate measurements that can be related to existing networks—for example, snowpack measurements or the use of bulk-sampling instruments with no power requirements—may be required to determine changes in atmospheric deposition at high elevations. In addition to problems in measurement of wetfall at high elevations, other depositional mechanisms, such as dry deposition, fog, and rime ice, are almost unevaluated by direct methods at high elevations in the region. Although hypotheses presented in this chapter indicate that such forms of atmospheric deposition are unlikely to affect most watersheds, these hypotheses need to be verified by some direct measurements.

The same problems of access that hinder measurement of atmospheric deposition at high elevations also hinder measurement of aquatic chemistry during periods of snowmelt or intense rainfall. Most of the data about the chemistry of lakes and streams are restricted to sampling during midsummer through early fall. This period is not as likely to indicate early stages of acidification as is the period of snowmelt, for example. In particular, legal restrictions on mechanized access to wilderness areas severely hamper determination of acidification or watershed processes that are important to acidification of these areas.

Only a few watersheds in the Rocky Mountain region have been studied with respect to watershed processes that are important to acidification. Regional data about soil-exchange chemistry, weathering reactions, groundwater chemistry, in-lake processes, and hydrologic flow paths are almost nonexistent. These same data are even less common in watersheds typical of those sensitive to acidification. No catchment-size areas exist in which experimental manipulation has been done in sensitive watersheds.

To prevent acidification of sensitive watersheds it will be necessary to determine how much effect a given change in atmospheric emissions from a particular source or group of sources will have on atmospheric deposition. The comparative effects of local, regional, or extraregional anthropogenic or natural sources in controlling the chemistry of atmospheric deposition are not known. Coordinated study of transport, mixing, reaction, and deposition of emissions is needed before reliable predictive models can be developed, calibrated, and verified for the region.

To be most useful, study of effects of changes in atmospheric emissions on atmospheric deposition needs to be coordinated with study of the processes important to the transport and fate of atmospheric deposition in watersheds. Such watershed studies need to represent the variety of geographic, geologic, and hydrologic ranges of conditions common to watersheds sensitive to acidification. In particular, the likely response of mineral weathering, cation release from exchange sites, vegetative growth, bacterial metabolism, and algal metabolism to changes in deposition of nitrogen, sulfur, and hydrogen ions needs to be studied. Watershed and in-lake processes need to be incorporated in realistic predictive models both to enable prediction of effects from proposed sources and to guide effective monitoring of response to present sources.

V. Summary

Published studies about individual lakes and watersheds, as well as regional surveys, indicate that lakes in the Rocky Mountain region are not chronically acidic. Data generally are unavailable to determine whether episodic acidity occurs, the magnitude of the occurrence, and the relative effects of natural and anthropogenic causes. Researchers have hypothesized that decreases in pH of about 0.2 unit have occurred in lakes along the Colorado Front Range between the late 1930s and 1979; however, confirmation of this hypothesis is impossible with the present data.

Many lakes in the Rocky Mountain region have only small concentrations of alkalinity to buffer any future increase in the deposition of acid. Lakes that have minimum alkalinity typically occur on any of several slow-weathering bedrock types common to the region. Within the bedrock types commonly associated with lakes that have minimal alkalinity, the primary controls on alkalinity seem to be related to the time of contact between water and rock and to the surface area exposed to such contact. Loss of alkalinity, caused by atmospheric deposition of acidity, has been hypothesized for lakes along the Colorado Front Range and in west-central Colorado; however, confirmation of this hypothesis is impossible with the present data. Maximum loss of alkalinity for lakes in two areas of Colorado is estimated to be 10 µeq/L or less.

Comparison of lake chemistry among lakes within the Colorado Front Range, Wind River Range, and Bitterroot Range enables determination of the relative importance of several sources, sinks, and processes that affect lake chemistry and acidification. Wetfall is the primary control on both chloride and sulfate concentrations in all three areas. Comparison of chloride concentrations in lakes and wetfall indicates that evapotranspiration is small in most lakes that have alkalinity less than 200 µeq/L. Comparison of sulfate concentrations in lakes and wetfall indicates that processes that add to or remove sulfate from the sulfate concentration supplied by wetfall rarely change lake-sulfate concentrations by more than 50% of wetfall-sulfate concentration. Dryfall, which is difficult to measure directly, is small compared to wetfall contributions for lakes that have alkalinity less than 200 µeq/L.

The maximum acidification of lakes can be estimated from present wetfall pH and from the increase of lake-sulfate concentration by processes other than wetfall-sulfate contribution. Maximum acidification amounts are about 10 µeq/L for the Colorado Front Range, about 16 µeq/L for the Wind River Range, and about 9 µeq/L for the Bitterroot Range. Maximum acidification is less for most lakes in each of these areas because processes other than wetfall-sulfate contribution are negligible for most lakes.

Data generally are inadequate to calibrate and verify models that might be used to predict the effects of changes in atmospheric deposition of acidity and associated constituents on lake chemistry. The few watersheds for which sufficient data exist to calibrate such models are inadequate to represent the variety of conditions that occur in the Rocky Mountain region. Research watersheds, in which experimental manipulation of catchment-size areas is done to verify predictions, do not exist for systems sensitive to acidification.

Physical and legal restrictions on access and operations hamper the collection of data that are essential to evaluating watershed processes, measuring lake chemistry during periods of ice cover and snowmelt, and even measuring the quantity and chemistry of atmospheric deposition. Even if legal restrictions are eased, operational difficulties will require alternative approaches to data collection required for determination of chemistry and of processes. Because wilderness areas comprise much. of the water resources that are sensitive to acidification, alternative instrumentation and approaches especially are needed for these areas.

References

Baron, J., S. A. Norton, D. R. Beeson, and R. Herrmann. 1986. Canadian Journal of Fisheries and Aquatic Sciences 43/7:1350–1362.

Chappell, W. R., R. R. Meglen, G. A. Swanson, L. A. Taylor, R. J. Sistko, R. B. McNelly, E. L. Hartman, M. L. Rottman, R. W. Klusman, and T. D. Margulies. 1986. *Acidification status of Colorado lakes—part II: chemical classification*. Univ. of Colo. at Denver. 96 p.

Chappell, W. R., R. R. Meglen, G. A. Swanson, L. A. Taylor, R. J. Sistko, R. B. McNelly, and R. W. Klusman. 1985. *Acidification status of Colorado lakes—part I: chemical classification*. Univ. of Colo. at Denver, 70 p.

Eilers, J. M., P. Kanciruk, R. A. McCord, W. S. Overton, L. Hook, D. J. Blick, D. F. Brakke, P. Kellar, M. E. Silverstein, and D. H. Landers. 1986. *Characteristics of lakes in the western United States—volume II: data compendium for selected physical and chemical variables*. EPA-600/3-86/054B., USEPA, Washington, D.C.

Ellis, B. K., J. A. Stanford, G. R. Gregory, and L. F. Marnell. 1986. Open-File Report, Flathead Lake Biological Station, Polson, Mont. 59 p.

Epstein, C. B., and M. Oppenheimer. 1987. Nature 323(6085):245–247.

Galloway, J. N., and E. B. Cowling. 1978. Air Pollution Control Association Journal 28/3:229–235.

Gherini, S. A., L. Mok, R. J. Hudson, G. F. Davis, C. W. Chen, and R. A. Goldstein. 1985. Water, Air, and Soil Pollution 26:425–459.

Gibson, J. H., J. N. Galloway, C. L. Schofield, W. McFee, R. Johnson, S. McCarley, N. Dise, and D. Herzog. 1983. *Rocky Mountain acidification study*. FWS/OBS-80/40.17. U.S. Fish and Wildlife Serv. 137 p.

Grant, M. C., and W. M. Lewis, Jr. 1982. Tellus 34:74–88.

Harte, J., G. P. Lockett, R. A. Schneider, H. Michaels, and C. Blanchard. 1985. Water, Air, and Soil Pollution 25:313–320.

Hendry, G. R., J. N. Galloway, S. A. Norton, C. L. Schofield, P. W. Shaffer, and D. A. Burns. 1980. *Geological and Hydrochemical Sensitivity of the Eastern United States to Acid Precipitation*. Brookhaven Natl. Lab., Upton, N.Y., 71 p.

Huebert, B. J., R. B. Norton, M. J. Bollinger, D. D. Parrish, C. Hahn, Y. A. Bush, P. C. Murphy, F. C. Fehsenfeld, and D. L. Albritton. 1983. *In* R. Hermann and A. I. Johnson, eds. *Acid rain—a water resource issue for the 80's*, 17–23. AWRA International Sym., Denver.

Kelly, T. J., and D. H. Stedman. 1980. Science 210:1043.

Landers, D. H., J. M. Eilers, D. F. Brakke, W. S. Overton, R. D. Schonbrod, R. E. Crowe, R. A. Linthurst, J. M. Omernik, S. A. Teague, and E. P. Meier. 1986. *Characteristics of lakes in the western United States—volume I: population descriptions and physico-chemical relationships*. EPA600/3-86/054a, USEPA, Washington, D.C.

Lewis, W. M., Jr. 1982. Limnology and Oceanography 27/1:167–172.

Lewis, W. M., Jr., and M. C. Grant. 1979. Ecology 60:1093–1097.

Lewis, W. M., Jr., and M. C. Grant. 1980. Science 207:176–177.

Lewis, W. M., Jr., M. C. Grant, and J. F. Saunders, III. 1984. Water Resources Research 20/11:1691–1704.

National Atmospheric Deposition Program. 1982. *Distribution of surface waters sensitive to acid precipitation: a state-level atlas.* NADP Tech. Report IV, Univ. of Maine, Orono, Maine, 72 (unnumbered) p.

Omernik, J. M., and G. E. Griffith. 1986. *Total alkalinity of surface waters—a map of the western region.* EPA-600/D-85-219., USEPA, Corvallis, Ore.

Oppenheimer, M., C. B. Epstein, and R. E. Yuhnke. 1985. Science 229:859–862.

Schneider, R. 1984. *In Acid rain in the Rocky Mountain West,* 28–34, Colo. Dept. of Health, Denver, Colo.

Smith, R. A., and R. B. Alexander. 1983. U.S. Geol. Surv. Circ. 910. 12 p.

Stuart, S. 1984. *In Air quality and acid deposition potential in the Bridger & Fitzpatrick Wildernesses,* 130–173. U.S. For. Serv., Ogden, Utah.

Turk, J. T., and D. B. Adams. 1983. Water Resources Research 19/2:346–350.

Turk, J. T., and D. C. Campbell. 1984. *In Air quality and acid deposition potential in the Bridger & Fitzpatrick Wildernesses,* 268–274. U.S. For. Serv., Ogden, Utah.

Turk, J. T., and D. C. Campbell. 1987. Water Resources Research 23/9:1757–1761.

Influence of Airborne Ammonium Sulfate on Soils of an Oak Woodland Ecosystem in the Netherlands: Seasonal Dynamics of Solute Fluxes

N. van Breemen,* P.M.A. Boderie,* and H.W.G. Booltink*

Abstract

This chapter is based on three years of hydrochemical monitoring at three different sites in an oak woodland ecosystem in the Netherlands. Atmospheric inputs, mainly dry deposited ammonium sulfate, are high, and amount to 2 to 3 kmol $ha^{-1}yr^{-1}$ of $(NH_4)_2SO_4$. Intensive nitrification in the acidic soils (pH 3.5–4) is the main source of H^+ ions in these soils, which in turn causes mobilization of Al^{3+}, Ca^{2+}, and Mg^{2+} ions.

By combining measurements of solute concentrations and hydrologic modeling to evaluate seasonal hydrology and changes in the storage of solutes, the variation in time and depth of solute transfer between the solid phases and the soil solution has been calculated. Concentrations of nitrate and the cations Al^{3+}, Ca^{2+}, and Mg^{2+} are generally highest in summer, mainly as a result of evaporative concentration of solutes. Mobilization and leaching of most cations, however, takes place mainly during the fall. During summer, soil acidification is less and confined to a shallow surface layer. Contrary to the other cations, K^+ is weathered most strongly in summer, presumably under the influence of strong depletion by root uptake.

Strong soil acidification in the fall appears to be due to a combination of increased net nitrification and increased drainage, both of which are obviously related to decreased physiological activity of the tree vegetation.

I. Introduction

In 1980 a monitoring program was set up to study the effects of acid atmospheric deposition on the chemistry of soils under oak in the east-central part of the Netherlands. Soon we discovered at the research site that (1) atmospheric deposition is high and is dominated by ammonium sulfate, (2) soil acidification is intense, mainly as a result of nitrification of ammonium, even under very acid conditions, and (3) levels of dissolved Al^{3+} in the soil solution are high (van Breemen et al., 1982; Mulder et al., 1987). Annual nutrient and proton budgets

*Department of Soil Science and Geology, Agricultural University, P.O. Box 37, NL-6700 AA Wageningen, The Netherlands.

and detailed annual N and Al budgets for the period April 1981 to April 1984 have been published elsewhere (van Breemen et al., 1986; van Breemen et al., 1987; Mulder et al., 1987). Here we summarize those results and present information about the seasonal dynamics of the production and consumption of major cations and anions and of H^+, at various depths in three plots in an oak woodland ecosystem.

The findings reported here may be relevant for most forest areas in the Netherlands, where deposition of ammonium sulfate, mainly from dry deposited NH_3 (volatilized from manure produced along with the widespread intensive animal husbandry) and SO_2 (mainly from industrial sources) is high. In view of growing evidence for increasing nitrogen loadings of European forest and moorland systems, however, the results reported here may be of more than regional interest.

II. Environmental Setting, Materials, and Methods

The study area is an old 3.2 ha woodland dominated by *Quercus robur* and *Betula pendula*. Trees were coppiced until 1939, but the vegetation has been left largely undisturbed since then. The undergrowth consists mainly of *Maianthemum bifolium, Pteridium aquilinum, Holcus mollis, Rubus idaeus,* and *Anemone nemorosa*.

The soil parent material consists of sandy to loamy Pleistocene sediments of the Rhine River, locally with inconspicuous (30 to 50 cm high) river dunes. Differences in topography, in intensity of soil-forming processes (decalcification of the originally calcareous sediments, clay illuviation, and later accumulation of iron conveyed by groundwater), and in human action have resulted in strong variations in soil properties. Three plots characterized by soils with an acidic, sandy root zone (0 to 50 cm deep) were selected for hydrochemical monitoring. Soils are classified as coarse loamy, mixed, acid, mesic Aeric Haplaquept (A); sandy, mixed, acid, mesic Umbric Dystrochrept (B); and mixed, acid, mesic Aquic Udipsamment (C). The soils have a well-developed organic forest floor, 2 to 5 cm thick. Some relevant soil properties are given in Tables 7-1 and 7-2. Further details on the sites are given by van Breemen and associates (1988).

At each of three $10 \times 20 \text{ m}^2$ plots represented by the soils A, B, and C, canopy throughfall and soil solutions were sampled and analyzed chemically for over 3 years. Precipitation in the open field was sampled at two sites outside the woodland at 100 to 200 m from the monitoring plots. Aboveground water (rain and throughfall) was collected via 400 cm^2 funnels of high-pressure polyethylene into 5-L, opaque polyethylene bottles, sampled fortnightly and analyzed in pooled monthly samples. Seven throughfall collectors were placed 120 cm above the land surface. Deposition fluxes into the bulk collectors used here may be 10 to 30% higher than those in wet-only collectors. Soil solutions were sampled from Soil Moisture Equipment Type 1910 high-flow porous ceramic cups within one day of evacuating the cups. At each plot cups were placed at depths of 10, 20, 40, 60, and

Table 7-1. Classification and short descriptions of the soils at the three monitoring plots.

Site soil classification	Short soil description
A Coarse loamy, mixed, acid, mesic Aeric Haplaquept	Strong brown, acid fine sandy loam surface (0 to 50 cm) on fine sandy subsoil abruptly below 80-cm depth
B Sandy, mixed, acid, mesic Umbric Dysto-chrept	Yellowish-brown, acid loamy sand to sand (0 to 80 cm), slightly podzolized in the surface 10 cm; elevation about 30 cm above the other two soils
C Mixed, acid, mesic Aquid Udipsamment	Yellowish-brown, medium to coarse, moderately acid (pH 4) sandy soil, presumably distributed by ditching and leveling in past centuries.

Reproduced, by permission, from van Breemen et al., 1988.

90 cm, with duplicates at 10, 40, and 90 cm. Chemical analysis was done by IR spectrophotometry (organic and inorganic dissolved C), ion chromatography (Cl^-, NO_3^-, and SO_4^{2-}) spectrophotometry (NH_4^+, $H_2PO_4^-$, Al^{3+}, and SiO_2), atomic absorption spectrophotometry (Ca^{2+}, Mg^{2+}, Mn^{2+} and Fe), and flame emission (Na^+ and K^+). Organic anion concentrations were estimated from discrepancies between summed equivalent concentrations of analyzed cations and anions. Further details on sampling and on pretreatment and storage of samples are given by van Breemen and co-workers (1988); analytical procedures are outlined by Begheyn (1980). Concentrations and fluxes of ions are presented as equivalent ionic concentrations, that is concentrations per unit charge of the ion considered (e.g. $mol(+/-)$ m^{-3}), which are equal to the now obsolete equivalents (e.g. eq m^{-3}).

Vertical fluxes of solutes aboveground and through the soil were quantified by multiplying solute concentrations with measured (aboveground) or estimated (below-ground) water fluxes. Monthly integrated water fluxes in unsaturated conditions in the soil at the depths of the ceramic sampling cups were quantified using the dynamic simulation model SWATRE (Belmans et al., 1983). The model was calibrated and validated using hydraulic conductivity measurements on undisturbed soil samples and hydraulic potentials at intervals of 10-cm depth, measured weekly in the field (van Grinsven et al., 1987). Potential evapotranspiration was estimated by an adapted Penman method. Hydraulic conductivity was calculated from the soil-water retentivity curves according to the van Genuchten (1981) method. Simulated pressure heads could be fitted to field observations after reduction of conductivities by a factor of between 6 and 60. Coefficients of variation for chemical budgets were determined by sensitivity analysis of the simulated soil-water flux, and by measuring spatial, temporal, and analytical variability of the solute concentrations. Coefficients of variation of solute fluxes

Table 7-2. Selected chemical and physical data about the three soils of the study area.

Site depth	C	N	Free Fe_2O_3 mass fraction (%)	$CaCO_3$	Clay	pH H_2O	pH KCl	CEC (mmol (+/−) kg^{-1})	Base saturation (%)
A 5-0[a]	42	1.9	—	—	—	—	—	—	—
0-8	3.1	0.2	5.2	0.0	11	3.7	3.4	50	3
8-20	0.9	0.07	6.3	0.0	11	3.6	3.4	46	1
20-30	0.4	0.03	11.6	0.0	14	3.7	3.6	—	—
30-40	0.3	0.02	1.1	0.0	15	4.2	3.7	42	28
40-50	0.1	0.01	1.7	0.0	7	4.7	3.8	31	63
50-60	0.1	0.01	0.7	0.0	5	4.5	4.0	29	67
65-80	0.2	0.01	0.9	0.1	5	6.1	5.9	33	91
80-100	0.1	0.02	0.3	17.4	2	7.1	8.1	20	100
B 5-0[a]	47	2.1	—	—	—	—	—	—	—
0-4	6.6	0.18	2.2	0.0	7	3.8	3.0	145	20
4-8	3.0	0.07	2.4	0.0	8	3.7	3.2	62	8
8-20	2.5	0.08	2.3	0.0	7	4.0	4.0	31	3
20-30	1.5	0.03	2.5	0.0	7	4.1	4.2	20	2

30–40	0.5	0.02	2.9	0.0	6	4.2	4.3	15	3
40–48	0.3	0.01	2.7	0.0	7	4.2	4.2	16	1
48–60	0.2	0.01	1.0	0.0	4	4.3	4.2	11	2
60–100	0.1	0.01	0.4	0.0	1	4.5	4.3	7	5
100–120	0.2	0.01	0.9	0.1	6	6.8	6.1	39	100
C 4–0[a]	41	1.8	—	—	—	—	—	—	—
0–10	2.3	0.14	0.7	0.0	2	3.8	3.8	30	5
10–15	0.3	0.02	0.5	0.0	2	4.1	3.8	15	3
15–25	0.2	0.02	0.7	0.0	1	4.2	3.8	14	2
25–35	0.2	0.01	0.6	0.0	2	4.4	3.9	13	2
35–45	0.4	0.02	4.6	0.0	7	4.3	4.0	19	1
45–60	0.5	0.02	4.6	0.0	7	4.2	4.1	18	1
60–70	0.1	0.01	1.4	0.0	5	4.4	4.0	15	4
70–80	0.1	0.01	2.4	0.0	5	4.3	3.9	17	13
80–90	0.1	0.01	0.8	0.0	1	4.7	4.0	13	4
90–100	0.1	0.01	0.6	0.0	2	4.7	4.1	13	16

[a]Organic forest floor.
Reproduced by permission from van Breemen et al., 1988.

for Al and NO_3 ranged from 10 to 30% between the 10- and 90-cm depth, with spatial variability as the most important source of uncertainty (van Grinsven et al., 1987).

For any compartment (= horizontal soil layer), conservation of mass dictates that for a given solute i, the following relationship exists between the supply from the overlying layer ($J_{i,in}$), the leaching to the lower adjacent layer ($J_{i,out}$), the production or consumption of solid components by precipitation or dissolution and mineralization (including ion-exchange reactions), PC_i, uptake by plant roots U_i, and changes in the pool of the solute dS_i, over a certain period of time:

$$dS_i = J_{i,in} - J_{i,out} - PC_i - U_i \qquad (1)$$

Our annual budgets presented earlier (van Breemen et al., 1986, 1987, 1988) were calculated for hydrological years (April to April), over which period dS_i can be taken as nil. This allows the calculation of PC_i from the solute fluxes, J_i, and the uptake by plants, U_i. To quantify the monthly production or consumption of solutes, however, changes in storage of solutes and the distribution of nutrient uptake by the vegetation over time need to be considered. These were estimated for the soil layers from the top of the forest floor to 10-cm depth, from 10 to 20 cm, 20 to 40 cm, 40 to 60 cm and 60 to 90 cm depth. For Al and Si, inputs from canopy throughfall or from the forest floor were not known, and were assumed zero, so that the uppermost soil layer considered was 0 to 10 cm. The PC_i values were calculated as follows:

Using mean monthly data on the volumetric water content, θ, provided by the dynamic simulation model, the change in the pool of solutes over a monthly period, dS_i, was calculated from:

$$dS_i = [\theta_{t+1/2} - \theta_{t-1/2}] \cdot C_{i,t} \cdot d \qquad (2)$$

where $C_{i,t}$ is the measured solute concentration at time t (in months), and d is the thickness of the layer.

Total annual uptake by plants was estimated as the sum of annual tree litter, leaf leaching, and annual net biomass increase of trees.

Tree litter (except branches exceeding 2 cm in diameter) was sampled monthly to fortnightly from 0.5 m^2 cone-shaped plastic collectors at 0.5 m above the soil surface (six replicates per plot), oven-dried at 70°C, pooled to annual samples, and analyzed chemically. The number of stems of different tree species and their dbh (diameter at breast-height, 1.30 m) per 100 m^2-square plots were measured in 1977 (Minderman, 1981) and in 1985 for this study. From January to March 1985, above-ground tree biomass was estimated using a stratified sampling strategy. At each of the plots A, B, and C, five trees with a dbh similar to the weighted-mean dbh of the trees at the plot in question were selected. Each plot was considered as one stratum or cluster (Cunia, 1981). Bole sections (2 m long), branches (1.5 to 4.0 cm in diameter), and twigs (<1.5 cm Ø) were collected separately, weighed, and sampled for analysis (Rennie, 1966). Bark from two heights within each tree was analyzed. Stemwood of the same age was sampled from each bole section. All samples were dried at 70°C, weighed, ground, and analyzed for N (colorimetrically with Na-salicylic acid), P (colorimetrically with Al-molybdate), Na, K, and Ca (atomic emission spectrometry), Mg (atomic adsorption spectrometry), and S (ICP). For analytical details, see Houba and associates (1985). De Visser (1986) estimated standing crops per plot for the years 1977 and 1985, using the fitted parameters of the regression of mass with dbh, assuming identical allometric functions after crown closure in 1952. Net biomass production for 1981 to 1984 was calculated, assuming equal annual growth. Root biomass was estimated as 13% of the aboveground biomass (Duvigneaud and Denaeyer-de Smet, 1970).

The monthly element uptake from each soil layer was estimated from the total annual uptake by plants (as reported by van Breemen et al., 1986), by assuming element uptake to be proportional to water uptake for transpiration. The water uptake per soil layer was calculated by the hydrologic simulation model according to van Grinsven and associates (1987). For NH_4^+ and NO_3^- the relative contributions to the total uptake of N are unknown, and monthly uptake per N species could not be estimated. For H_4SiO_4 the uptake by the vegetation was unknown too. For those solutes only the values of $PC_i + U_i$, that is, the combined effect of production or consumption, uptake, and mineralization (plus, for N, nitrification and denitrification) has been calculated. For nitrogen, the corresponding values can be used to quantify the contribution of all processes involving N to the production and consumption of H^+ (Van Breemen et al., 1983):

$$H^+ \text{ produced} = (PC_{NH_4} + U_{NH_4}) - (PC_{NO_3} + U_{NO_3}) \qquad (3)$$

It should be stressed that a negative value for PC_i for solutes other than NH_4^+ and NO_3^- includes mineralization from organic matter, in addition to dissolution and/or desorption processes involving solid phases in the soil.

In the absence of relevant data for the organic forest floor, monthly storage differences in this layer were ignored. The H^+ production due to nitrogen transformations has been estimated for the forest floor plus the 0 to 10 cm mineral soil, according to equation 3. Throughfall fluxes were used for the values of $J_{i,in}$ to the forest floor. For Al and Si, the forest floor was ignored, and inputs from the forest floor into the mineral soil were assumed to be zero.

III. Results and Discussion

A. Atmospheric Inputs

Table 7-3 shows the 3-year means for amounts of rainfall and throughfall (i.e. canopy drip) and for solute concentrations in bulk precipitation and in throughfall. In both rain and throughfall water, the dominant ionic constituents are, in order of quantitative importance: (1) NH_4^+ and SO_4^{2-}, (2) Na^+ and Cl^-, and (3) NO_3^-. Concentrations in throughfall are 2 to 4 times higher than in bulk precipitation. This increase in concentration is only partly due to evaporation of intercepted water and must be attributed mainly to dry deposition plus canopy leaching.

Throughfall fluxes of nitrogen and sulfur are very high, in the order of 50 to 60 kg ha^{-1}yr^{-1} for both elements. About two-thirds of this can be attributed to dry

Table 7-3. Flux-weighted means of pH, and of molar concentrations (C_{org}), or ionic equivalent concentrations (mmol $(+/-)$ m^{-3}) in precipitation and in throughfall water at the three monitoring plots from April 1981 to 1984.

	pH	C_{org}^{a}	K	Na	Ca	Mg	Fe	Mn	NH$_4$	Cl	NO$_3$	SO$_4$	H$_2$PO$_4$	Σ+	Σ−	mm^{b}
Bulk precipitation																
	4.56	204	7.1	124	41	30	2.5	0.85	177	127	76	185	1.7	410	390	706
Throughfall																
A	4.80	1474	197	205	128	87	3.6	12.0	474	260	172	575	12.7	1122	1020	551
B	4.69	1560	209	195	117	84	3.0	6.1	542	254	185	542	16.3	1177	997	551
C	4.62	1279	159	172	105	81	2.8	8.1	421	228	153	433	15.6	973	830	554

[a]C_{org} refers to total dissolved organic carbon.
[b]mm refers to mean annual precipitation for this period.
Reproduced, by permission, from van Breemen et al., 1988.

deposition of SO_2 and NH_3 on tree canopies, as is discussed below. Concentrations of most solutes are still much higher in stemflow than in throughfall. However, stemflow volumes are small (equivalent to about 10 mm/year), and stemflow inputs amount to less than 5% of the total input of nitrogen and sulfur.

Concentrations of other solutes are lower, but the increase in their concentration when rainwater passes the canopy is in the same order or even higher than in the case of the bulk solutes NH_4^+, SO_4^{2-}, Na^+, Cl^-, and NO_3^-. By comparing differences in throughfall composition between the vegetation period (15 April to December 1) and the leafless period, van Breemen and associates (1988) concluded that leaching of nutrients from leaves is the main cause of increased concentrations of K^+, Ca^{2+}, Mg^{2+}, Fe^{2+}, and $H_2PO_4^-$, which were distinctly higher in summer than in winter. By contrast, Na^+, Cl^-, NH_4^+, and SO_4^{2-} show the highest enrichment in throughfall during winter, indicating that they are largely of atmospheric origin and are mainly deposited in dry form. Na^+ and Cl^- are probably derived from cyclic sea salt, which is conveyed in larger quantities during winter than during summer. Most of the NH_4^+ and SO_4^{2-} is probably derived from NH_3 (volatilized from manure produced in intensive animal husbandry) and from SO_2 of industrial origin. Anthropogenic emissions of ammonia in Europe have, on the average, increased by at least a factor of 2.3 since 1870, as a result of increased livestock production (Asman, 1987). In the Netherlands, where animal production has expanded most, NH_3 emissions have increased 4- to 5-fold. Adema and coworkers (1986) discusses the mechanism by which NH_3 and SO_2 mutually influence each other, leading to increased dry deposition of both gases as $(NH_4)_2SO_4$, formed stoichiometrically. Throughfall NO_3^- is higher in summer than in winter. Because it is unlikely that leaves leak NO_3^-, this difference must probably be attributed to (1) higher dry deposition of HNO_3 and NO_x in summer, and/or (2) partial nitrification of dry deposited NH_x in summer.

B. Uptake by Plants

Table 7-4 shows, for each of the three plots, the annual fluxes of elements in leaf litter (means for May 1980–1981 and May 1981–1982), the fluxes of elements leached from the tree canopy, and the annual net uptake in wood, bark, and roots. Canopy leaching was assumed to be zero for Na^+, Cl^-, NO_3^-, NH_4^+, and SO_4^{2-} and assumed to be equal to throughfall input for the other elements. The annual net uptake was obtained from dbh data for all trees per plot, using the excellent regressions ($r > 0.9$) between dbh and fresh weights of stems, branches, and twigs of sampled trees. Branch litter fall during this period was neglected. For all elements except N in the three acid soil plots, we assumed a zero net accumulation in the forest floor, so that summation of these three fluxes gives the total annual uptake of elements from the soil by trees (*total uptake*). From data by Minderman (1981) and recent measurements of the N pools in the forest floor (Winkels, 1985), we estimated that 0.7 kmol $ha^{-1}yr^{-1}$ of N accumulate in the forest floor. Only a small fraction of the annual uptake from the soil is stored in woody parts and bark (*net uptake*); most is returned to the soil as leaf litter (especially Ca, Mg, Mn, S, P, and N) and canopy leachate (especially K).

Table 7-4. Element cycling through the tree vegetation at each of the monitoring plots. Values are equivalent ionic fluxes in $kmol(+/-) ha^{-1} yr^{-1}$, assuming valences of 1 for N (NO_3^- or NH_4^+) and P ($H_2PO_4^-$), 2 for S, and 3 for Al.

	H	K	Na	Ca	Mg	Mn	Al	Cl	S	P	N
A. litter	—	0.65	0.11	1.75	0.57	0.23	0.24	0.14	0.46	0.18	7.54
Leaf leach.	-0.62	1.03	—	0.31	0.27	0.06	—	—	—	0.06	—
Net uptake	—	0.07	0.00	0.09	0.03	0.00	0.00	—	0.00	0.01	0.36
Total uptake	—	1.75	0.11	2.15	0.87	0.29	0.24	0.14	0.46	0.25	7.20
B. litter	—	0.91	0.12	2.23	0.71	0.12	0.25	0.14	0.30	0.26	9.50
Leaf leach.	-0.49	1.10	—	0.25	0.25	0.03	—	—	—	0.08	—
Net uptake	—	0.25	0.02	0.37	0.12	0.00	0.00	—	0.00	0.03	1.19
Total uptake	—	2.26	0.14	2.85	1.08	0.15	0.25	0.14	0.30	0.37	10.0
C. litter	—	0.75	0.11	1.84	0.71	0.18	0.15	0.14	0.24	0.22	8.10
Leaf leach.	-0.32	0.84	—	0.21	0.24	0.04	—	—	—	0.08	—
Net uptake	—	0.29	0.02	0.41	0.13	0.00	0.00	—	0.00	0.03	1.34
Total uptake	—	1.88	0.13	2.46	1.10	0.22	0.15	0.14	0.24	0.33	8.74

Slightly modified from van Breemen et al., 1988.

C. Chemical Composition of the Soil Solution

The chemistry of the soil solution can be considered as a mirror of the chemical conditions in the soil as seen through the plant roots. Moreover, changes in soil solution chemistry with time reflect seasonal chemical processes, and can be used to calculate the mass transfer involved in such processes. Soil solution concentrations at various depths are shown in Table 7-5. The most noticeable differences between the composition of throughfall and that of the soil solution are the marked decrease in the concentration of ammonium, the concomitant increase in the concentration of nitrate, and the appearance of aluminum as the dominant cation in the soil solution.

D. Soil Acidification and H^+ Budgets

Acidification or alkalinization of soils occurs through H^+ transfer processes involving vegetation, soil solution, and soil minerals. A permanent change in the acid-neutralizing capacity of the inorganic soil fraction ($ANC_{(s)}$), that is, soil acidification ($\Delta ANC < 0$) or soil alkalinization ($\Delta ANC > 0$), results from an irreversible H^+ flux. This irreversible H^+ flux can be caused either by direct proton addition or depletion, by different mobility of components of the $ANC_{(s)}$, or by a permanent change in redox conditions. The contributions of acidic atmospheric deposition, nitrogen transformations, deprotonation of CO_2 and of organic acids and protonation of their conjugate bases, assimilation of cations and anions by the vegetation, weathering or reverse weathering of minerals, and stream output to changes in the $ANC_{(s)}$ can be quantified by means of H^+ budgets.

The measurements required to compile H^+ budgets over relatively long periods (so that dS_i can be taken as nil) should include: the flux of solutes entering and leaving the system, net assimilation (or mineralization) of solutes by living and dead biomass, net quantities of potentially acidic gases entering or leaving the system, and net adsorption (c.q. precipitation) or desorption (c.q. dissolution) of solutes in the soil.

Fluxes of quantitatively important cations (Ca^{2+}, Mg^{2+}, K^+, Na^+, NH_4^+, H^+, Al^{3+}, and Mn^{2+}) and anions (Cl^-, SO_4^{2-}, NO_3^-, HCO_3^-, $H_2PO_4^-$, and organic anions) must be monitored in incoming precipitation and outflowing drainage water. For the elements taken up by the vegetation, only Ca, Mg, K, Na, S, and P need to be considered. Nitrogen uptake can be considered under the heading N transformations, as discussed below.

The amount of HCO_3^- ($+ \times 2CO_3^{2-}$) and organic anions transported from a system may be equated to proton dissociation from CO_2 and organic anions resulting in weathering and the associated depletion of soil $ANC_{(s)}$.

The net proton flux attributed to nitrogen transformations can be determined from net output of ammonium less net output of nitrate to the system. A negative value for this quantity suggests that nitrogen transformations are generating protons to the ecosystem; a positive value suggests that protons are being consumed from the ecosystem. This expression is valid regardless of the amount of N transported to and from the ecosystem as neutral gases (NO_x, N_2, NH_3).

Table 7-5. Dissolved organic C (C_{org}), pH, SiO_2 (mmol m^{-3}), and concentrations of ionic solutes (mmol ($+/-$) m^{-3}) in soil solutions from various depths at each of the three plots.

Depth	pH	C_{org}	SiO_2	K	Na	Ca	Mg	Al	Fe	Mn	NH_4	Cl	NO_3	SO_4	H_2PO_4	HCO_3	Organic anions
Plot A																	
0	3.60	3370	86	371	245	480	235	77	19	66	384	305	996	709	31.0	—	108
10	3.59	4793	417	293	248	778	311	592	34	75	239	322	1441	846	3.79	—	153
20	3.55	3408	506	228	255	572	335	985	17	64	71	297	1667	826	1.11	—	107
40	3.81	2847	580	113	353	621	353	2063	4	80	50	430	1902	1353	1.79	—	101
60	4.60	1860	695	30	491	2387	888	587	2	118	27	776	2067	1504	4.59	—	91
90	7.05	1968	422	10	494	4786	1506	61	2	2	36	572	2369	1441	1.45	2322	142
Plot B																	
0	4.20	6655	100	401	210	314	165	29	31	16	1096	310	870	695	65.4	—	280
10	3.50	5537	323	310	229	407	195	570	31	20	185	237	1231	699	4.74	—	169
20	3.51	4384	380	326	216	549	243	946	32	22	151	253	1623	7325	5.73	—	135
40	3.95	2943	314	208	239	486	250	1589	5	20	105	349	1492	934	1.57	—	111
60	3.98	3682	323	218	257	508	234	1902	6	17	80	345	1685	993	0.47	—	141
90	3.99	3068	401	120	363	465	249	2144	4	14	42	539	1373	1533	0.43	—	118
Plot C																	
0	3.84	11793	191	425	196	354	224	58	47	30	974	236	1178	564	55.1	—	242
10	3.85	4975	424	223	213	294	201	585	23	21	161	259	717	719	2.7	—	180
20	3.82	4031	425	218	199	201	201	475	24	18	160	230	583	585	1.81	—	44
40	4.04	3983	482	159	225	281	368	704	10	21	91	276	769	681	0.95	—	157
60	4.12	1862	429	127	224	331	366	753	2	42	32	273	768	850	0.89	—	76
90	4.14	2282	481	67	323	344	484	1054	2	39	28	471	670	1174	0.46	—	94

Values are means of all available data between April 1981 and April 1984 (10- to 90-cm depth) or between February 1983 and April 1984 (0-cm depth). The number of samples involved is 30 to 40 for the 20- and 60-cm depths, about 60 for the 10- and 40-cm depths, and about 90 for the 90-cm depth.

Reproduced, by permission, from van Breemen et al., 1988.

For the theory behind soil acidification and H^+ budgets, reference is made to van Breemen et al. (1983, 1984) and de Vries and Breeuwsma (1987). As has been discussed in greater detail elsewhere (van Breemen et al., 1987; Mulder et al., 1987), nitrification accounted for a major part of all soil acidification. H^+ budgets for the soil down to 90 cm depth (Table 7-6) show that nitrification plus uptake of ammonium accounted for the production of 3 to 7.4 kmol ha^{-1}yr^{-1} of H^+. In soils B and C, most of this acidification was neutralized by dissolution of Al^{3+}. In the surface soil at plot A, Al was mobilized too, but essentially all of it was precipitated again at greater depth. Precipitation of Al^{3+} is due to the presence of a calcareous subsoil and is associated with the solubilization of Ca^{2+}. Soil acidification due to nitrification in the 10-cm surface soil alone was higher than by the net nitrification throughout the soil profile (e.g. at A, 9.1 against 7.4 kmol$(+)$ha^{-1}yr^{-1}). The high acidification in the surface soil was partly alleviated at greater depth by removal of nitrate. Presumably, most of this nitrate was removed by plants and microorganisms and, perhaps, in part by denitrification. At plot A, however, leaching of nitrate below the root zone still exceeded atmospheric throughfall input indicating (1) that atmospheric N inputs were even higher than throughfall inputs, for example, due to direct assimilation of atmospheric NH_3 by tree leaves, or (2) that the pool of soil N decreased by mineralization in excess of uptake. At plot C most of the atmospheric N was assimilated and soil acidification due to N transformations was relatively small. Low net biomass accumulation at plot A (van Breemen et al., 1986) is probably the main cause for the high rate of nitrate utilization at plot A.

Aluminum was relatively important in neutralizing strong acid inputs, due to the high acid loading and the low rate of acid neutralization by basic cations in our soils. Most H^+ from atmospheric deposition and nitrification of atmospheric NH_4^+ resulted in the dissolution of mineral Al (mainly Al-hydroxide):

$$Al(OH)_3 + 3H^+ \longrightarrow Al^{3+} + 3H_2O$$

Over the years, considerable amounts of Si were removed along with Al from the surface soil layers, which suggests that at least part of the Al originated from dissolution of Al-silicate minerals. At the low pH values of our soil solutions (annual mean pH 3.27 to 4.26) dissolved Al in the mineral soil was mainly in aquo Al^{3+} form at all depths. Highest Al^{3+} concentrations (up to 4 mol$(+)$m^{-3}) were reached in summer when nitrification was most intense. During high discharge in winter and spring, the export of Al^{3+} with drainage water from the acidic soils peaked and solute concentrations decreased.

E. Seasonal Changes in Solution Chemistry and in Solid-Solution Interactions

The conclusions presented above were based on annual budgets and on seasonal changes in concentrations in the soil solution. Seasonal changes could not be interpreted directly in terms of processes taking place, because they result from several simultaneous processes: interactions between the solution and solid

Table 7-6. Annual H^+ budgets for the soils (0- to 90-cm depth) of each of the plots A, B, and C, based on measurements in three hydrological years (April 1981–April 1984). Units are kmol (+/−) ha^{-1} yr^{-1}.

	H^+ sources						H^+ sinks					
	External		Internal					Internal				
								Weathering				
	H^+	N	N	CO_2	Biomass	Weathering[b]	Sum	Sum	Al	Other[a]	Biomass	Output	Δ ANC
A	0.7	5.8	1.7	3.4	0.2	0.7	12.5	12.6	0.2	12.4	0.0	0.0	−11.9
B	0.6	3.9	0.0	−0.2	0.8	0.8	5.9	6.3	3.8	2.3	0.0	0.2	−5.3
C	0.4	3.0	0.0	0.0	0.9	0.7	5.0	5.1	2.6	2.4	0.0	0.1	−4.3

[a] Cation dissolution (excluding Al) + anion precipitation.

[b] Anion dissolution + cation precipitation.

Reproduced, by permission, from van Breemen et al., 1988.

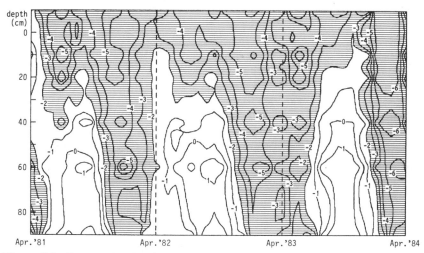

Figure 7-1. Contour plot showing the distribution of vertical water fluxes with time and depth in one of the three plots (A) studied. Contour intervals are 1 cm month^{-1}. Downward flows exceeding 2 cm month^{-1} have been shaded.

phases, uptake by plants, mineralization, evaporative concentration of solutes in summer, and dilution by infiltrating rain and snowmelt in winter. The strongly seasonal hydrology is illustrated in Figure 7-1, which is derived from the hydrological model calculations. Figure 7-1 shows that net drainage takes place throughout the soil profile in the winter periods and only in the upper parts of the root zone throughout the year. In the summer period, moreover, capillary rise from the groundwater (generally present at 100 to 140 cm below the soil surface) reverses the flow in the lower part of soil profile.

To discriminate between the various processes that influence the composition of the soil solution at monthly intervals, we estimated the contribution of the various processes as outlined in the methods section.

The results are presented as contour diagrams that show the production $(-)$ or consumption $(+)$ of various solutes by the solid phases as a function of soil depth in cm below the boundary of the forest floor and the mineral soil and of time between April 1981 to April 1984 (Figures 7-2 to 7-10).

Caution is needed in interpreting the contour plots. The errors in the calculated values of PC_i are unknown and cannot be evaluated precisely with the information available. Although errors in the annual solute fluxes (due to errors in the water budget, and spatial, temporal, and analytical variability in the solute concentrations) increase with depth from 10% (10 cm depth) to 30% (90 cm), errors in annual chemical budgets $(J_{i,in} - J_{i,out})$ were 10 to 30% at the surface, but ranged from 40 to 200% in the deeper soil layers (van Grinsven et al., 1987). Errors in the calculated PC_i may be even larger. An impression about the reliability of the contour plots can be obtained from the plots for Cl^- (Figure 7-2). Because interaction between Cl^- and the solid phases of the soil is probably insignificant,

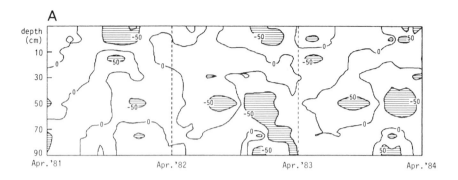

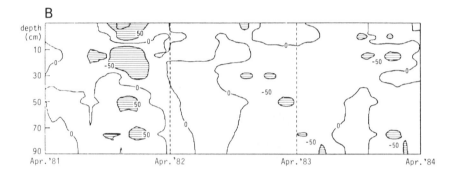

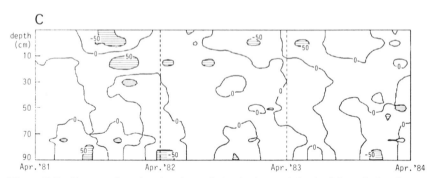

Figure 7-2. Contour plots showing the variation in time and depth of the calculated mass transfer between solid phase and the soil solution for Cl^- in soils in each of the three studied plots (mmol$(-)$ m^{-3} month^{-1}). Dark shading indicates release of Cl^- into solution exceeding 50 mmol$(-)$ m^{-3} month^{-1}.

PC_{Cl} values should be close to zero at all times. In order to merit serious consideration in terms of solid-solution interaction, patterns of production and/or consumption of other solutes therefore need to be either more strongly expressed, or more reproducible on a seasonal basis, than those of Figure 7-2. Generally PC_{Cl} values are lower than 50 mmol$(-)$ m^{-3}month^{-1} (about one-third of the annual flux of Cl^- through a 1 m^{-2} cross-section of soil) throughout the monitoring period and show no consistent pattern with time or depth. The distinctly positive PC_{Cl} values sometimes observed in winter at plot A and plot B are probably artifacts due to errors of various kinds. Errors of the same relative magnitude can be expected to mar similar plots for other solutes.

Contour diagrams showing the net production and/or consumption for several of the major solutes (NH_4^+, NO_3^-, Al^{3+}, Ca^{2+}, Mg^{2+}, Na^+, K^+, SO_4^{2-}, and H_4SiO_4) are discussed in some detail only in case of plot A (Figures 7-3 and 7-4). Next the most important differences between plot A and plots B and C are dealt with briefly (Figures 7-5 through 7-7).

A striking aspect revealed by the contour diagrams for plot A is the distinct seasonality of the appearance or disappearance of most solutes by solid-solution interactions. For NO_3^-, Al^{3+}, Ca^{2+}, Mg^{2+}, K^+, Na^+, SO_4^{2-}, and H_4SiO_4, the seasonal patterns are consistent, and the peak values are in the order of the annual net downward fluxes. Therefore, we consider these production and/or consumption patterns real, and not due to various errors, as those for Cl^- in Figure 7-2. Patterns for ammonium (Figure 7-3) show only a weak seasonality. Moreover, ammonium is consumed only: essentially all the ammonium entering from the atmosphere is removed from solution in the surface soil throughout the year. Nitrate (Figure 7-4) is produced throughout the year at the same depths where ammonium disappears, suggesting that ammonium is nitrified more or less continually. Quick nitrification of ammonium from atmospheric input at plot A has indeed been demonstrated by field experiments involving ^{15}N (Stams et al., 1989). From annual budgets it is clear that nitrate production in the surface soil exceeds consumption of atmospheric ammonium (van Breemen et al., 1987). So part of the nitrate formed in the surface soil must be derived from organic N, presumably leaf litter. At greater depths nitrate is produced too, but only during

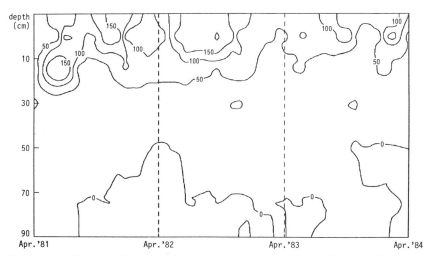

Figure 7-3. Variation in the production and/or consumption of dissolved NH_4^+, not corrected for uptake by the vegetation ($PC_i + U_i$), over time and depth in plot A (mmol($+$) $m^{-3}month^{-1}$).

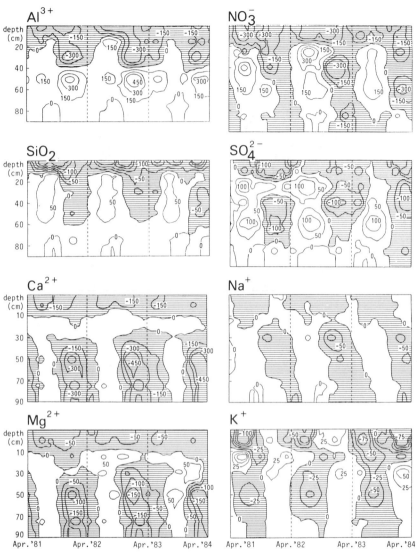

Figure 7-4. Variation in the production and/or consumption of dissolved Al^{3+}, Ca^{2+}, Mg^{2+}, SO_4^{2-}, Na^+, K^+ (PC_i), and of that for NO_3^- and H_4SiO_4, not corrected for uptake by the vegetation ($PC_i + U_i$), over time and depth in plot A (mmol(+/−) m^{-3}month^{-1}).

winter. In the absence of aqueous NH_4^+, the source of the nitrate formed at greater depths is probably mineralized and nitrified organic N, in this case mainly from decaying roots. In summer, however, nitrate disappears from solution at depths below 10 cm. Because contours for NO_3^- and NH_4^+ indicate the net effects of production and/or consumption plus uptake and/or mineralization, this removal of nitrate can be explained by root uptake of the vegetation.

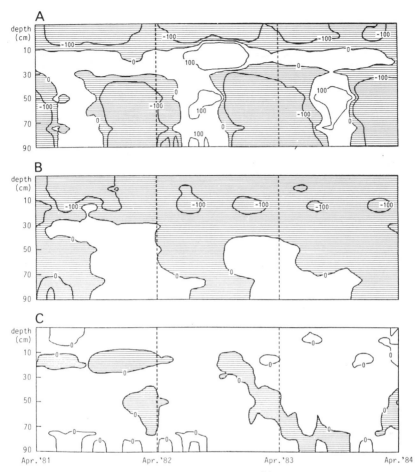

Figure 7-5. Comparison of PC_i contour diagrams of Ca^{2+} for plots A, B, and C (mmol($+$) m^{-3}month^{-1}).

The pattern of release of Al^{3+}, Ca^{2+}, Mg^{2+}, SO_4^{2-}, and H_4SiO_4 into the soil solution is rather similar to that of NO_3^-. Two zones of production can be distinguished. First, in the forest floor and the top 10 cm of the mineral soil, these solutes are released throughout the year, but mainly in summer, at a relatively low intensity. Second, at greater depth, they are produced mainly in winter and generally in larger quantities than in summer. The production of these solutes must be due to a combination of mineralization and mineral dissolution. For Ca^{2+} and Mg^{2+}, the shallow and the deeper soil layers, where these solutes are produced, are distinctly separate. The release of Ca^{2+} and Mg^{2+} in and just below the forest floor can be attributed mainly to mineralization. In the subsoil, however, release of solutes is mainly dissolution of minerals and/or desorption. For Al^{3+}, Ca^{2+}, Mg^{2+}, SO_4^{2-}, and H_4SiO_4, solid-solution transfers during summer in the subsoil

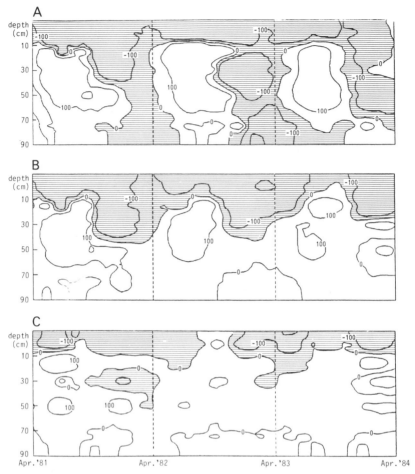

Figure 7-6. Comparison of ($PC_i + U_i$) contour diagrams of NO_3^- for plots A, B, and C ($mmol(+)\ m^{-3}month^{-1}$).

are small and tend toward precipitation or sorption. So the high summertime concentrations of Al^{3+} noted earlier are the result of evaporative concentration, not of enhanced dissolution. In addition to sorption, the disappearance of dissolved sulfate must be due to other processes, which may include microbial incorporation of sulfate in organic compounds (Mitchell et al., 1986). As would be expected for an element that undergoes little biocycling, there is practically no release of Na^+ from the forest floor. Presumably as a result of mineral weathering, Na^+ is released, in low quantities and during winter only, from the subsoil.

In the mineral soil, the release pattern of K^+ is the opposite of that of the other solutes. It is released from the mineral soil during summer, and shows a tendency to precipitate (or adsorb) in winter. Of the nutrients considered so far, K is the only one that is taken up by the trees in larger quantities than it is supplied by inputs

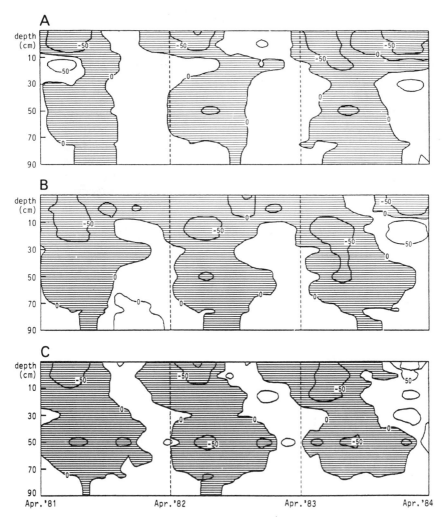

Figure 7-7. Comparison of PC_i contour diagrams of K^+ for plots A, B, and C (mmol(+) $m^{-3}month^{-1}$).

from the atmosphere (Tables 7-3 and 7-4). Therefore, it is likely that weathering of K-bearing minerals, induced by nutrient uptake, is the cause of the release during summer. Weathering or desorption of Al^{3+}, Ca^{2+}, Mg^{2+}, Na^+, and H_4SiO_4, on the contrary, seems to be related mainly to acidification associated with the formation of nitrate via nitrification and, to a much lesser extent, of sulfate.

Release of Al^{3+}, Ca^{2+}, and Mg^{2+} occurs simultaneously but at different depths. Aluminum is released from the 0 to 40-cm surface soil, and all of it is removed again from solution (during the winter season, shortly after it was released into solution) between 50- and 70-cm depth. This precipitation or sorption of Al^{3+} coincides with a release of Ca^{2+} and Mg^{2+}. Release of Ca^{2+} and Mg^{2+}, however, is not confined to the zone where Al^{3+} precipitates, but continues to greater depths. Below a depth of 75 cm, the soil contains 10 to 20% calcite (with about 7 mol % Mg), which is probably the source of the Ca^{2+} and Mg^{2+} released in winter. The following processes explain the observations on Al^{3+}, Ca^{2+}, and Mg^{2+} described above.

Acidic (pH 3.5 to 4) water with Al^{3+} and NO_3^- as the dominant solutes, percolating downward in late fall and winter, encounters somewhat less acidic soil (pH 4 to 5) with some adsorbed Ca^{2+} and Mg^{2+} at 40- to 60-cm depth (Table 7-2). As a result, Al^{3+} is exchanged for adsorbed Ca^{2+} and Mg^{2+}, which take over the role of Al^{3+} in accompanying NO_3^- on its way down the profile. As the solution reaches the calcareous subsoil, the pH further increases sufficiently for any CO_2 present to become an efficient proton donor. This eventually results into dissolution of calcite:

$$CaCO_3 + CO_2 + H_2O \longrightarrow Ca^{2+} + 2\,HCO_3^- \qquad (4)$$

Magnesium is released similarly from the (Mg)calcite lattice. The fact that the highest rates of release of Mg^{2+} are about 25% of those of Ca^{2+}, while the Mg/Ca ratio in the calcite in our soil is less than 10% (van Breemen et al., 1988), may indicate that Mg^{2+} is released preferentially over Ca^{2+}.

The pool of exchangeable Ca and Mg between 40- to 60-cm depth is in the order of 40 kmol(+)/ha. At the ambient rate of supply of Al^{3+} (about 5 kmol(+)/ ha · yr), such a pool could be exhausted within a decade. However, the pool of exchangeable bases is probably recharged occasionally by upwelling of groundwater containing $Ca(HCO_3)_2$. In 6 years of monitoring between 1980 and 1987, two such events were recorded. In February 1985 and January 1987, the pH of the soil solution at 60-cm and 45-cm depth rose from 4.5 to near neutrality, respectively, and the concentration of dissolved Ca^{2+} increased simultaneously.

A major difference between plot A and plots B and C is the absence of a shallow calcareous subsoil at B and C. So, the abrupt disappearance of Al^{3+} and the simultaneous appearance of Ca^{2+} and Mg^{2+} in the subsoil of plot A is not observed in plots B and C (shown only for Ca in Figure 7-5). Otherwise, the temporal patterns are quite similar, with release of solutes from the solid phase mainly coinciding with the period of net nitrate formation and strong drainage of water during the fall. Net nitrate formation, however, decreases from A to B to C (Figure 7-6), and mobilization of the cations accompanying NO_3^- decreases accordingly. This is in line with the annual N budgets, which show that N cycling is tighter at plot C than at B, and much more so than at A (van Breemen et al., 1987). At all locations some solutes (SO_4^{2-}, Al^{3+}, SiO_2) tend to disappear from solution during summer. However, K^+ is released in summer only (Figure 7-7).

F. Seasonality of Soil Acidification

Figure 7-8 shows the variation with time and depth of net removal of cations (base cations plus aluminum) from the solid phases of the soil, minus that of anions (Cl, SO_4^{2-}, and $H_2PO_4^-$), expressed as kmol(+) ha^{-1} month^{-1}, or, in other words, of

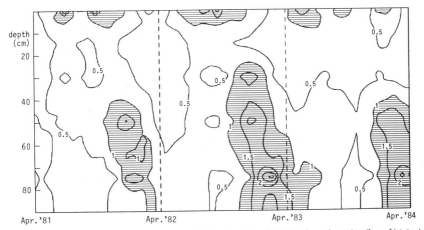

Figure 7-8. Variation with time and depth of $-\Delta ANC$ in plot A (kmol($+/-$) ha^{-1}month^{-1}).

$-\Delta ANC$. The change in the acid-neutralizing capacity of the soil, or ΔANC, can be seen as the capacity function indicating soil acidification (van Breemen et al., 1984). Soil acidification appears to be a seasonal process, which is strongest during summer and early fall in the surface soil and peaks during fall and winter in the subsoil. Table 7-7 shows the values of ΔANC per soil layer for winter and summer separately. The small (5%) differences between the annual ΔANC values in Tables 7-6 and 7-7 are due to the different calculation methods used. In the

Table 7-7. Mean values of ΔANC (kmol ($+$) ha^{-1} (half year^{-1}) for summer (April 1 to October 1) and winter, per soil layer, for each of the three plots studied.

	Depth (cm)	A	B	C
Summer	0–10	−5.1	−3.6	−3.1
	10–20	1.0	0.3	0.4
	20–40	2.1	1.6	0.6
	40–60	2.5	1.7	0.9
	60–90	0.3	0.1	0.1
	0–90	+0.8	0.1	−1.1
		−0.6[a]	−0.1[a]	−1.0[a]
Winter	0–10	−4.4	−2.9	−2.2
	10–20	−0.6	−2.1	0.0
	20–40	−2.8	−1.5	−1.0
	40–60	−0.9	0.5	0.7
	60–90	−4.8	0.3	−0.5
	0–90	−13.5	−5.7	−3.0
		−8.6[a]	−5.0[a]	−1.5[a]
Annual	0–90	−12.6	−5.5	−4.0

[a]The contribution of ammonium uptake and nitrification to ΔANC over 0- to 90-cm depth.

surface layers, soil acidification is stronger in summer than in winter, and the reverse is true in winter. Below the 10-cm depth at plots A and B, there is even an increase in ANC during summer. This increase in ANC is the effect of precipitation plus sorption of cations (Ca, Mg, and Al) under the influence of relatively dry conditions in the soil during summer.

As has been discussed in detail elsewhere (van Breemen et al., 1984), soil acidification is not necessarily reflected directly by a change in pH or a production or consumption of free H^+. Net production of free H^+ (which, in soil solutions such as these with essentially zero alkalinity, reflect acidification of the *soil solution*) is mainly confined to the forest floor and the upper 10 cm of the soil (Figure 7-9). Comparison between Figures 7-9 and 7-10 indicates that the production of free acid in the surface soil is associated with net production of H^+ due to consumption of NH_4^+ and formation of NO_3^-. Practically all of this free H^+ is consumed again somewhat deeper in the mineral soil. However, particularly during peak water flow in winter, appreciable amounts of free H^+ appear to be drawn down into the mineral soil to about 50-cm depth before neutralization takes place (compare Figures 7-1 and 7-9). This neutralization cannot be traced to a particular process, but must be due to a combination of (1) H^+-cation (mainly Al^{3+}) exchange in weathering or sorption reactions, (2) sorption of SO_4^{2-}, and (3) net uptake of NO_3^-.

G. Synthesis

The results discussed above suggest that in the soils studied here dissolution and leaching of base cations and aluminum is strongly seasonal and takes place mainly

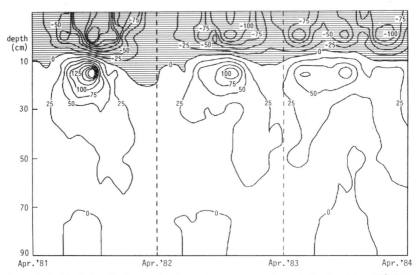

Figure 7-9. Variation in the production and/or consumption of dissolved free H^+ (PC_i) over time and depth in plot A (mmol(+) m^{-3}month^{-1}).

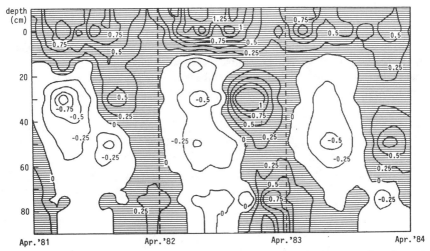

Figure 7-10. Variation over time and depth of the production and/or consumption of a total H^+ as a result of transformations involving NO_3^- and NH_4^+ in plot A (mmol$(+/-)$) m^{-3}month^{-1}).

in fall and in winter. In summer, however, most solutes tend to be transferred from the solution to a solid phase by precipitation or adsorption. Removal of dissolved substances mainly in the fall and winter period is obviously a result of increased drainage. Because precipitation is rather evenly distributed over the year, the increased drainage is due to decreased evapotranspiration. The coincidence of dissolution and increased drainage is not so obvious, however. This point is discussed in some detail below.

Increased drainage of water and the ensuing dilution of the soil solution could enhance the rate of mineral dissolution, that is, by increasing the degree of undersaturation of the solution with respect to weatherable minerals. At the same time, however, anions (or soluble proton donors that can be transferred to anions) must be available to release and accompany those cations. In slightly acid to near neutral environments, CO_2 from the usually large pool of soil CO_2 is able to fulfill the role of a proton donor and lead to the formation of bicarbonate solutions. If, in addition, the rate of mineral dissolution is high enough, the rate of weathering by CO_2 can be governed completely by the rate of water drainage, as has been shown for calcareous soils by van Breemen and Protz (1988). In the course of their development, the pH of our soils has become too low for CO_2 to act as a proton donor. Helped by increased atmospheric deposition of N and S, nitric and sulfuric acid have taken over that role. Increased dissolution and desorption of sulfate as a result of dilution of the soil solution during the fall could explain the appearance of SO_4^{2-}. In the absence of sorbed or solid nitrate, however, such a process cannot account for the appearance of dissolved NO_3^-, which is the dominant anion. Increased nitrification, decreased nitrate uptake, or any combination thereof must therefore explain the increased net production of nitrate during the fall. Although conditions for nitrification itself probably do not improve when soil temperatures fall from 12 to 14°C around July to 5 to 8°C in October (van Breemen et al., 1988), the increased supply of ammonia from decaying organic matter may increase over that period as a result of root turnover and may enhance nitrification. Decreased consumption of NO_3^- by plants and soil microbes during the fall, however, may be more important in causing increased net nitrification. Comparison of the values of ΔANC with the acidity produced in nitrogen transformations (Tables 7-6 and 7-7), shows that net nitrification is the main process involved in mobilizing the cations and in acidifying the soil.

In conclusion, increased mineral dissolution and increased drainage of water are not, or only partially, linked by a direct causal relationship. Rather, decreased activity during the fall of the dominant primary producers of the ecosystem, the trees, causes a simultaneous increase in the drainage of water and of the availability of (nitrified) nitrogen, both leading to increased weathering and leaching.

It cannot be excluded that some of the seasonal patterns described here are an effect of the methodology applied. In summer the soil solution is not necessarily homogenous: the dryest pockets of soil may contain solutes at very high concentrations, in water that is not in hydrological contact with the sampling cups. These solutes would appear in the water sampled by the cups only after dilution during the fall and would remain "invisible" in summer. Such a course of events would be observed in our approach as an apparent removal from solution during summer and an apparent dissolution during the fall. The contour diagrams for Cl^-, however, indicate that this happens to a very limited extent at the most, suggesting that the patterns described here really reflect solid-solution interactions.

IV. Future Research Directions

The analysis presented here shows that seasonal patterns in solute fluxes in soils are very pronounced and need to be considered if the long-term processes of soil acidification are to be well understood. The data needed to evaluate seasonal processes are even more extensive than that required for "simple" annual element or H^+ budgets. In this study, various assumptions had to be made to arrive at estimates for fluxes that could not be quantified in a more direct way, for example, atmospheric deposition and nutrient uptake. More detailed work would be necessary to check the validity of these assumptions.

Before generalizing the conclusions about seasonal patterns from studies as presented here, many years of data are needed. As has been emphasized by others (Likens et al., 1977; Matzner, 1987; Ulrich, 1987) long-term shifts in element cycling of forest ecosystems require decades of monitoring. The establishment or continuation of such long-term monitoring programs is therefore an urgent task.

References

Adema, E. H. 1986. *In* H. F. Hartmann, P. Heeves, and J. Hulskotte, ed. *Proc. 7th World Clean Air Congress, Sydney, Australia*, vol 2, 1–8.

Asman, W. A. H. 1987. *Atmospheric behaviour of ammonia and ammonium.* PhD thesis, Agricultural University, Wageningen, The Netherlands, 173 p.

Begheyn, L. Th. 1980. *Methods of chemical analysis for soils and waters.* Int. Rept., Dept. Soil Science and Geology, Agric. Univ. Wageningen, The Netherlands.

Belmans, C., J. G. Wesseling, and R. A. Feddes. 1983. J Hydrol 63:271–286.

Cunia, T. 1981. *Cluster sampling and tree biomass tables construction. In Kyoto Biomass Studies.* IUFRO workshop proceedings, Maine, Orono., 3–15.

de Visser, J. B. 1986. *Interactions between soil, vegetation and atmospheric deposition at the research plots of the RIVM/LH/EC/RIN project.* Rept. 02-01. Dutch Priority Programme on Acidification, Dept. Soil Science and Geology, Agric. Univ. Wageningen, The Netherlands, 88 pp.

de Vries, W., and A. Breeuwsma. 1987. Water, Air and Soil Poll 35;293–310.

Duvigneaud, P., and S. Denaeyer-de Smet. 1970. *Biological cycling of minerals in temperate deciduous forests. In* D. E. Reichle, ed. *Analysis of temperate forest ecosystems.* Springer Verlag, Berlin.

Houba, V. J. G., I. Novozamsky, J. J. Van der Lee, W. Van Vark, and E. Nab. 1985. *Chemische analyse van Gewassen.* Report of the Dept. of Soil Science and Plant Nutrition, Agricultural University, Wageningen, The Netherlands.

Likens, G. E., F. H. bormann, R. S. Pierce, J. S. Eaton, and N. M. Johnson. 1977. *Biogeochemistry of a forested ecosystem.* Springer-Verlag. 146 p.

Matzner, E. 1987. *Der Stoffumsatz zweier Waldökosysteme im Solling.* Habilitationsschrift, Faculty of Forestry, University of Göttingen, BRD. 254 p.

Minderman, G. 1981. *Bodemkundige en bosecologische gegevens van het eikenhakhoutbosje Hackfort in de gemeente Vorden.* 4 volumes. Inst. for Nature Management, Arnhem, The Netherlands.

Mitchell, M. J., M. B. David, D. G. Maynard, and S. A. Tellang. 1986. Can J Forest Res 16:315–320.

Mulder, J., J. J. M. Van Grinsven, and N. Van Breemen. 1987. Soil Sci Soc Amer J 51:1640–1646.

Rennie, P. J. 1966. Commonwealth Forestry Review 45:119–127.

Stams, A. J. M., H. W. G. Booltink, I. J. Lutke-Schipholt, B. Beemsterboer, J. R. W. Wottiez, and V. van Breemen. 1989. ^{15}N Field study on the fate of atmospheric ammonium in an acid forest soil. Submitted for publication. Soil Sci Soc Amer J.

Ulrich, B. 1987. *Stability, elasticity and resilience of terrestrial ecosystems with respect to matter balance. In* E. D. Schulze and H. Zwölfer, ed. Ecological Studies 61:11–49. Springer–Verlag, Berlin.

van Breemen, N., P. A. Burrough, E. J. Velthorst, H. F. Van Dobben, T. De Wit, T. B. Ridder, and H. F. R. Reijnders. 1982. Nature 299:548–550.

van Breemen, N., J. Mulder, and C. T. Driscoll. 1983. Plant and Soil 75:283–308.

van Breemen, N., C. T. Driscoll, and J. Mulder. 1984. Nature 307:599–604.

van Breemen, N., P. H. B. De Visser, and J. J. M. Van Grinsven. 1986. J Geol Soc London 143:659–666.

van Breemen, N., J. Mulder, and J. J. M. Grinsven. 1987. Soil Sci Soc Amer J 51:1634–1640.

van Breemen, N., and R. Protz. 1988. *Rates of calcium carbonate weathering in soils.* Can J Soil Sci (in press).

van Breemen, N., W. J. F. Visser, Th. Pape. 1988. *Biogeochemistry of an oak woodland affected by acid atmospheric deposition.* Agr. Res. Repts 930, PUDOC, Wageningen, The Netherlands.

van Genuchten, M. Th. 1981. Soil Sci Soc Amer J 44:892–898.

van Grinsven, J. J. M., N. van Breemen, and J. Mulder. 1987. Soil Sci Soc Amer J 51:1629–1634.

Winkels, H. 1985. *Kwantitatieve aspecten van de koolstof en stikstof huishouding van de strooisellaag in de gronden van het Hackfortprojekt.* MSc thesis, Agricultural University, Wageningen, The Netherlands.

Acidic Deposition: Case Study Scotland

F.T. Last*

Abstract

The history of atmospheric pollution in Scotland can be traced back to the beginning of the thirteenth century, when the first coal charter was granted.

Annual average rural atmospheric concentrations of both SO_2 and NO_2 are typically 5 $\mu g\ m^{-3}$, with ozone episodes on about 10 days per year in central and eastern areas. Amounts of annual dry deposition, 0.4 g S m^{-2} and 0.2 g N m^{-2}, respectively, are exceeded by wet deposition, 1.6 g S m^2 and 0.6 g N m^{-2}, with additional contributions from cloudwater. This pollution climate, dominated by wet deposition, is similar to that in parts of southern Norway and Sweden but differs from those in central England, eastern Belgium, and much of the Federal Republic of Germany.

The appearance of beech (*Fagus sylvatica*), oak (*Quercus* spp.), and Scots pine (*Pinus sylvestris*), but not Norway spruce (*Picea abies*) and Sitka spruce (*Picea sitchensis*), deteriorates towards the north and west of Britain. In 1986 and 1987 there were progressively increasing losses of conifer needles, but surveys made by the Forestry Commission have not identified the cause(s) of these losses. Correlations suggest that prevailing concentrations of SO_2 and NO_2 may even be enhancing needle retention.

Stemflow of "old" stands of alder (*Alnus glutinosa*), oak (*Q. petraea*), Scots pine, Norway spruce, and Sitka spruce was always more acidic than incident precipitation. In contrast, throughfall in alder, oak, and Sitka spruce was less acidic than incident rain during summer and autumn but not at other times. Throughfall was never as acidic as stemflow. The contribution of foliar leachates is still questionable. Unlike that of other species, foliage of Norway spruce and Sitka spruce did not decrease net concentrations of ammonium and nitrate in throughfall.

Although pollutants have been shown to be capable of affecting soil processes, there is, although suspected, no unequivocal field evidence to show that they have exacerbated the natural acidity of soils in Scotland.

In contrast, there is substantive evidence from diatom reconstructions to show that atmospheric pollutants have been implicated in the acidification of some

---

*Department of Soil Science, The University, Newcastle-upon-Tyne NE1 7RU, UK.

freshwaters clustered in Galloway, southwest Scotland. Affected lochs and associated streams drain "acidification-sensitive" basins with slowly weathering base-poor bedrock: they are subject to heavy rainfall and have therefore been receiving large inputs of deposited acidity. In addition to diatoms, acidification has altered populations of freshwater invertebrates and decreased, if not eliminated, populations of fish, notably salmonids. Afforestation with conifers has exacerbated the acidity of freshwaters in acidification-sensitive locations, as in west-central Scotland.

During the last decade amounts of annually deposited acidity have decreased, and at the same time there is evidence indicating that the acidification of some lakes has been halted if not reversed.

I. Introduction

What is *Scotch?* Perhaps, a seemingly irrelevant question but during the 1910s and 1920s the word *Scotch* ("the motor spirit of most excellent quality") was applied to one of the major fossil fuels; it was the label found on pumps in Scotland selling petrol refined in West Lothian, to the west of Edinburgh (Last, 1978). It was a reminder of the pioneering work of the Ninth Earl of Dundonald who in 1781 had patented a crude process for producing oil (and tar and pitch) from coal. He was followed by James Young, a Glaswegian-born chemist nicknamed "Paraffin Young," who developed a two-part process—distillation and the treatment (refining) of the distillate—to refine oil from the mineral torbanite and also from waxy and resinous shales (Cook, 1971). His development, which was patented in 1850, foreshadowed and facilitated the development of the Pennsylvania oil industry, which was started 9 years later when Drake struck oil in Titusville in 1859.

To Scots interested in the use of fossil fuels, the refining of oil is a relatively recent development remembering that the monks of Newbattle Abbey, near Dalkeith south of Edinburgh, were granted a charter to extract coal sometime between 1210 and 1217. As suggested by McNeill (1883), "The monks were not slow to put its (the beautiful black stone) heat-producing properties to the test." Originally it was taken from surface outcrops, as had been the case with peat, the traditional source of fuel.

In setting the scene for the Scottish case study and, because of the scientific and political interest in the reversal of water acidification as, and when, amounts of acidifying pollutants are lessened, it is of interest to refer to events in 1661. John Evelyn (1661), an exile from Britain who lived in Paris, was particularly concerned with the dirty and undignified condition of London, which at the time, the mid-1600s, largely depended on mines in Northumbria, northeast England, for its supplies of coal. Having earlier referred to

> this horrid Smoake which obscures our Churches, . . . and corrupts the Waters, so as the very Rain and refreshing Dews which fall in the several seasons, precipitate this impure vapour,

he continued,

> Not therefore to be forgotten, is that which was by many observed, that in the year (1661) when Newcastle was besieged and blocked-up in our late Wars, so as through the great Dearth and Scarcity of Coales. . . . Divers Gardens and Orchards planted even in the very heart of London. . . . were observed to bear such plentiful and infinite quantities of Fruits, as they never produced the like either before or since.

As it happens, it was the Scots who "besieged and blocked-up" Newcastle (the shipping outlet for Northumbria) and forced Londoners to witness the benefits of emission control: Many conclusions can be drawn from Evelyn's narrative, not the least being the very long lead time that is usually needed before a warning signal is given the attention that it merits; the Clean Air Acts were introduced into Britain in 1956 and 1968, that is, about 300 years later.

II. The Environment of Scotland

Scotland has an area of about 8.1 million ha and lies between latitudes 55°N in the south (corresponding with Svendborg in Denmark, Schefferville in Quebec and Hazelton in British Columbia) and 61°N in the north (Figure 8-1). Its geology is complex with rocks ranging in age from Precambrian to Tertiary. The greater part of Scotland north of the Highland Boundary Fault (Figure 8-2) is underlain by granites and Precambrian crystalline metamorphic rocks (gneisses and schists). South of the Southern Upland Fault, the gently rounded Southern Uplands are formed from Ordovician and Silurian shales and graywackes with some granite intrusions. Between the Southern Upland and Highland Boundary Faults is the relatively low-lying but physiographically diverse central part of Scotland, the Midland Valley. It is an ancient rift valley with considerable areas having Devonian and Carboniferous sedimentary rocks (red and gray sandstone) and also igneous rocks of the same periods (Cameron and Stephenson, 1985; Kinniburgh and Edmunds, 1986).

The soils overlying these rocks reflect the nature of the rocks, the influences of weather, and the activities of man. In the wet and cool uplands of Scotland, where precipitation regularly and substantially exceeds evapotranspiration, there is an abundance of acidic soils, including a variety of orthods (podsols), histosols (peats), humic aquepts (peaty gleys) and entisols (rankers) (Figure 8-2). Predictably, strongly acid freshwaters are likely to be found where these acidic soils overlie granite and acid igneous rocks, most metasediments, grits, and other types of rock encompassed by the high-susceptibility category in the classification of solid geology evolved in relation to the occurrence of acid waters (Table 8-1).

By applying Indicator Species Analysis (Hill et al., 1975) to extant map-readable information concerning a range of environmental attributes (climatic, "topographical," geological, and others associated with human artifacts), R. G. H. Bunce generated 32 land classes encompassing the range of habitats found in Great Britain (GB). Twenty-seven of the GB land classes occur in Scotland, with land

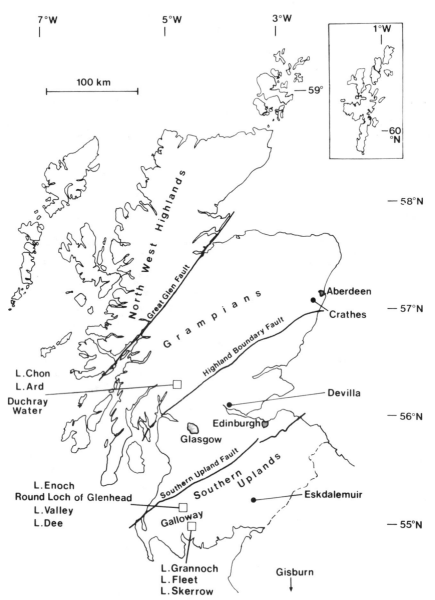

Figure 8-1. Scotland: key locations in relation to the study of the nature and impacts of acidic deposition.

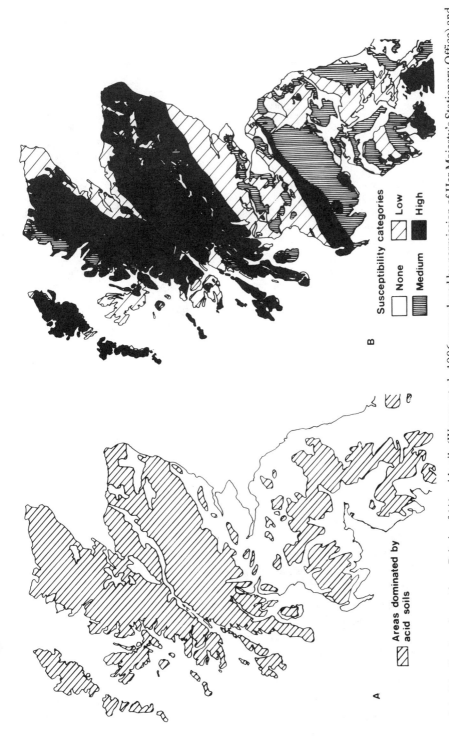

Figure 8-2. Distributions in northern Britain of (A) acid soils (Warren et al., 1986; reproduced by permission of Her Majesty's Stationery Office) and (B) different rock types influencing the susceptibility of surface groundwaters to acidic deposition (Kinniburgh and Edmunds, 1986).

Susceptibility categories

None
Low
Medium
High

B

Areas dominated by acid soils

A

Table 8-1. Categories used in the UK to classify the sensitivity to acidification of groundwaters related to different types of bedrock

Sensitivity category	Types of bedrock	Buffer capacity and/or impact on groundwaters
1	Granite and acid igneous rock, most metasediments, grits, quartz sandstones, and decalcified sandstones, some Quaternary sands/drift	Most areas susceptible to acidification, little or no buffer capacity, except where significant glacial drift
2	Intermediate igneous rocks, metasediments free of carbonates, impure sandstones and shales, coal measures	Many areas could be susceptible to acidification; some buffer capacity due to traces of carbonate and mineral veining
3	Basic and ultrabasic igneous rocks, calcareous sandstones, most drift and beach deposits, mudstones and marlstones	Little general likelihood of acid susceptibility except very locally
4	Limestones, chalk, dolomitic limestones and sediments	No likelihood of susceptibility; infinite buffering capacity

Edmunds and Kinniburgh, 1986.

classes 17 to 24 (upland) and 25 to 32 (northern) dominating; classes 30 to 32 occur only in Scotland (Bunce and Last, 1981). In total, 54% of Scotland is regarded as upland, with land class 22, the most widespread in Scotland, having the following characteristics:

87% of its area is at altitudes ranging from 199 to 488 m
A slope of 8°
A mean minimum January temperature of $-0.2°C$
A mean number of days with snow falling of 51.6
A mean daily bright sunshine of 4.4 h

The estimates suggest that peatlands account for substantial areas of land classes 21, 24, 28, 30, and 32, with forests and woodlands concentrated in classes 17, 19, 21, 22, and 28, which have fairly widely varying proportions of different soil types, including dystrochrepts, histosols, humic aquepts, and orthods (Table 8-2). Because of our interest in the acidifying influences of atmospheric pollutants, it is of significance that calcareous soils, with infinite buffering capacities, are not associated with land classes 17 to 32.

   In the past, research workers concerned with acid waters have tended to focus on pollutants reaching their targets by wet deposition, whereas those concerned with the growth of plants (nonwoody and woody) have veered towards gases (sulfur dioxide, oxides of nitrogen, and ozone) reaching their targets by dry deposition. They have tended to overlook the concepts propounded by Smith (1872) in his book *Air and Rain: The Beginnings of a Chemical Climatology* and more recently by Benarie (1980), who referred to *air pollution climate,* an umbrella term encompassing a variety of pollutants whether reaching their targets by wet or dry deposition (Last et al., 1986). In developing the pollution climate concept, Fowler and associates (*in litt.*) examined pollutant gas and rainfall chemistry data for many of the countries of midlatitude Europe, specifically taking note of the influences of altitude. In the Federal Republic of Germany (FRG) they were concerned with the influences of altitude on the amplitude of diurnally changing concentrations of ozone (large concentrations sometimes persist longer at high, than at low, altitudes; Reiter and Kanter, 1982), but in Scotland there is greater concern for effects of altitude on the occurrence of cloud droplets—"occult deposition." By separating rainwater from cloudwater during a number of rain events, Unsworth and Crossley (1987) showed that cloudwater can be much more acidic and contain larger concentrations of particles than rainwater (Table 8-3).

With the available data, Fowler and associates identified the locations of distinct pollution climates:

Location of pollution climate 1: southern areas of FRG, including Baden-Württemberg and Bavaria, and most of Switzerland
Location of pollution climate 2: eastern Belgium, the Nordrhein-Westfalen region of the FRG, western Czechoslovakia, and parts of central England
Location of pollution climate 3: northwest Britain (Figure 8-3)

The dominant pollutants and/or deposition pathways in these climates appear to be:

Pollution climate 1: ozone, particularly at high altitudes, with significant inputs of S and N, primarily by wet deposition. Mists contribute significant amounts of sulfate, nitrate, and hydrogen ions.
Pollution climate 2: SO_2 and NO_2 with a significant number of summer ozone episodes. Dry deposition dominates inputs of S, N, and acidity
Pollution climate 3: Wet deposition of acidity, SO_4^{2-} and NO_3^- and large concentrations of these ions in intercepted cloudwater

In climate 3, in NW Britain, average annual atmospheric concentrations of both SO_2 and NO_2 are typically 5 $\mu g\ m^{-3}$; their dry deposition gives annual inputs of 0.4 g S m^{-2} and 0.2 g N m^{-2}. These amounts are greatly exceeded by the wet deposition of nonmarine SO_4^{2-} and NO_3^- approximately to 1.6 g S m^{-2} and 0.6 g N m^{-2}. Additionally, with the frequent occurrence of mists, there is a predictable extra contribution from cloudwater containing concentrations of SO_4^{2-}, NO_3^-, H^+, and NH_4^+, ranging from 50 to 2,000 $\mu eq\ L^{-1}$. Thus, in Scotland, part of northwest Britain, inputs are dominated by wet deposition (Figure 8-4). The data for 1983 suggest that wet deposition accounted for more than 75% of the total sulfur deposited in much of west and southwest Scotland, compared with less than 50% in most of England (Irwin and Williams, 1988).

Within Scotland ozone concentrations greater than 75 ppb (volume) have not been measured in high-altitude, remote locations, but in lowland, more populous

Table 8-2. Predicted occurrence of different soil types in G.B. land classes occurring in Scotland (% of the area of each land class in Scotland with the different soil types).

Soil types

GB land classes in Scotland	Dystro-chrept / Brown earth	Aquic dystro-chrept / Gleyed Brown earth	Aquept / Gley	Eutro-chrept / Calcareous Brown earth	Rendoll / Rendzina	Eutric aquept / Calcareous gley	Hapl-orthod / Brown podsolic	Humic aquept / Peaty gley	Aquod / Peaty podsol	Orthod / Podsol	Entisol / Ranker	Lithic udorthents / Brown ranker	Lithic udorthents / Peat ranker	Histosol / Peat	Aquentic Hapl-orthod / Gleyed podsol	Spodic entisol / Podsolic ranker	Spodic aquept / Podsolic gley
1	52.5	15.0	25.0	—	—	—	—	—	—	7.5	—	—	—	—	—	—	—
5	32.5	7.5	32.5	10.0	—	—	—	10.0	—	2.5	—	—	—	5.0	—	—	—
7	48.7	7.7	10.2	2.6	12.8	—	5.1	—	2.6	2.6	—	5.1	2.6	—	—	—	—
8	45.0	—	50.0	—	—	10.0	—	—	—	—	5.0	—	—	—	—	—	—
9	35.0	15.0	27.5	—	2.5	—	2.5	5.0	—	7.5	—	—	—	—	—	—	—
10	22.5	12.5	52.5	—	—	—	2.5	—	—	5.0	—	—	—	—	—	—	—
11	35.0	7.5	47.5	7.5	—	—	—	7.5	—	—	2.5	—	—	—	—	—	—
13	30.0	2.5	42.5	—	—	—	10.0	—	—	—	2.5	—	—	—	—	—	—
14	27.5	27.5	27.5	7.5	—	7.5	2.5	2.5	—	5.0	—	—	—	—	—	—	—
15	35.0	7.5	27.5	—	2.5	—	12.5	—	—	—	5.0	—	—	—	—	—	—
16	35.0	7.5	35.0	—	—	2.5	5.0	2.5	—	—	5.0	—	—	—	—	—	—
17	52.5	2.5	5.0	—	—	—	15.0	7.5	—	12.5	—	—	2.5	5.0	—	—	—
18	10.0	—	12.5	—	—	—	20.0	7.5	2.5	7.5	2.5	2.5	2.5	15.0	—	—	—
19	17.5	7.5	7.5	—	—	—	7.5	2.5	12.5	17.5	7.5	—	10.0	17.5	—	—	—
20	30.0	2.5	35.0	—	—	—	5.0	2.5	10.0	7.5	2.5	—	5.0	7.5	—	—	—
21	7.5	—	2.5	—	—	—	5.0	20.0	—	—	—	2.5	—	40.0	2.5	—	—
22	10.0	2.5	12.5	—	—	—	7.5	32.5	5.0	10.0	2.5	—	—	15.0	2.5	—	—
23	—	—	—	—	—	—	5.0	10.0	15.0	—	—	2.5	12.5	35.0	—	—	—
24	5.0	—	7.5	—	—	—	2.5	12.5	17.5	15.0	7.5	2.5	5.0	30.0	—	—	—
25	38.5	20.5	33.3	—	—	—	2.6	—	17.5	7.5	5.1	—	—	—	—	—	2.5
26	38.4	10.3	30.8	—	—	—	10.3	2.6	—	—	—	2.5	2.6	2.5	—	—	—
27	55.0	5.0	22.5	—	—	—	10.0	—	—	—	—	5.0	—	—	—	—	—
28	22.5	—	37.5	—	—	—	5.0	2.5	5.0	2.5	2.5	—	—	20.0	—	—	—
29	20.0	—	5.0	—	—	—	—	7.5	7.5	5.0	10.0	—	—	32.5	—	5.0	—
30	7.5	—	—	—	—	—	—	2.5	10.0	2.5	2.5	2.5	15.0	57.5	—	—	—
31	27.5	7.5	5.0	—	—	—	—	2.5	5.0	17.5	7.5	—	15.0	27.5	—	2.5	—
32	7.5	7.5	5.0	—	—	—	—	7.5	7.5	—	—	2.5	7.5	55.0	—	—	—

Bunce and Last, 1981.

Table 8-3. The particulate content and pH of rain and cloudwater collected at sites near Edinburgh, Scotland.

Date	Wind direction	Periods of collection (days)	Types of samples	Volumes collected (ml)	pH	Content of particles (mgl^{-1})
Dunslair Heights						
6 Feb. 85	SE	1	Cloudwater	169	2.79	25
Castlelaw						
24 April 85	NW	5	Cloudwater	139	3.96	50
24 April 85	NW	5	Cloudwater + rain	372	4.09	43
24 April 85	NW	5	Rain	71	4.40	3
15 May 85	S	9	Cloudwater	2500	3.38	102
15 May 85	S	9	Cloudwater + rain	2500	3.44	89
15 May 85	S	9	Rain	513	3.87	10

Dunslair Heights (altitude, 602 m) is about 30 km south of Edinburgh; Castlelaw (altitude, 408 m) is about 10 km southwest of Edinburgh.

Unsworth and Crossley, 1987.

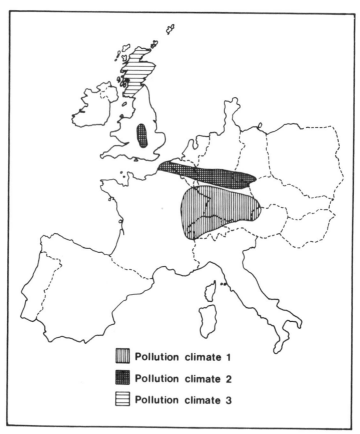

Figure 8-3. The location in midlatitude Europe of three distinct pollution climates (Last et al., 1986).

parts of central and eastern Scotland they are not uncommon (50 h yr^{-1} or about 10 days with ozone episodes). Concentrations of sulfur dioxide and oxides of nitrogen are largest in the east-to-west central belt encompassing the valleys of the rivers Clyde and Forth. In contrast there is a gradient of acidity in rainfall from the west (average pH 4.7 to 4.4) to the east coast (pH 4.4 to 4.0). Cape and associates (1984) found that concentrations of H$^+$, NO$_3^-$, and nonmarine SO$_4^{2-}$, but not NH$_4^+$, systematically increased from the northwest to the south and east (H$^+$, from 10 to 40 μeq L^{-1}; NO$_3^-$, from 10 to 30 μeq L^{-1}; nonmarine SO$_4^{2-}$, from 15 to 60 μeq L^{-1}). Although it is incorrect to assume that the chemistry of rain falling at different altitudes in the same region is the same—in different circumstances, they may be either larger or smaller (see Unsworth and Crossley, 1987)—the deposition of all ions and also acidity (Figure 8-5) was greatest in the mountainous areas of the south and west where amounts of rainfall were largest ($\sim$2000 to 3000 mm yr^{-1}): these areas include (1) the west-central Highlands of Scotland, where the Duchray and Loch Chon catchments are to be found, and (2) the southwest of

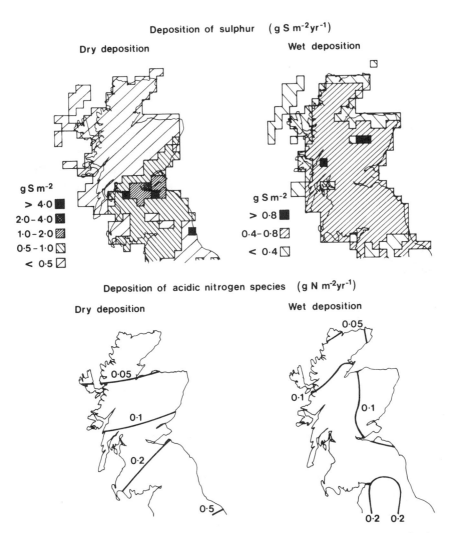

Figure 8-4. Modeled wet and dry deposition of sulfur and acidic nitrogen species in northern Britain (Barrett et al., 1987).

Scotland (including Galloway) where the lochs investigated by Battarbee and his team are located (see section on changes in aquatic ecosystems).

As happens elsewhere, the deposition of acidity is highly episodic. Near Edinburgh 30% of the total acidity deposited involves only 2.7% of rain events (Barrett et al., 1983). As suggested by Davies and associates (1986), this distribution reflects, to a very considerable extent, patterns of weather. They examined rainfall at Eskdalemuir in southern Scotland and separated rain events by the prevalent daily weather types (W = westerly; A = anticyclonic;

Figure 8-5. Mean annual amounts of wet deposited acidity in northern Britain, 1981–1985 (Barrett et al., 1987).

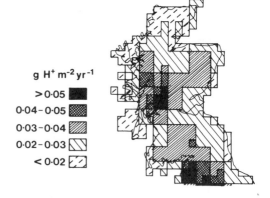

g H⁺ m⁻² yr⁻¹

$g\ H^+\ m^{-2}\ yr^{-1}$

> 0.05
0.04 – 0.05
0.03 – 0.04
0.02 – 0.03
< 0.02

c = cyclonic; Lamb, 1972). Rain coincident with anticyclonic weather is likely to be sparse but nevertheless important for acidic deposition. Rain at Eskdalemuir coincident with cyclonic weather is likely to be heavy, to have originated in England and nearby parts of continental Europe, and to be relatively acidic; westerly weather is similarly heavy but, in contrast to cyclonic weather, is more commonly associated with less heavily polluted maritime trajectories. In 1984 Davies and co-workers reported the deposition in the Cairngorm Mountains of black, acid snow (pH 3.0) containing large amounts of particulate matter ($\sim$0.01 g L^{-1} of melted snow) and having a substantial carbonaceous component, in addition to much more SO_4^{2-}, Cl^- and NO_3^- than cleaner snow deposited either before or afterwards (Table 8-4). Because they are morphologically and chemically distinctive, the assemblages of particles, derived from different sources, enable the conjectured trajectories of relevant airmasses to be confirmed with a reasonable degree of probability. As with many strongly acidic rain and/or snow events in Scotland, the relevant air mass of this particular black snow event traveled northwards from industrial England. It is thought that black snow events occur more frequently than is usually recognized.

Table 8-4. Composition (ppm) of black snow that fell on the Cairngorm Mountains, Scotland, on 20 Feb, 1984, compared with that of "less dirty snow" collected on 12 other occasions in the period January to April 1984.

	pH	SO_4^{2-}	Cl^-	NO_3^-	NH_4^+	Ca^{2+}	Mg^{2+}	Na^+	K^+
				Concentrations (ppm) of different ions					
Black snow	3.0	19.8	14.8	20.9	0.3	0.63	0.52	4.52	0.39
Other snows	3.5–4.7	0.27–4.86	0.15–10.23	0.11–3.52	ND[a]	ND[a]	0.02–0.64	0.07–4.45	ND[a]

[a]ND = not determined.

Davies et al., 1984. Copyright © 1984 Macmillan Magazines Ltd. Reprinted by permission.

III. Effects on Forested Ecosystems

Since 1910, the proportion of coniferous trees in new plantings has greatly increased. In 1910 the totals for coniferous and broad-leaved trees in Scotland were 38,000 ha and 24,000 ha, respectively; thereafter, the areas of new plantings have totaled 740,000 ha and 38,000 ha. Within the broad-leaved category, plantings of beech (*Fagus sylvatica* L.) and oak (*Quercus robur* L. and *Q. petraea* [Matt.] Lieblein) have declined, and those of birch (*Betula* spp.) have increased; among the conifers, the importance of Sitka spruce (*Picea sitchensis* Bong. Carr.) and Lodgepole pine (*Pinus contorta* Douglas ex Loud.) has greatly increased (Table 8-5). As a result of these changes, the age distributions of the different species differ markedly: in 1980 most oaks and beeches were more than 70 years old, whereas the majority of lodgepole pine and Sitka spruce were less than 20 years old, a difference of significance when considering forest health.

In Britain, starting in 1984, the Forestry Commission has annually assessed forest health. In doing so, it has adopted a system of stratified random sampling in which the strata relate to (1) altitude (above or below 244 m), (2) amounts of annual rainfall (more or less than 1,000 mm), and (3) amounts of dry deposited sulfur (more or less than $2g \ S \ m^{-2}yr^{-1}$) (Binns et al., 1985). In 1984 and 1985, surveys of needle losses and discoloration were limited to stands, 30 to 50 years old, containing Sitka spruce (economically the most important conifer in Britain), Norway spruce (*Picea abies* L. Karst.; included because this species is of major importance in continental Europe), and Scots pine (*Pinus sylvestris* L.; the most important native conifer). In 1986 these procedures were extended to include 33 stands of Norway spruce more than 60 years old; in 1987 the Forestry Commission added stands of beech and oak (*Quercus petraea* and *Q. robur,* and their hybrids), while also increasing the number of stands of Scots pine and Sitka spruce to include some nearly 100 years old. To comply with the regulations of the Commission of the European Communities, members of the Forestry Commission now assess trees at the intersections of a 16×16 km grid in addition to those designated by its own more resource-effective system of stratified sampling. Like its counterparts in continental Europe, the Forestry Commission has concentrated, but not exclusively, on two aspects, namely, crown density (previously designated *needle loss*) and the discoloration of needles or leaves (including the browning or yellowing of current, and older, needles) (Anon, 1986). The published data for 1986 and 1987 include the outcome of stepwise multiple regression analyses made with a variety of independent variables including age of tree, altitude and slope of site, presence of recent stand thinning, amounts of rain, dry deposition of sulfur suggested from a "model," nitrogen deposition, grid east, and grid north (Innes et al., 1986; Innes and Boswell, 1987a and 1987b).

Even allowing for the need to be cautious about the effects of observer-error, changes in crown densities (needle retention) are greater than might have been predicted. Rates of needle loss, noted in Scots pine, Norway spruce, and Sitka spruce in 1985 and 1986, continued into 1987 (Figure 8-6), whereas the extent of

Table 8-5. Approximate areas (ha) in Scotland, when censused in 1980, of mainly broad-leaved and coniferous trees. Details are also given for the three most widespread species of both types of trees.

| | Dates of planting | | | | | | | | | | |
	Pre-1861	1861–1900	1901–1910	1911–1920	1921–1930	1931–1940	1941–1950	1951–1960	1961–1970	1971–1980	TOTAL
Broad-leaved trees											
Beech	4,600	3,100	420	110	320	260	260	460	600	410	10,500
Birch	7	220	650	1,200	1,900	2,600	5,600	2,200	1,400	700	16,600
Oak	5,000	6,400	2,300	350	360	1,200	120	550	170	140	16,500
Total (including other broad-leaved species)	12,300	19,300	6,400	2,900	6,300	6,000	8,700	5,900	5,200	2,800	
Coniferous trees											
Lodgepole pine	Nil	Nil	Nil	16	280	630	610	17,500	39,700	45,100	103,900
Scots pine	4,900	9,900	4,500	3,900	12,200	10,700	14,300	44,700	29,000	10,300	144,400
Sitka spruce	47	120	93	170	3,900	11,300	21,200	49,800	108,700	169,300	364,600
Total (including other coniferous species)	5,300	11,900	6,900	6,300	25,300	37,400	52,700	160,600	212,400	247,600	
Total of broad-leaved and coniferous trees	17,600	31,200	13,300	9,200	31,600	43,400	61,400	166,500	217,600	250,400	

Forestry Commission, 1983.

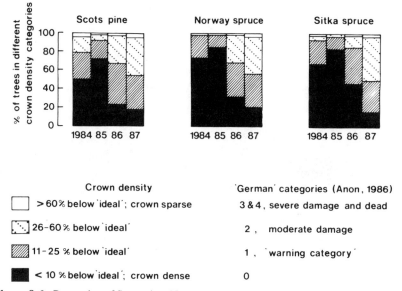

Figure 8-6. Proportion of Scots pine, Norway spruce, and Sitka spruce in different crown density categories when recorded in Britain between 1984 and 1987 (Innes and Boswell, 1987a).

needle discoloration (yellowing and browning) in 1987 was either the same or less than in 1986.

On using correlation and regression analyses, also principal component analyses, to associate changes in crown densities with environmental variables, Innes and his colleagues found, as already known with Scots pine (White, 1982) that the appearances of this species, beech, and oak deteriorate towards the north and west of Britain (Figure 8-7). The crowns of Scots pine and oak were significantly less dense at locations with heavy rainfall and low air temperatures; there was also an inverse relation with altitude. In contrast, the crowns of Norway spruce and Sitka spruce were not significantly influenced by geographical location, rainfall, and/or temperature. Instead, there were significant relationships with age, crown density being less in old than in young stands.

In making their surveys, Innes and his colleagues correctly stress that they have been assessing tree health and not pollution damage. In addition to showing that health is sometimes related to geographical location, age, and weather variables, they obtained other evidence suggesting that health may be influenced by atmospheric pollutants. Whereas there were indications of inverse relationships between crown densities of Scots pine, beech, and oak and mean annual atmospheric concentrations of ozone, the evidence in relation to SO_2 and oxides of nitrogen suggests that the small amounts of these pollutants that now prevail in forested areas of Britain may be beneficial. Roberts (1984) and Abrahamsen (1984) might suggest that this result could have been anticipated; there are many

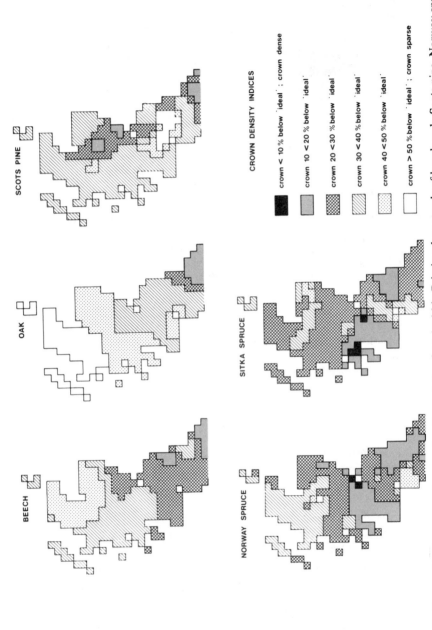

Figure 8-7. Geographical patterns of crown density variation found within Britain when stands of beech, oak, Scots pine, Norway spruce, and Sitka spruce were recorded July to September 1987 (Innes and Boswell, 1987b).

instances where small concentrations of sulfur and nitrogen pollutants have been beneficial. However, Nilsson (1986) might argue that the favorable link between crown density and atmospheric pollutants is the first sign of incipient nitrogen saturation. But what should be concluded from the surveys made by the British Forestry Commission? Innes and associates (1986) suggested that the series of regression analyses had not greatly helped the search for the causes of changes in crown densities. This may be so, but they have highlighted our profound ignorance of events in "normal" conditions, which clearly vary cyclically whether measured diurnally, seasonally, annually, or over longer periods of time. There could be many reasons for Innes and his colleagues' inability to detect key influences. The key influences may have been omitted from the Forestry Commission's system of stratification. Differences in weather from year to year may be of crucial significance, particularly as it is known that changes in some aspects of the environment, for example, light intensity and temperature, greatly affect plant responses to atmospheric pollutants (Mansfield et al., 1986). Conversely, exposure during spring and summer to dry deposited pollutants, at concentrations that neither cause blemishes nor significantly alter rates of shoot growth, can significantly increase the sensitivity of trees to unseasonably early autumn frosts. Colleagues working at Lancaster University and the Bush station of the Institute of Terrestrial Ecology found that intermittent exposures between June and August to concentrations of ozone that did not affect shoot extension increased the sensitivity of Sitka spruce needles to simulated frosts in November, that is, 10 weeks later (Lucas et al., 1988). Similar observations have been made by Brown and co-workers (1987) working with Norway spruce.

In forests, rain (incident precipitation) comes into contact with the canopies of trees, layers of scrub, ground vegetation, and litter on route to underlying soils. Inevitably, therefore, its chemistry is likely to be modified, the changes being effected by a variety of processes (Parker, 1983). As water passes over leaves and branches, its content of solutes and suspended material is likely to be modified by epiphytic microbes, by dry deposited pollutants, by the foliar absorption of ammonium and nitrate, and by the efflux of base cations, particularly potassium. Whereas the few available data suggest that conifers of all ages "acidify" stemflow, Miller (1984) indicated that stands of young conifers, unlike those of older specimens, do not acidify throughfall. Ralph (1986), working with Sitka spruce, showed that this differential effect on throughfall was a reflection of physiological processes associated with aging trees and not simply a changing balance of current-, first-, second-, and latter-year needles. Stevens (1987) suggested that the change occurred when stands of Sitka spruce were between 25 and 30 years old, but other observations presented by Miller and associates (1987) and Reynolds and co-workers (1989) suggest that the age at which the switch occurs is, to some extent, site dependent.

Recently Cape and Brown (1986a; 1986b) studied the influences on throughfall and stemflow of six tree species—two deciduous broad-leaved (oak; *Q. petraea* and alder; *Alnus glutinosa* L. Gaertn.), one deciduous conifer (European larch; *Larix decidua* Miller), and three evergreen conifers (Scots pine, Norway spruce, and Sitka spruce)—at three locations in northern Britain—Crathes in northeast Scotland, Devilla Forest in central Scotland, and Gisburn in northwest England. Gisburn and Devilla, both with annual mean concentrations of 8.0 to 9.0 ppb V each of SO_2 and NO_2, were more polluted than Crathes (2.0 to 3.0 ppb V SO_2 and NO_2); relatively small concentrations of ozone were recorded at all three sites. Irrespective of time of year and species of tree (they were all at least 22 years old), stemflow was always more acidic than incident precipitation, that of European larch was usually more acidic than that of Scots pine, which in turn was more acidic than that of Sitka spruce and Norway spruce, and the least acidic were oak and alder (Figure 8-8). The degree of stemflow acidification attributable to oak and alder was minimal in late summer/early autumn. The relatively large effect of European larch, irrespective of the presence or absence of needles, suggests that bark, in addition to foliage, plays an important role in ion exchange and/or leaching processes. In contrast to their stemflow, throughfall from oak, alder, and Sitka spruce was less acidic than incident rain during summer and autumn (July through November). At other times their throughfall was more acidic than incident precipitation but much less acidified than stemflow. Throughfall from Norway spruce, Scots pine, and European larch was always acidic, with the deciduous larch having the largest effect even, as with stemflow, in the absence of needles during winter (Cape and Brown, 1986a). When interpreting these data, it is necessary to recognize that different tree species have different effects on interception losses and on the subsequent apportionment of the remaining incident precipitation between stemflow and throughfall (Parker, 1983). Stemflow from European larch contributed 3.6% to the aggregate volume of stemflow plus throughfall at Devilla, compared with 20.8% collected under Scots pine; at Gisburn the percentages for alder, oak, Scots pine, and Norway spruce were 11.0%, 11.8%, 15.7%, and 10.2%, respectively. In terms of acidity, stemflow from European larch contributed 9.7% to the total at Devilla, compared with 48.0% for Scots pine. At Gisburn, stemflow from the deciduous alder and oak contributed about 25% to total acidity, whereas that from the evergreen conifers represented 14.0 to 20.7% of the total (Cape and Brown, 1986b).

In general, ion balances (cations and anions) were in good agreement (Figure 8-9), suggesting that most of the acidity was associated with inorganic acids, an outcome, in keeping with observations made by Skiba and associates (1986), who, in working with simulated acid rain, found that it increased the leaching of cations from both Sitka spruce and heather (*Calluna vulgaris* L. Hull), the sum of the cations (Mg^{2+}, Ca^{2+}, Mn^{2+}, K^+, and Na^+) being linearly related to the uptake of H^+ by foliage. This relationship suggests that the transfer was subject to cation exchange. At pH 3.5, leaching was greater with rain acidified with H_2SO_4 than with either HNO_3 or an equinormal mixture of H_2SO_4 and HNO_3. When examining the aggregate composition of throughfall and stemflow in their experiments with alder, oak, Scots pine, Norway spruce, and Sitka spruce, Cape and Brown (1986b) found that potassium, calcium, and magnesium were "leached" from all five species (at each of three sites), whereas ammonium and nitrate were absorbed except by the two species of *Picea* spp. (Figure 8-9). Not

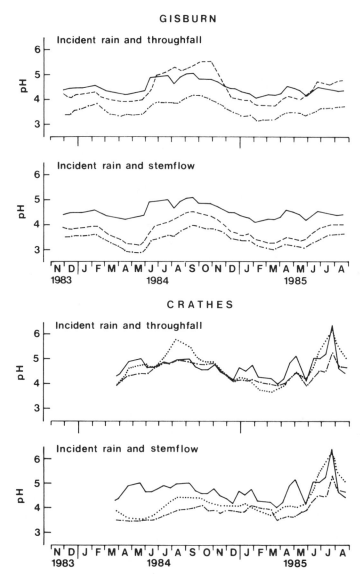

Figure 8-8. Influences of alder (– – – – – –), Scots pine (— · — · — · —), and Sitka spruce (---------------) on the pH of throughfall and stemflow at two sites, Gisburn and Crathes, in northern Britain (————, incident precipitation) (Cape and Brown, 1986a).

surprisingly, amounts of water-soluble organic carbon were consistently and appreciably increased by factors ranging from ×1.5 to ×17.8.

These results add to our knowledge of the chemistry of foliar runoff, a subject pioneered by Tukey (1971), his father,, and S. H. Wittwer, and of great concern to Ulrich (1983). At one extreme it is suggested that the enrichment of throughfall

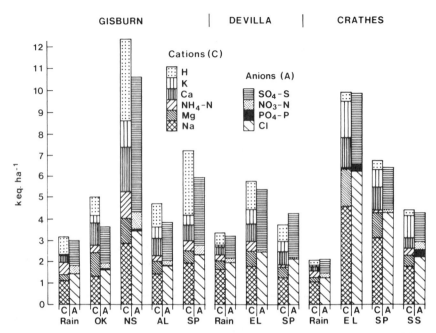

Figure 8-9. Amounts of different cations (*C*) and anions (*A*) in rain and throughfall + stemflow collected in 1985 and 1986 from stands of alder (*AL*), oak (*OK*), European larch (*EL*), Scots pine (*SP*), Norway spruce (*NS*), and Sitka spruce (*SS*) at three sites in northern Britain (Crathes, Devilla, and Gisburn) (Cape and Brown, 1986b).

and/or stemflow is largely attributable to the acquisition of "elements" originally deposited on plant surfaces from the atmosphere; at the other extreme, the acquisition of solutes is attributed to ions derived, by one means or another, from within plants. Whatever proves to be the correct balance of these and other processes (see Cape et al., 1987), the acidification of throughfall and stemflow should be regarded as a stress on terrestrial ecosystems immediately impacting on soil.

A consideration of natural processes inevitably leads to the conclusion that soils in most relatively cool areas, where precipitation regularly and substantially exceeds evapotranspiration, such as the uplands of Scotland, will attain pHs in the range of 4 to 5. This being so, recognizing that (1) "Accelerated acidification of soil by acid deposition at a distance from point sources is by its nature extremely difficult to detect" (Bache, 1984) and (2) if the base cations leached as a consequence of deposited acidity could be replaced by geochemical weathering, the base status of soil would not deteriorate and pHs would remain unaltered (Wood and Bormann, 1977), has it been possible to detect (in Scotland) an additional intensifying effect attributable to atmospheric pollutants? As far as I know, strongly argued evidence of soil changes in Scotland attributable to atmospheric pollutants has not been presented. In their review, Unsworth and associates (1988) question the influence of pollutants on the acidity of hill peats, a soil formation of importance in Scotland. By examining events in comparable, but unpolluted, situations elsewhere in the world, notably New Zealand, they concluded that "pH shifts in Britain caused by moderate amounts of pollutants may generally be modest. However, pollutant inputs may substantially reduce the timescale over which the ultimate degree of soil acidification is attained." However, it is desirable to be mindful of the observations made by Hallbacken and Tamm (1986) in southwest Sweden, where the pollution climate, unlike those of central England, eastern Belgium, most of the FRG, western Czechoslovakia, and near Sudbury, Ontario, Canada (Hutchinson and Whithby, 1977), is similar to that over much of Scotland. With detailed observations of 90 profiles made in 1927 and 1982 to 1983, Hallbacken and Tamm detected an increase in acidity of 0.3 to 0.9 pH units in all soil horizons irrespective of tree species (beech, oak, and Norway spruce). Having considered and/or made an allowance for the effects of age (years after establishment), Hallbacken and Tamm concluded that it was "difficult to explain (the observed acidification) without assuming an influence of acid deposition." Abrahamsen (1984) suggested that acid deposition may have three soil-mediated effects: "(i) a fertilizer effect caused by the deposition of N, and possibly, under specific conditions, also of S (—this has already been mentioned—) (ii) an acidification effect caused by increased leaching of base cations and (iii) an Al toxicity effect in cases where soil acidity is increased." Nobody doubts that acidic inputs can accelerate the rate of cation leaching, but there is a lack of objective data showing the magnitude of losses of this sort in Scottish conditions (see Cresser et al., 1988). Dighton and associates (1986) have shown that the repeated application of simulated acid rain at pH 3.0 can decrease the number of root tips per seedling and, at the same time, alter the type of fungus forming mycorrhizal associations with Scots pine. In microcosm experiments using SO_2, Ineson and Wookey (1988) found that as little as 15 ppb V SO_2 could significantly decrease the rate of microbial respiration associated with conifer litter decomposition and, by inference, affect nutrient cycling, including the accelerated loss of calcium and magnesium. Although most of these experiments have been done in conditions that are too far removed from reality, there is little doubt that acid deposition may increase the acidity of soils but the extent of this change and its significance to tree growth in Scottish conditions remains to be resolved. Perhaps it is necessary to adopt a bolder approach by following the example of the Rain Project, in which Norwegian colleagues working in the field either exclude or add acid rain to minicatchments (Wright, 1987).

IV. Changes in Aquatic Ecosystems

Despite the absence of unequivocal evidence to show that acidic deposition has changed the pH of soils in Scotland or altered their rates of change, there are many biological indicators of freshwater acidification. For some time it has been known that the acidity of water at breeding sites restricts the distribution of some northern temperate amphibians (Gosner and Black, 1957). During 1986 Cummins and Ross studied the development of 42 spawn clumps produced by the common frog (*Rana temporaria* L.) in natural locations, mostly overlying granite rock, near to Loch Fleet, which, as shown by Battarbee and co-workers (1985a), has become "acidified" since the onset of the Industrial Revolution. By plotting the mortality of developing embryos (%) against the most acid pH recorded at the different breeding sites, Cummins and Ross found that mortality was directly related to increasing acidity, with pH 3.5 being critical (Figure 8-10). Although embryos in the outer layers of spawn clumps were killed at pH 3.40, those within the clumps continued to develop, probably because the clumps themselves afforded a degree of protection. Increasing acidity was, as expected, linked with increasing concentrations of monomeric aluminum, which, in laboratory experiments, were shown to decrease rates of tadpole growth with associated delays in metamorphosis, a series of effects that may increase risks of predation losses as ephemeral ponds disappear (by drainage and/or evaporation).

Although the uncontrolled loss of amphibia is reprehensible, popular interest in acidification is more active when stocks of fish are at risk. Brown and Turnpenny (1988), when examining the fishing records of Loch Fleet in Galloway, found that about 100 brown trout (*Salmo trutta fario* L.) were caught per year between 1935 and 1950; thereafter, numbers decreased, despite three attempts at restocking. Characteristically, and as described by Muniz and Leivestad (1980), fish caught during the period of decline were, probably as a result of decreased competition, appreciably larger (900 g) than before (220 g) (Figure 8-11). In this example there is clear evidence of a progressive change with time, which Anderson and

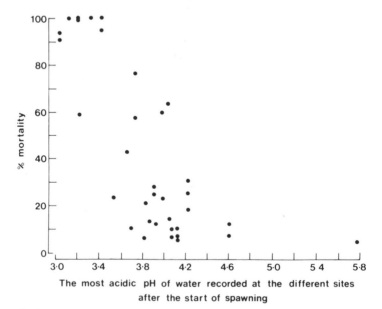

Figure 8-10. Relation between breeding site pH in the Cairnsmore of Fleet, a granitic area of Galloway, Scotland, and the mortality of embryos developing within spawn clumps of the common frog (Cummins and Ross, 1986).

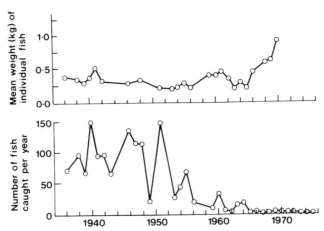

Figure 8-11. Numbers and mean weights of brown trout caught with rod and line in Loch Fleet, Galloway, Scotland between 1936 and 1975 (Brown and Turnpenny, 1988).

co-workers (1986) suggested was, at least in part, attributable to the partial afforestation of the Loch Fleet catchment. However, this change in land use, starting in the 1960s, cannot account for the gradually decreasing numbers of fish caught during the 1950s. Instead, it is more likely that the changes reflect the impacts of acidic deposition on the granitic catchment of Loch Fleet, whose waters lack buffering capacity. With this relationship in mind, Maitland and his colleagues (1986) surveyed a range of lochs in granitic and nongranitic areas of Scotland, taking note of acidic deposition and land use (freshwaters in nongranite areas, the controls, were selected as close as possible to their granitic counterparts so as to minimize differences attributable to weather and amounts of acidic inputs; additionally, attempts were made to match the sizes and other characteristics of each pair of lochs).

Within Scotland, Maitland and associates estimated that 1,536 of 3,788 lochs (Smith and Lyle, 1979) have basins overlying granite, and only 377 of the granite basins were larger than 1 ha in surface area. Of the latter group of 377, 279 were entirely contained by granitic bedrock. Fifty of the latter group of lochs were studied, plus six "fringe" lochs whose granitic catchments included small proportions of other types of bedrock and 27 control lochs whose catchments were outwith areas of granite. Two series of observations were made. One was concerned with water chemistry (pH, conductivity, and concentrations of calcium, magnesium, aluminum, PO_4-P, NO_3-N, NH_4-N, chloride, and SO_4-S), and the other with the occurrence of fish, leeches (*Hirudinea*), and snails (*Gastropoda*). Fish in lochs were trapped with gill nets, identified, measured (length and weight), and sexed before samples were taken of scales, flesh, and stomach contents; those in major inflow and outflow streams were caught with standard electrofishing procedures, before being subsequently returned. These observations of the different lochs, which from historical sources were mostly known to have had fish in the past, showed that all of the control lochs had fish whereas 27% of those overlying granite are now fishless; data from inflow and outflow streams were in broad agreement with this generalization. When the presence and absence data were superimposed on plots of pH against calcium concentrations, it was immediately apparent that virtually all of the fishless lochs were grouped on the acidified side of the empirical curve used by Henriksen (1979) to separate acidified from nonacidified lakes in Norway. Additionally it was apparent that only one area of Scotland, Galloway in the southwest, had significant numbers of lochs affected by acidification, namely, those associated with two blocks of granite (Doon and Cairnsmore). On studying lochs on these blocks of granite with surface areas greater than 1 ha, 6 of 23 were found to be fishless; in another 5, the condition of the fish (tail deformities, the lack of recruitment, etc.) suggested acid stress. Some of the control lochs overlying slates and schists had fish, although chemically they appeared to be acidified.

Although progressively declining populations of fish may suggest continuing freshwater acidification, how can independent proof be obtained, bearing in mind the necessity for information about events in the past? Some of the most elegant pollution research in this regard has depended upon sediment accumulations of diatoms and knowledge of the conditions in which present-day populations of those diatoms flourish. Battarbee (1984) and his colleagues have studied the assemblages of diatoms in lake sediments in Galloway, southwest Scotland, where the underlying bedrock is granitic and where there are very large inputs of deposited acidity. Although there are often very good relationships between water acidity and different types of diatoms, whether in open water or attached to plants, stones, sand grains, and mud surfaces, acidity may additionally be correlated with other important attributes of water quality, such as ionic strength and nutrient availability (see Patrick, 1977). In the first instance, Battarbee accepted Hustedt's (1937–39) characterization of diatoms occurring in waters naturally differing in pH:

1. Alkalibiontic: occurring at pHs >7
2. Alkaliphilous: occurring at pH 7 with widest distribution at pH >7
3. Indifferent ("circumneutral"): equal occurrences on both sides of pH 7
4. Acidophilous: occurring at pH 7 with widest distribution at pH <7
5. Acidobiontic: occurring at pH <7, optimum distribution at pH 5.5 or pH <5.5

Thereafter, for each sample he calculated index B, as recommended by Renberg and Hellberg (1982):

$$\text{index B} = \frac{\% \text{ indifferent} + (5 \times \% \text{ acidophilous}) + (40 \times \% \text{ acidobiontic})}{\% \text{ indifferent} + (3.5 \times \% \text{ alkaliphilous}) + (108 \times \% \text{ alkalibiontic})}$$

knowing that Renberg and Hellberg had found, when examining the diatom assemblages occurring in 30 lakes in Scandinavia, that

$$\text{pH} = 6.40 - 0.85 \log \text{index B} \ (r^2 = 0.91)$$

with values of pH having standard errors of ±0.3 of a pH unit.

With this procedure, and knowing that dead diatoms are well preserved in acid conditions, Battarbee examined the diatoms in lake sediment cores that were frozen in situ, so reducing the mixing of diatoms during disturbance, and in which the different layers within cores were dated using ^{210}Pb.

The six lochs in Galloway studied by Battarbee and co-workers are situated either entirely or partly on granite bedrocks; the catchments of three (Loch Enoch, Loch Valley, and Round Loch of Glenhead) have remained undisturbed (and therefore unafforested), while the others (Loch Grannoch, Loch Dee, and Loch Skerrow) have been plowed and afforested (see Wells et al., 1986; Battarbee et al., 1985a). For those that have remained undisturbed, the diatom reconstructions imply a consistent increase in acidity of about one pH unit since 1850. As data for Round Loch of Glenhead indicate, populations of circumneutral taxa, *Anomoeoneis vitrea* (Grun.) Ross and *Achnanthes microcephala* (Kütz.), decreased rapidly at the onset of acidification in 1859 to be followed by other circumneutral taxa that were replaced by populations of acidophilous species such as *Tabellaria flocculosa* (Roth.) Kütz. and *Eunotia veneris* Kütz., which in turn were beginning to be replaced by acidobiontic taxa such as *Tabellaria binalis* (Ehr.) Grun. In the disturbed catchments, developments have been less consistent. Acidification in Loch Grannoch was delayed compared to events in other unafforested catchments (acidification was first discernible in 1920 to 1930), but by 1960 its diatom assemblage was dominated by *T. binalis* and *T. quadriseptata* Knudsen (as in Loch Enoch and Loch Valley). Since then, the picture has been clouded as a result of catchment plowing and consequent increased soil erosion. In Loch Dee and Loch Skerrow, acidification has been less intense, probably because their catchments include substantial areas of nongranitic rocks. Nonetheless, their populations of circumneutral species of *Cyclotella* had decreased greatly by the end of the nineteenth century. Because afforestation cannot be implicated, the changes that have occurred in Loch Enoch, Loch Valley, and Round Loch of Glenhead are likely to be associated with the deposition of atmospheric pollutants. As pointed out by Rosenquist (1978), however, changes in water quality can also reflect changes in the ways in which the associated catchments are managed, for instance, changes in grazing intensities and/or the possibly increased growth of "acidifying" heather. However, the evidence from sediment cores from Loch Enoch (which has not been afforested) does not support this contention. At this site heather has become less abundant while pollen grains of grasses have become more abundant. Instead, the increase in acidity is associated with increases in copper, zinc, lead, and soot (carbonaceous particles) (Darley, 1985), all indicators of increased industrial activity (Figure 8-12).

This was the picture until recently, but observations made since 1981 of two of the undisturbed lochs, Round Loch of Glenhead and Loch Enoch, suggest that their acidification has been halted. In fact, analyses of sediment cores taken from Loch Enoch in 1982 and 1986 and from the Round Loch of Glenhead in 1981, 1984, and 1986 suggest that these lochs have become less acidic (Battarbee et al., 1988). This result is in keeping with observations made at nearby Loch Valley (pH 4.47 in 1978 and 1979; pH 4.75 in 1984 to 1986; Harriman and Wells, 1987) and in other moorland catchments in Scotland (Flower et al., 1988).

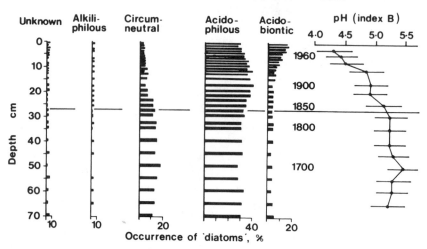

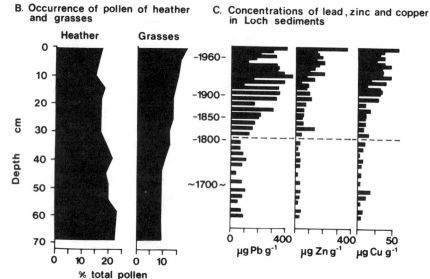

Figure 8-12. Historical records in sediments of events in Loch Enoch, Galloway, Scotland, an unafforested catchment overlying granitic bedrock (Battarbee et al., 1985a and 1985b).

Although the diatom data indicate that lochs, in catchments overlying slowly weathering granitic and other similar bedrocks in areas where large amounts of acidity are deposited, can be acidified, they give no indication of the within-season variations in water acidity. Having concluded that waters susceptible to acidification are likely to be in high-rainfall, steeply sloping areas with granitic or similar bedrock, Cresser and associates (personal communication) analyzed the outflows from two lochs meeting these requirements in the Grampian Mountains of northeast Scotland. The data for the outflows from Dubh Loch and Loch Muick give an indication of the extent of seasonal variations and show the increases in acidity coincident with abnormally wet weather in November 1984 (Figure 8-13), an effect comparable to the acid flushes recorded elsewhere in Britain and during snowmelt in southwestern Norway (Henriksen et al., 1984).

In similar situations Harriman and Morrison (1980 and 1982) have found that afforestation, notably with Sitka spruce and also including Scots pine, lodgepole pine, and Norway spruce can exacerbate the situation. They observed a set of streams in the Duchray and Loch Chon catchment areas overlying schistose grits, graywackes, slate, and phyllite, in the west of central Scotland where the mean weighted pH of bulk precipitation was, from 1973 to 1979, about pH 4.4. In the event H^+ was significantly correlated with excess sulfate (total sulfate minus marine sulfate), the deposition of acid pollutants invariably was associated, as already mentioned, with air masses crossing Scotland from southerly or easterly directions, where major sources of emission are located.

Within the Duchray area, three of the streams drained unafforested moorland minicatchments (the controls), and six drained minicatchments that had been afforested for varying numbers of years; in the Loch Chon area, two control streams and one draining a minicatchment had been afforested for about 30 years. From stream samples taken in 1979, usually at intervals of 2 weeks, Harriman and Morrison found that streams draining afforested minicatchments were always more acid than their nearby control-stream counterparts, the effect of forests about 25 years old being appreciably greater than that of a stand 4 years old. Afforestation increased the frequency of pH events more acid than 5.0 and 4.4, these changes being associated with larger concentrations of SO_4^{2-} and Cl^- and the increased mobilization of aluminum and manganese. The concentrations of

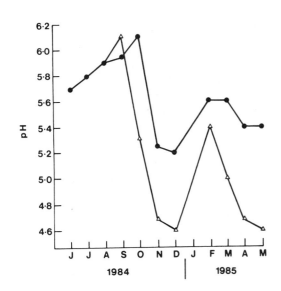

Figure 8-13. Seasonal changes in the pH of water leaving two freshwater lochs—Dudh Loch ($\triangle$) and Loch Muick ($\bullet$)—in the Grampian Mountains, NE Scotland (Cresser et al., personal communication).

aluminum increased from a typical maximum of 100 μg L^{-1} up to 350 μg L^{-1}, and those of manganese went from 30 μg L^{-1} in the control streams, draining nonafforested minicatchments, to 90 μg L^{-1}. As in Galloway, changes in water quality were matched by alterations to the aquatic biota. Although the biomass of invertebrates remained fairly constant, mayflies (*Ephemeroptera*) were precluded by increasing acidity from assemblages that in the control streams also included a variety of worms, stone flies (*Plecoptera*), and caddis flies (*Trichoptera*).

V. Discussion

The pollution literature contains many apparent disagreements and inconsistencies. For example, Cape and associates (1987) concluded from a study of the annual sulfate budgets in a Scots pine forest in central Scotland that about 50% of the "excess" sulfate found in throughfall represented leachings from foliage ("excess" = concentration in throughfall minus concentration in incident precipitation). In contrast, Lindberg and co-workers (1986) suggested, from observations of a mixed forest in eastern USA, that there was little or no evidence of leaching, a point of view agreeing with Skeffington (see Unsworth and Fowler, 1988). However, is it possible that these deductions differ because observations were being made on ecosystems developing in discernibly different pollution climates? It is probably true that most research workers concerned with the fate and effects of atmospheric pollutants have usually had unduly single-minded and narrow views of their particular problems. Teams concerned with acid waters have tended to focus on the fate and effects of atmospheric pollutants reaching their targets by wet deposition, and those concerned with the effects of sulfur dioxide, oxides of nitrogen, and ozone on the growth of plants (nonwoody and woody) have veered toward pollutants reaching their targets by dry deposition. By developing the pollution climate concept, it has been possible to show that the mixture of pollutants encountered in Scotland is distinct from that occurring in central England, eastern Belgium, the Nordrhein-Westfalen region of the Federal Republic of Germany, and so on. More importantly, the identification of pollution climates facilitates, as a matter of principle, a more objective approach to comparisons and contrasts, a process that can be carried further by objective classifications of catchments (using sensitivity characteristics; Table 8-1) and climates (Jones and Bunce, 1985). In doing so, it becomes obvious that conditions in much of Scotland have affinities with some of those in southern Norway (Wright et al., 1980) and are unlike those, for instance, in much of the FRG and western Czechoslovakia. As a result, it is not surprising that the main area of concern in Scotland about atmospheric pollutants is closely allied to that in southern Norway, namely, impact on freshwater ecosystems; it differs from that in FRG, where the impact on the growth of trees is foremost.

Within Scotland there is evidence of seasonally varying changes in the quality of streams, of historical changes in the quality of lochs (and their inflow and outflow streams) associated with the deposition of pollutants since the onset of the

Industrial Revolution, and also of increasing acidity where water drains from catchments and/or watersheds planted with conifers. Although these occurrences are predictably found in poorly buffered waters draining watersheds that overlie granite bedrock (also schistose grits and graywackes), can the magnitude of these changes be predicted? Together Cosby, Whitehead, Neal, and Neale have modeled changes in surface water chemistry, assuming that water chemistry is determined by (1) the deposition of atmospheric pollutants, (2) the rates and amounts of mineral weathering, and (3) exchanges in soil and soilwater. In their model MAGIC (model acidification of groundwater in catchments), they further assume that "the vertical stratification of soils is unimportant or, equivalently, that all water reaching the stream contacts, and has its chemical quality determined by, a single layer within the soil column" (Cosby et al., 1986). Despite, or possibly because of, the simplicity of these assumptions, they estimated, when simulating pH changes in the Dargall Lane moorland subcatchment of Loch Dee (Figure 8-1) from historical records of sulfate deposition (Barrett et al., 1983) that water acidity increased significantly from 1860 onward (Figure 8-14), a deduction in keeping with that made from diatom records of nearby lochs (Battarbee et al., 1985a). On extending their analysis to include effects of afforesting the Dargall Lane subcatchment, they progressively included the influences of (1) increasing evapotranspiration losses from 16 to 30% when replacing moorland vegetation by trees and (2) additional inputs following afforestation of sea salt (30%) and scavenged pollutants (40%) (Figure 8-14). As found by Harriman and Morrison (1980), calculations made by Neal and associates (1986) suggest that afforestation is likely to increase the acidity of poorly buffered streams, the increased inputs of sea salts and pollutants being individually and in combination greater than that of increased evapotranspiration.

In their model linking hydrological and chemical processes operating in snow cover, near surface soils, and in groundwater, Seip and co-workers (1985) controlled the flow between soil and groundwater compartments with a percolation factor (P). With the introduction of "piston flow" between soil and groundwater compartments, the effect of P is restricted to the regulation of discharge chemistry; it does not affect total amounts of discharge. By altering P so that the baseflow contribution to streams decreased from 89 to 52%, with a corresponding increase from 11 to 48% in the contribution of surface flows, Whitehead and associates (1986) found that the simulated concentrations of H^+ and Al^{3+} were significantly increased (Figure 8-14), as happens during spates. The changes are not linear and, furthermore, they are of greater importance as amounts of deposited pollutant increase.

In Scandinavia, attention has been increasingly focused on critical loads that "will not cause chemical changes leading to long-term harmful effects on the most sensitive ecological systems" (Nilsson, 1986). For acid deposition, critical loads have been separately argued and defined for forest soils, groundwater, and surface water. Accepting a variety of caveats, it is thought that loads in excess of 15 kg H^+ km^{-2} yr^{-1} are expected to "maintain pressure" on forest soils and freshwater ecosystems in Scandinavian conditions. The question of what the critical load(s)

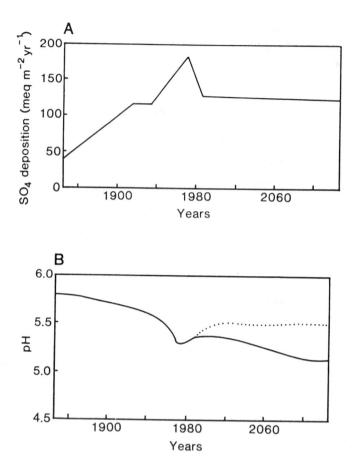

Figure 8-14. Simulations of the chemistry of waters draining areas overlying graywacke, shale, granite, schist, and/or gneiss-type bedrocks. (A), (B), and (C): data relevant to the Dargall Lane moorland subcatchment of Loch Dee, Galloway, Scotland (Neal et al., 1986). (A) Historical amounts of sulfate deposition to 1984 with constant amounts shown thereafter. (B) Effects of different amounts of sulfate deposition from 1984 onwards on stream pH. ———, pH when amounts of sulfate deposition from 1984 onwards are maintained at 1984 levels. ·········, pH when amounts of sulfate deposition are progressively decreased by 50% between 1984 and 2000. (C) Effects of afforesting the Dargall Lane moorland subcatchment with conifers. ———, pH of stream draining unafforested subcatchment, assuming sulfate deposition remains constant from 1984 onwards. ·········,

would be in Scotland is as yet unanswered; however, Unsworth and his colleagues (1988) think it should be accorded high priority in research programs.

Knowledge of critical loads would add substance to debates about reversibility. How soon after lessening or abating the emission of pollutants will the water quality (including pH) of acidified lochs and lakes return to its pre-Industrial Revolution condition? Fowler and Cape (1985) discussed the factors interrelating

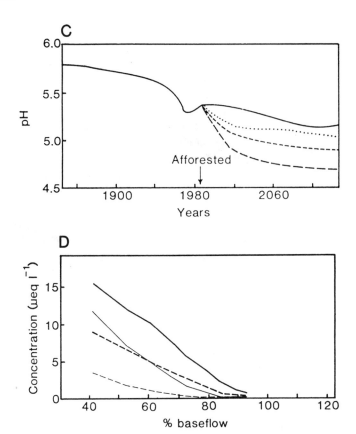

effects of increased evapotranspiration ("interception loss") after afforestation; losses increased from 16% (moorland vegetation) to 30% after afforestation. ------, effects of increased evapotranspiration losses and 30% additional sea salt inputs (by scavenging) following afforestation. – – – –, effects of increased evapotranspiration losses and additional inputs (by scavenging) of (i) sea salts (30%) and (ii) atmospheric pollutants (40%). (D) Hypothetical examination of the effects of varying the proportions of *baseflow* (groundwater) and *quickflow* (water draining near surface soils) on concentrations of H^+ (——,——) and Al^{3+} (– – –,– – –) in a stream flowing into Harp Lake, Ontario assuming two rates of sulfur deposition—2g Sm^{-2} yr^{-1} (——,– – –) and 6g Sm^{-2} yr^{-1} (——,– – –) (Whitehead et al., 1986).

the emission and deposition of atmospheric pollutants—atmospheric transport, dry deposition, atmospheric chemistry, and the occurrence of rain. Instead of relying upon models to give conjectural solutions, they preferred to rely, more pragmatically, upon actual data. They noted, however, that Fisher and Clark (1985) had concluded that the relationship between emissions and deposition was approximately linear for distant receptors. This deduction seems to be supported by the data given by Fowler and Cape (1985): "For the United Kingdom, emissions of sulphur dioxide decreased by 25% between 1978 and 1982 and if 80% of the total sulphur (wet and dry) deposition originates in the UK, the deposition of acidity, which is closely linked with sulphur deposition, should have decreased by a similar amount if the processes are linear." Continuous records of rainfall chemistry at eight sites in Scotland and northern England show systematic decreases in SO_4^{2-} (Figure 8-15) and H^+ concentrations, the decreases of the latter amounting to 50% in the period 1978 to 1983. There is therefore a strong relationship between emissions and deposition, but caution is needed when inferring causality because systematic changes in weather could be of significance. Furthermore, cognizance should be taken of an increasing proportion of NO_3^- to SO_4^{2-}, with an expected continual increase in emissions of oxides of nitrogen attributable to vehicle emissions (Scottish Development Department, 1984).

It would therefore be of interest to judge how a reduction of acidic inputs will be reflected in changes in the quality of water in lochs and streams, knowing that eight stages of acidification and recovery have been identified (Galloway et al., 1983). Although Cosby (1987) suggested, when using the MAGIC model, that recovery response times are likely to be measured in decades, he and his colleagues have shown that recovery might begin as soon as amounts of acidic deposition lessen (Neal et al., 1986). Like Dillon and associates (1986), who observed freshwaters near Sudbury, Ontario, Battarbee and co-workers (1988) have found a discernible improvement in Galloway lochs in less than 10 years, a matter of intense interest when judging the effects of other ameliorative procedures (such as the application of lime), as is being attempted in Loch Fleet (Howells, 1987). However, much remains to be done to reconcile predictions and actuality; more needs to be learned of the mechanisms involved in acidification and its reversal. This conclusion applies with similar aptness to the mechanisms involved in damage done to plants,

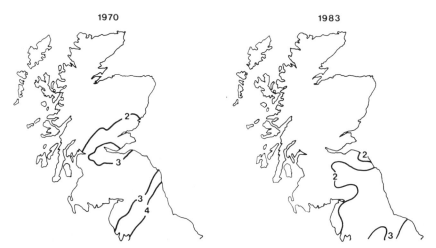

Figure 8-15. Changes in the modeled total amounts (wet + dry) of sulfur annually deposited in northern Britain in 1970 and 1983 (g $Sm^{-2}yr^{-1}$) (Barrett et al., 1987).

particularly where, as in Scotland, the prevalent ambient concentrations of atmospheric pollutants, if not beneficial as has been suggested in relation to crops of oilseed rape *Brassica napus* L., are likely to cause chronic, not acute, damage.

VI. Future Research Directions

In discussing acidic deposition in Scotland it has been possible to identify five areas of further enquiry that should be accorded priority:

1. *The development of two-way matrices embracing the "pollution climate" and "land class" concepts.* With these matrices it should be possible to examine more effectively the interrelations of events related to the acidification of soils and freshwaters or the incidence of forest decline whether in different continents, in different countries of the same continent, or in different locations within a country. To achieve this objective, the derivation of pollution climates is dependent upon networks (1) monitoring atmospheric concentrations of SO_2, NO_x, ozone, ammonia . . . and (2) making direct measurements of occult, in addition to wet, deposition and hopefully, in due course, dry deposition. The objective system of land classification devised by Jones and Bunce (1985) from map readable data primarily concerned with daylengths, amounts of sunshine and precipitation, including snow, and atmospheric temperatures would be greatly enhanced if geochemical attributes, related to the acid sensitivity of soils and catchments, were added.
2. Process studies. Although predictions from some models concerned with the movement and chemistry of soil solutions, also their impacts on the acidity of freshwaters, agree reasonably closely with events deduced from paleoecological studies, it is nevertheless desirable to verify the validity of the mathematical functions used in these models; the functions commonly integrate complexes of soil and freshwater processes.
3. *Soil changes and critical loads.* Nilsson (1986) defined a critical load as "The highest load that will not cause chemical changes leading to long-term harmful effects on the most sensitive ecological systems." For surface waters it can be modified to read "The highest load that will not lead in the long-term (within 50 years) to harmful effects on biological systems, such as decline and disappearance of natural fish populations" whereas for forest soils it was suggested that "The soil should be protected from a long-term chemical change, due to anthropogenic impact, which cannot be compensated by natural soil processes." To aid the development of a rational approach to the moderation of problems associated with atmospheric pollutants it is essential to develop the critical load concept which is akin to the dose-response relationships of classical toxicology. A development of this sort is likely to provide, coincidentally, information on changes in soil properties since the onset of the Industrial Revolution.
4. *Reversal of acidification.* Wright (1987) has outlined two challenging experiments being done in the southern half of Norway. In essence, attempts are being

made to apply Koch's postulates—the bulwark of pathology whether human, veterinary, or plant (Last, 1987)—to the eludication of factors contributing to the acidification of freshwaters. In one of the experiments minicatchments are being "protected" (by exclusion) from acidic precipitation whereas in the other, minicatchments are receiving augmented acidic inputs. Preliminary results suggest that these experiments might profitably be repeated elsewhere but not at the expense of detailed observations at sites where ambient loads of atmospheric pollutants lessen as a result of political decisions to restrict emissions. The opportunity of observing the effects of hopefully progressively decreasing emissions of pollutants should be seized.

5. *Forest decline*. The existence of forest decline in Scotland is arguable but what is not arguable is our profound ignorance of the responses of trees to biotic stresses whether related to weather (drought, unseasonal frosts, and the like) and/or pollutants including occult deposition. The investigation of stress physiology should be accorded priority.

6. *The investigation of the effects of soil acidity on the growth of heather*. Since ordering my priorities I have attended a symposium—Acidification in Scotland 1988—organised by the Scottish Development Department. As a result I have reason to add this sixth priority, involving the plant for which Scotland is famous. Surprisingly, M. Cresser of the University of Aberdeen has reason to believe that some of the hill peats in Scotland have acidified to an extent which is restricting the growth of acid-tolerant heather. If this is true of heather, the effects on other acid-tolerant species in the seminatural habitats of Scotland is an issue that should be addressed.

Acknowledgments

In addition to the authors I wish to acknowledge the willingness of the following publishers to grant permission to reprint, in part or whole, Figure 8-2 and Table 8-1, British Geological Survey, Keyworth, UK; Figure 8-2, Controller of Her Majesty's Stationery Office; Figure 8-3, J.G.R. Stevens, publisher of *Span*; Figures 8-4, 8-5, and 8-15 (Crown Copyright), Warren Spring Laboratory, Stevenage, UK; Figures 8-6 and 8-7, and Table 8-5 (Crown Copyright), Forestry Commission, Edinburgh, UK; Figures 8-8, 8-9, and 8-10 (data ex Community R & D programme), Commission of the European Communities, Brussels, Belgium; Figure 8-11, Imperial College Centre for Environmental Technology, London, UK; Figure 8-12 and Table 8-4, (ex *Nature*) Macmillan Magazines Ltd., London, UK; Figure 8-14, Elsevier Science Publishers B.V., Amsterdam, the Netherlands; Table 8-2, Edinburgh Centre of Rural Economy, UK; and Table 8-3, Blackwell Scientific Publications Ltd., Oxford, UK.

References

Abrahamsen, G. 1984. Phil Trans R Soc Lond B305:369–382.
Anderson, N. J., R. W. Battarbee, P. G. Appleby, A. C. Stevenson, F. Oldfield, J. Darley,

and G. Glover. 1986. *Palaeolimnological evidence for the acidification of Loch Fleet.* Palaeoecology Research Unit Working Paper no. 17, Department of Geography, University College, London.

Anon. 1986. *Waldschäden in der Bundesrepublik Deutschland: Ergebnisse der Waldschädenserhebung 1986.* Schriftenreihe des Bundesministers für Ernährung, Landwirtschaft und Forsten, Reihe A: Angewandte Wissenschaft 334.

Bache, B. W. 1984. Phil Trans R Soc Lond B305:393–407.

Barrett, C. F., D. H. F. Atkins, J. N. Cape, J. Crabtree, T. D. Davies, R. G. Derwent, B. E. A. Fisher, D. Fowler, A. S. Kallend, A. Martin, R. A. Scriven, and J. G. Irwin. 1987. *Acid Deposition in the United Kingdom 1981–1985.* Warren Spring Laboratory, Stevenage.

Barrett, C. F., D. H. F. Atkins, J. N. Cape, D. Fowler, J. G. Irwin, A. S. Kallend, A. Martin, J. I. Pitman, R. A. Scriven, and A. F. Tuck. 1983. *Acid Deposition in the United Kingdom.* Warren Spring Laboratory, Stevenage.

Battarbee, R. W. 1984. Phil Trans R Soc Lond B305:451–477.

Battarbee, R. W., R. J. Flower, A. C. Stevenson, V. J. Jones, R. Harriman, and P. G. Appleby. 1988. Nature 332:530–532.

Battarbee, R. W., R. J. Flower, A. C. Stevenson, and B. Rippey. 1985a. Nature 314: 350–352.

Battarbee, R. W., R. J. Flower, A. C. Stevenson, and B. Rippey. 1985b. *In Report of the Acid Rain Inquiry, Edinburgh, September 1985,* 89–95. Scottish Wildlife Trust, Edinburgh.

Benarie, M. M. 1980. *In Special environmental report no. 14 WMO technical conference on regional and global observation of atmospheric pollution relative to climate,* 393–398. WMO Secretariat, Geneva.

Binns, W. O., D. B. Redfern, K. Rennolls, and A. J. A. Betts. 1985. *Forest Health and Air Pollution 1984 Survey.* Forestry Commission R & D Paper no. 142. Forestry Commission, Edinburgh.

Brown, D. J. A., and A. W. H. Turnpenny. 1988. *In* M. R. Ashmore, J. N. B. Bell, and C. Garretty, eds. *Acid Rain and Britain's Natural Ecosystems,* 87–96. Centre for Environmental Technology, Imperial College, London.

Brown, K. A., T. M. Roberts, and L. W. Blank. 1987. New Phytol 105:149–155.

Bunce, R. G. H., and F. T. Last. 1981. *In Edinburgh Centre of Rural Economy Annual Report 1980–81.*

Cameron, I. B., and D. Stephenson. 1985. *British Regional Geology—The Midland Valley of Scotland,* 3d ed. Her Majesty's Stationery Office, London.

Cape, J. N., and A. H. F. Brown. 1986a and 1986b. *The effects of different tree species on the amount and chemical composition of precipitation passing through the canopy.* Interim and Final Reports to the Commission of the European Communities Contract ENV774 UK (H). Commission of the European Communities, Brussels.

Cape, J. N., D. Fowler, J. W. Kinnaird, I. S. Paterson, I. D. Leith, and I. A. Nicholson. 1984. Atmos Environ 18:1921–1932.

Cape, J. N., D. Fowler, J. W. Kinnaird, I. A. Nicholson, and I. S. Paterson. 1987. *In* P. J. Coughtrey, M. H. Martin, and M. H. Unsworth, eds. *Pollutant Transport and Fate in Ecosystems,* 155–169. Blackwell Scientific Publications, Oxford.

Cook, F. M. 1971. Chemy Ind 71:585–589.

Cosby, B. J., P. G. Whitehead, and R. Neale. 1986. J Hydrol 84:381–401.

Cosby, B. J. 1987. *In* H. Barth, ed. *Reversibility of acidification,* 114–125. Elsevier Applied Science, London and New York.

Cresser, M. S., R. Harriman, and K. Pugh. 1988. *In* M. R. Ashmore, J. N. B. Bell, and C.

Garretty, eds. *Acid Rain and Britain's Natural Ecosystems,* 55–70. Centre for Environmental Technology, Imperial College, London.

Cummins, C. P., and A. Ross. 1986. *Effects of acidification of natural waters upon amphibia.* Final Report to the Commission of the European Communities EV3V907 UK(H). Commission of the European Communities, Brussels.

Darley, J. 1985. *Particulate soot in Galloway Lake sediments: its application as an indicator of environmental changes and as a technique for dating recent sediments.* B.Sc. Dissertation, University College, London.

Davies, T. D., P. W. Abrahams, M. Tranter, I. Blackwood, P. Brimblecombe, and C. E. Vincent. 1984. Nature 312:58–61.

Davies, T. D., P. M. Kelly, P. Brimblecombe, G. Farmer, and R. J. Barthelmie. 1986. Nature 322:359–361.

Dighton, J., R. A. Skeffington, and K. A. Brown. 1986. *In* V. Gianinazzi-Pearson and S. Gianinazzi, eds. *Physiological and Genetical Aspects of Mycorrhizae,* 739–743. Institut National de la Recherche Agronomique, Paris.

Dillon, P. J., R. A. Reid, and R. Girard. 1986. Water, Air and Soil Pollut 31:59–65.

Edmunds, W. M., and D. G. Kinniburgh. 1986. J Geol Sci 143:707–720.

Evelyn, J. 1661. *Fumifugium: or the inconveniencie of the aer and smoak of London dissipated.* Originally printed by W Godbid for Gabriel Bedel and Thomas Collins, London. Published (1976) in facsimile by the Rota, University of Exeter.

Fisher, B. E. A., and P. A. Clark. 1985. *In* C. de Wispelaere, ed. *Air pollution modelling and its application.* Plenum, New York.

Flower, R. J., R. W. Battarbee, J. Natkanski, B. Rippey, and P. G. Appleby. 1988. J Appl Ecol 25:715–724.

Forestry Commission. 1983. *Census of woodlands and trees, Scotland.* Forestry Commission, Edinburgh.

Fowler, D., and J. N. Cape. 1985. *In Report of the acid rain inquiry, Edinburgh,* September 1985, 30–45. Scottish Wildlife Trust, Edinburgh.

Fowler, D., J. N. Cape, D. Jost, and S. Beilke. 1987. *In litt.*

Galloway, J. N., S. A. Norton, and M. R. Church. 1983. Environ Sci Technol 17:541A–545A.

Gosner, K. L., and I. H. Black. 1957. Ecology 38:256–262.

Hallbacken, L., and C. O. Tamm. 1986. Scand J For Res 1:219–232.

Harriman, R., and B. Morrison. 1980. *In* D. Drabløs and A. Tollan, eds. *Ecological impact of acid precipitation,* 312–313. P.O. Box 61, 1432 Ås-NLH, Norway.

Harriman, R., and B. R. S. Morrison. 1982. Hydrobiologia 88:251–263.

Harriman, R., and D. E. Wells. 1987. *In* R. Perry, R. M. Harrison, J. N. B. Bell, and J. N. Lester, eds. *Acid rain: scientific and technical advances,* 287–292. Seiper, London.

Henriksen, A. 1979. Nature 278:542–545.

Henriksen, A., O. K. Skogheim, and B. O. Rosseland. 1984. Vatten 40:255–260.

Hill, M. O., R. G. H. Bunce, and M. W. Shaw. 1975. J Ecol 63:597–613.

Howells, G. 1987. *In* H. Barth, ed. *Reversibility of acidification,* 104–111. Elsevier Applied Science, London and New York.

Hustedt, F. 1937–39. Arch Hydrobiol (Suppl) 15 and 16.

Hutchison, T. C., and L. M. Whithby. 1977. Water, Air and Soil Pollut 7:421–438.

Ineson, P., and P. A. Wookey. 1988. Effects of sulphur dioxide on forest litter decomposition and nutrient release. *In* P. Mathy, ed. *Air Pollution and Ecosystems,* 254–260. D. Reidel, Dordrecht.

Innes, J. L., and R. C. Boswell, 1987a. *Forest health surveys 1987 part 1: results.* Forestry Commission Bulletin no. 74. Her Majesty's Stationery Office, London.

Innes, J. L., and R. C. Boswell. 1987b. *Forest health surveys 1987 part 2: analysis and interpretation.* Forestry Commission Bulletin no. 79. Her Majesty's Stationery Office, London.

Innes, J. L., R. Boswell, W. O. Binns, and D. B. Redfern. 1986. *Forest health and air pollution 1986 survey.* Forestry Commission R & D Paper no. 150. Forestry Commission, Edinburgh.

Irwin, J. G., and M. L. Williams. 1988. Environ Pollut 50:29–59.

Jones, H. E., and R. G. H. Bunce. 1985. J Environ Manage 20:17–29.

Kinniburgh, D. G., and W. M. Edmunds. 1986. *The susceptibility of UK groundwaters to acid deposition.* Hydrogeol Rep Br Geol Surv no. 86/3.

Lamb, H. H. 1972. *Geophysical memoir no. 116.* Her Majesty's Stationery Office, London.

Last, F. T. 1978. Trans Bot Soc Edinb 42 (Suppl):99–124.

Last, F. T. 1987. *In* L. Kairiukstis, S. Nilsson and A. Straszak, eds. *Forest decline and reproduction: regional and global consequences,* 63–77. International Institute for Applied Systems Analysis, Laxenburg, Austria.

Last, F. T., J. N. Cape, and D. Fowler, 1986. Span 29:2–4.

Lindberg, S. E., G. M. Lovett, D. D. Richter, and D. W. Johnson. 1986. Science 231:141–145.

Lucas, P. W., D. A. Cottam, L. J. Sheppard, and B. J. Francis. 1988. New Phytol 108:495–504.

McNeill, P. 1883. *Tranent and its surroundings.* Menzies, Edinburgh and Glasgow.

Maitland, P. S., A. A. Lyle, and R. N. B. Campbell. 1986. *The status of fish populations in waters likely to have been affected by acid deposition in Scotland.* Final Report to Department of the Environment Contract PECD 7/10/69. Department of the Environment, London.

Mansfield, T. A., W. J. Davies, and M. E. Whitmore. 1986. *In How are the effects of air pollutants on agricultural crops influenced by the interaction with other limiting factors?* 2–15. Directorate General XII, Commission of the European Communities, Brussels.

Miller, H. G. 1984. Phil Trans R Soc Lond B305:339–352.

Miller, H. G., J. D. Miller, and J. M. Cooper. 1987. *In* P. J. Coughtrey, M. H. Martin, and M. H. Unsworth, eds. *Pollutant transport and fate in ecosystems,* 171–180. Blackwell Scientific Publications, Oxford.

Muniz, I. P., and H. Leivestad. 1980. *In* D. Drabløs and A. Tollan, eds. *Ecological impact of acid precipitation,* 84–92. P. O. Box 61, 1432 Ås-NLH, Norway.

Neal, C., P. Whitehead, R. Neale, and J. Cosby. 1986. J Hydrol 86:15–26.

Nilsson, J. 1986. *Critical loads for sulphur and nitrogen.* Secretariat, Nordic Council of Ministers, Copenhagen.

Parker, G. G. 1983. Adv Ecol Res 13:57–133.

Patrick, R. 1977. *In* D. Werner, ed. *The biology of diatoms.* Botanical Monographs vol 13. Blackwell Scientific Publications, Oxford.

Ralph, G. A. 1986, *Acid rain induced crown leaching effect of needle position and tree age.* Unpublished Honours Thesis, Department of Forestry, University of Aberdeen.

Reiter, R., and H. J. Kanter. 1982. Arch Met Geoph Biokl Ser B, 30:191–225.

Renberg, I., and T. Hellberg. 1982. Ambio 11:30–33.

Reynolds, B., J. N. Cape, and I. S. Paterson. 1989. Forestry (in press).

Roberts, T. M. 1984. Phil Trans Roy Soc Lond B305:299–316.

Rosenquist, I. T. 1978. Sci Total Environ 10:39–49.

Scottish Development Department. 1984. *Acid rain: the Scottish dimension.* Scottish Development Department, Edinburgh.

Seip, H. M., R. Seip, P. J. Dillon, and E. de Grosbois. 1985. Can J Fish Aquat Sci 42:927–937.

Skiba, U., T. J. Peirson-Smith, and M. S. Cresser. 1986. Environ Pollut Ser B 11:255–270.

Smith, I., and A. Lyle. 1979. *Distribution of freshwaters in Great Britain.* Institute of Terrestrial Ecology, Abbots Ripton, Huntingdon (formerly at Cambridge).

Smith, R. A. 1872. *Air and rain: the beginning of a chemical climatology.* Longmans Green, London.

Stevens, P. A. 1987. Pl Soil 101:291–294.

Tukey, H. B., Jr. 1971. *In* T. F. Preece, and C. H. Dickinson, eds. *Ecology of leaf surface micro-organisms,* 67–80. Academic Press, London and New York.

Ulrich, B. 1983. *In* B. Ulrich and J. Pankrath, eds. *Effects of accumulation of air pollutants in forest ecosystems,* 33–45. D. Reidel, Dordrecht.

Unsworth, M. H., J. N. B. Bell, V. J. Black, M. S. Cresser, N. M. Darrall, A. W. Davison, P. H. Freer-Smith, P. Ineson, J. A. Lee, F. T. Last, T. A. Mansfield, M. H. Martin, H. G. Miller, T. M. Roberts, C. E. R. Pitcairn, and R. B. Wilson. 1988. *The effects of acid deposition on the terrestrial environ in the UK.* Her Majesty's Stationery Office, London.

Unsworth, H. M., and A. Crossley. 1987. *In* P. J. Coughtrey, M. H. Martin, and M. H. Unsworth, eds. *Pollutant transport and fate in ecosystems,* 125–137. Blackwell Scientific Publications, Oxford.

Unsworth, M. H., and D. Fowler. 1988. *In* P. Mathy, ed. *Air pollution and ecosystems,* 68–84. D. Reidel, Dordrecht.

Warren, S. C., B. W. Bache, W. M. Edmunds, H. J. Egglishaw, A. S. Gee, M. Hornung, G. D. Howells, C. Jordan, J. B. Leeming, P. S. Maitland, K. B. Pugh, D. W. Sutcliffe, D. E. Wells, J. N. Cape, J. C. Ellis, D. T. E. Hunt, R. B. Wilson, and D. C. Watson. 1986. *Acidity in the United Kingdom Fresh Waters.* Department of the Environment and Department of Transport Publication Sales Unit, South Ruislip, UK.

Wells, D. E., A. S. Gee, and R. W. Battarbee. 1986. Water, Air and Soil Pollut 31:631–645.

White, E. J. 1982. For Ecol Manage 4:225–245.

Whitehead, P. G., C. Neal, and R. Neale. 1986. J Hydrol 84:353–364.

Wood, T., and F. H. Bormann. 1977. Water, Air and Soil Pollut 7:479–488.

Wright, R. F. 1987. *In* H. Barth, ed. *Reversibility of acidification,* 14–29. Elsevier Applied Science, London and New York.

Wright, R. F., N. Conroy, W. T. Dickson, R. Harriman, A. Henriksen, and C. L. Schofield. 1980. *In* D. Drabløs and A. Tollan, eds. *Ecological impact of acid precipitation,* 377–379. P. O. Box 61, 1432 Ås-NLH, Norway.

Lange Bramke: An Ecosystem Study of a Forested Catchment

M. Hauhs*

Abstract

The forested catchment of Lange Bramke in the Harz Mountains (FRG) has been hydrologically monitored since 1948 when it was clear-cut. Today the catchment is covered by a 38-year-old Norway spruce (*Picea abies* Karst.) stand. Seepage water pathways through the catchment and ion cycling have been investigated in detail during the last 10 years. The results showed that seepage water through the root zone is vertical and unsaturated at the catchment slopes. The derived model of seepage water trajectories matches the observed chemical gradients between soil solution, headwater, groundwater, and runoff.

Acid deposition inside the valley of Lange Bramke is substantial with through-fall means of 45 kg ha^{-1}yr^{-1} as SO_4-S, 10 kg ha^{-1}yr^{-1} as NO_3-N, and 9.1 kg ha^{-1}yr^{-1} as NH_4-N. The fluxes of these ions are significantly higher at the wind-exposed ridge of the catchment. This additional input is probably due to cloudwater capture during winter months and amounts to 54% for SO_4, 69% for NO_3, and 37% for NH_4.

Soil acidification at Lange Bramke is driven by base cation (Ca and Mg) accumulation in the biomass and accelerated leaching of these ions due to acid deposition. The profiles of exchangeable Ca and Mg with depth down to 3.5 m can be interpreted as transient stages of the current base cation depletion. The net export of Mg from the soil has resulted in severe symptoms of Mg-deficiency in the forest stand that were first observed in 1982.

Today about one-third of the trees at Lange Bramke and the adjacent catchment of Dicke Bramke are affected by forest decline. This decline is manifested in two symptom types that show a spatially different distribution locally in the Bramke Valley and regionally for the Harz Mountains. Changes in soil solution and runoff chemistry occurred prior to and concurrent with the development of these decline symptoms. Among the changes are increased Al and decreased Ca concentrations over the 10 years of measurement. The fine root biomass in the mineral soil is low where nitrate leaching occurs. Nitrate export with runoff water from both catchments has increased dramatically over the last 10 years. These data provide empirical evidence for linkages among acid deposition, forest decline and water acidification, respectively.

*Institute for Soil Science and Forest Nutrition, Büsgenweg 2, D-3400 Göttingen, FRG.

I. Introduction

The assessment of acid deposition impacts on ion cyling in forest ecosystems and on streamwater chemistry is the main objective for this study. This aim requires the measurements of ion budgets and soil capacity factors such as base saturation. Additional information on intensity factors (e.g. toxic peak concentrations in precipitation, soil, and runoff water) is necessary to understand the possible changes in the biota. The monitoring setup at Lange Bramke is based on this concept.

The calibrated catchment of Lange Bramke (76 ha) is part of the upper Harz experimental basins (Wagenhoff and von Wedel, 1959). Investigations on the impact of clear-cutting on stream hydrology and soil erosion started in 1948. The 140-year-old spruce stand at Lange Bramke was clear-cut while the adjacent catchments of Wintertal (77 ha) and Dicke Bramke (32 ha) provided forested controls. These catchments now have a 40-year record of forest cover and hydrology (Table 9-1).

Beginning in 1977, major ions were measured in soil solution at Lange Bramke and in the runoff water from Lange and Dicke Bramke. Since 1981 major ions have also been analyzed in throughfall and bulk precipitation. In 1982 the symptoms of the widespread decline of Norway spruce stands in Germany appeared at the monitoring plots at Lange Bramke. We have a rare situation in which continuous measurements are available before and after forest decline became visible.

The catchment of Lange Bramke is located in the western Harz Mountains (Federal Republic of Germany), which range in height between 400 and 1,100 meters. The area is almost entirely forested (78,400 ha), and 78% of the forests consist of Norway spruce (*Picea abies* Karst.). Since about 500 AD the Harz has been settled for mining purposes, and by the 17th century large areas of the natural beech stands had already been deforested. Norway spruce was used for replanting during the following centuries, and the rotation period is about 100 years.

The geology is highly variable and includes material with a high silicate content, such as gabbro and diabase, as well as extremely poor quarzite. The Harz was not ice-covered during the last glaciation. The soils are developed in thick (2 to 4 m) residual layers that often had undergone some downslope transport. Most of the soils are well drained due to the coarse structure of the residual layers of saprolite.

Table 9-1. Percent of forest cover at three catchments in the Harz mountains during the past 40 years (changes are due to clearcutting and planting).

% forested	Lange Bramke	Wintertal	Dicke Bramke
1948	5	100	100
1965	65	80	95
1975	90	50[a]	95
1985	95	60	90[b]

[a]Mainly wind thrown in 1971.

[b]Dieback since 1980 at the northwestern wind-exposed ridge.

Surface waters from the Harz are important for the drinking water supply in northern Germany.

Today the Harz Mountains are a center of the forest decline in central Europe. The area is situated within the belt of high SO_2-gas concentrations across Europe (EMEP, 1987; Eliassen et al., 1988). A compilation of dry deposition estimates for SO_4 in European coniferous forest identifies the Harz region as heavily loaded by acid deposition (Hauhs et al., in prep.).

II. Methods

The catchment of Lange Bramke ranges in height from 535 to 700 m (Figure 9-1). The valley can be stratified into a north- and a southfacing subcatchment. The bedrock is a lower Devonian (Kahleberg) sandstone containing a mixture of quartz and muscovite (serecite) (Görz, 1962). Residual soils that may contain up to 60% stones have a total depth of about 3.5 m at midslope of the investigated

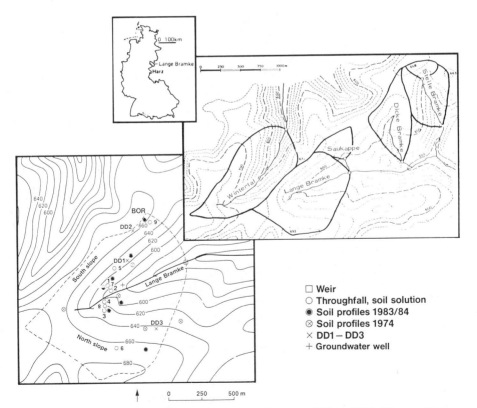

Figure 9-1. The Lange Bramke catchment (from Hauhs and Dise, 1988). The sampling design and periods are given in the text.

cross-sections. The soils are spodozols. The mean annual temperature is 5.8°C, and the precipitation averages 1,303 mm (1949–1977), falling in equal amounts in summer and winter months.

At Lange Bramke six monitoring plots are currently used inside the spruce stand (plots: 1, 2, 3, 7, 8, 9), and plot 4 is situated on a small clearing at the valley bottom (Figure 9-1). Plots 5 and 6 were used only for throughfall and soil water potential measurements from 1977 until 1980. Plots 1 to 8 were installed in 1977, and each site is equipped with nine throughfall sampler (summer) or five snow buckets (winter) (Meiwes et al., 1984a). Plot 9 was added in 1986 to test the possibility of a higher depositional load at the exposed ridge of the catchment. This plot has 12 replicates of throughfall sampler. The amount of throughfall is recorded weekly, and three samples from plots 1, 2, 3, 4, 7, and 8 are collected for chemical analysis (12 at plot 9). These samples are combined into monthly totals, and three adjacent collectors are mixed into one sample prior to analysis. By this procedure each month will be described by a total of five replicates from plots at the valley bottom and four replicates at the ridge. In addition snow depth and snow water equivalent are measured routinely at 30 locations inside the catchment.

Plots 1 to 4 are each equipped with eight porcelain-cup suction lysimeters (Meiwes et al., 1984a) at a depth of 80 cm. Laboratory and field experiments showed that this sampling design does not alter major ion chemistry at concentrations typical for Lange Bramke soils. At plot 9 we have four suction lysimeters at 40- and 80-cm depths respectively. The soil water samples are taken by application of 150-cm suction for one day. Since 1987 sampling has been done by continuous application of suction and includes also the winter months. Beginning in 1982, monthly totals of the soil water samples were combined into three replicates for each plot before analysis. The monthly samples from plot 9 (ridge, operating since 1986) are analyzed individually.

Runoff water at Lange Bramke and Dicke Bramke (since 1977) and spring water at Lange Bramke (1982 to 1985) were sampled weekly. A shallow groundwater well (Herrmann et al., 1987) in the valley bottom (at height 580 m, Figure 9-1) was occasionally sampled in 1982 and 1983. Analytical methods for water samples are described by Meiwes and co-workers (1984b) and did not change between 1977 and 1988 except for SO_4* and Al†.

In 1974, 24 soil samples (down to 100 cm) were taken by horizon from five profiles inside the spruce stand of the catchment (Geenen, 1976). In 1983 and 1984, two and seven profiles, respectively, inside the same stand were sampled (32 samples) down to 70 cm. These profiles are not identical with those of the first survey (Figure 9-1). Total biomass and ion contents in the humus layer are derived from five profiles at plots 1, 2, 3, and 5 and ten replicate profiles at plot 9. The samples were taken in September 1987 by a 20-cm-diameter core.

---

*Before 1981, $BaSO_4$ precipitation; after 1981, colorimetrically after American Health Association (1976).

†Before 1980, colorimetrically; after 1980, atomic adsorption spectrophotometrie.

In 1985, three boreholes were drilled down to the bedrock: one at the midslope position of each slope (south-facing: DD1; north-facing: DD3), and one at the exposed ridgetop of the south-facing slope (DD2) (Hauhs and Dise, 1988). Disturbed soil samples were taken from the core at about 30-cm intervals.

The soil samples from 1974 were analyzed for pH ($CaCl_2$) and exchangeable cations (K, Ca, Mg, Al, H; percolate: unbuffered 1N NH_4Cl) with the equivalent sum of these taken as CEC. The samples from 1983 to 1984 and the deep-drill samples were analyzed for the same constituents as the 1974 samples, plus exchangeable Na, Mn, and Fe. In the samples from 1983 to 1985 these ions were additionally analyzed in a saturation extract (Richards, 1954). For this method, water is added to the fresh soil sample until a suspension is reached. The water/soil ratio is kept below 0.8 ml/g^{-1}. This suspension is kept at room temperature for 24 h with occasional stirring. The solution is then suction-extracted and centrifuged, and the supernate filtered before analysis (Meiwes et al., 1984b).

The bulk density of the fine soil and the stone content were derived from soil samples taken in four depth intervals down to 120-cm depth. These samples were prepared as $30 \times 30 \times 30$ cm soil columns and coated in situ with an asbestos-fiber paste. These samples were also used for measurements of pF-curves and unsaturated hydraulic conductivity (Hauhs, 1985). The soil water potential at plots 1 to 8 was recorded by 45 mercury tensiometers at each plot (depths: 15, 25, 45, 60, and 105 cm with nine replicates, respectively). In 1980 this setup was replaced by five pressure-transducer tensiometer at plots 2 and 3, respectively. For the installation of these tensiometers we used locations out of the original setup that run close to averages at these depths (Hauhs, 1985). Tensiometer readings were taken manually each week until 1987, and then automatically by data logger at 15-minute intervals. In 1987 plot 9 was equipped with five tensiometers.

In 1984, 96 trees were felled for growth analysis. Twenty-four trees were taken within a distance of 100 m from plots 1 and 3. Height growth could be traced back to 1968 and diameter growth was quantified by measurement of tree-ring widths on four sectors at 2-m height. Needles were sampled from three shoots of each tree and classified by the degree of yellowing and missing needles.

The fine roots of trees were sampled in September 1987 at five plots (1, 2, 3, 5, 9) by undisturbed cores of 250 cm^3 down to 60 cm. At each plot, 10 profiles were sampled. This auger method is described by Murach (1988). From one soil sample of each of these profiles (depth 32 to 37 cm) a water extract was analyzed for Cl, NO_3, and SO_4.

III. Hydrology

The infiltration capacity of the humus layer at Lange Bramke exceeds rainfall intensities (Delfs et al., 1958; Herrmann et al., 1987). We used daily sums of throughfall (summer) or meltwater (winter) to calculate daily infiltration. To this end the rainfall rates of a continuously recording rainfall gauge at plot 3 were weighted with the weekly totals from the throughfall samplers. Snow cover development was calculated from six-hourly averages of air temperature and throughfall rates by a degree-day type model from Anderson (1973). This model was calibrated against measurements of soil water equivalents (Hauhs, 1985). Snow cover at plots 3 and 8 (north-facing slope) may contain up to 100 mm more water than at plots 1, 2, or 7 (south-facing slope). The snow cover at the north-facing slope usually lasts 1 to 2 weeks longer.

No lateral gradients in soil water potential occurred within the first 105 cm of depth among plots at the north and south-facing slopes (Figure 9-2). During the vegetation period, soil water potential at the south-facing slope always exceeded the levels in the rooting zone of the north-facing slope. The measurements with high time resolution since 1987 show a parallel development of water potential at representative plots of the two slope aspects during winter (Figure 9-3). With the onset of the vegetation period, soil water potential at plot 2 (south-facing slope) showed daily variation probably due to water uptake down to a 60-cm depth (Figure 9-3). At plot 3 (north-facing), this daily variation started later and is less pronounced.

The identification of seepage water pathways through the rooting zone of the spruce stand is essential to derive ion budgets. Two conditions must be met to calculate seepage water fluxes from our measurements at Lange Bramke: (1) seepage water through the root zone must follow vertical trajectories and (2) the lysimeters must collect representative seepage water. Hauhs (1985, 1986) describes a test of these hypotheses by a finite element model of water flow through a catchment cross-section. The model is based on transient, saturated and/or unsaturated-flow equations and calculates the daily distribution of water potential within and runoff from the cross-section. The sums of daily infiltration and an estimate of potential transpiration are used as input variables. The model was calibrated against runoff and the measured soil water potentials in 1977 for the undetermined saturated hydraulic conductivity of the flow region and run from 1978 to 1981.

Flow in the model cross section was vertical and unsaturated in at least the first 1.0 (2.0) meter of soil depth for all episodes from 1978 to 1981. The calculated runoff in 1981 (Figure 9-4) is entirely produced by translatory, subsurface flow returning from depths beyond 2 m. Other independent investigations of catchment hydrology at Lange Bramke using ^{18}O as tracer underlined the delayed nature of stormflow response (Herrmann et al., 1986). Only 11.5% of the annual runoff is derived from direct runoff (Herrmann et al., 1986). The high contribution from pre-event water to stormflow generation matches the observation that the unsaturated zone of flow at the slopes extends at least 1 to 2 m into the soil. The pathways at depths greater than 2 m, however, are less clear. The finite element model (Hauhs, 1986) uses a simple geometry of the flow along the bedrock interfaces in 2.5 to 3.5 m depth. In reality, an unknown fraction of this saturated, lateral-moving water will pass through fractures in the bedrock (Herrmann et al., 1987).

The finite element model is able to reproduce the observed differences in soil water potential between south- and north-facing slopes when the calculated yearly transpiration differs by about 100 to 130 mm (Figure 9-4). Such a difference is

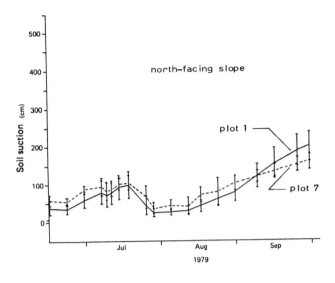

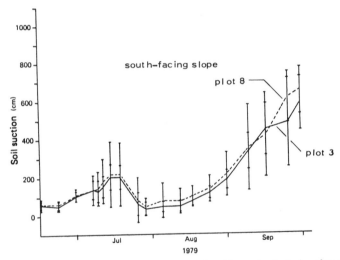

Figure 9-2. The time trends of soil water potential at 45-cm depth during the summer of 1979 at plots 1 (solid line) and 7 (dashed line) on the south-facing slope (upper figure) and at plots 3 (solid line) and 8 (dashed line) on the north-facing slope (lower figure). Means and standard deviations (n = 9) are given for each measurement. Note the difference in scale between the upper and lower figures.

unlikely to be caused by an insolation effect alone as transpiration above a closed forest canopy is not sensitive to net radiation at this spatial scale. (McNaughton and Jarvis, 1983; Hauhs, 1985). The rooting depth and the number of needles differs among slope aspects at Lange Bramke (see section V). It has been speculated that the lower transpiration rate at the north-facing plots is due to

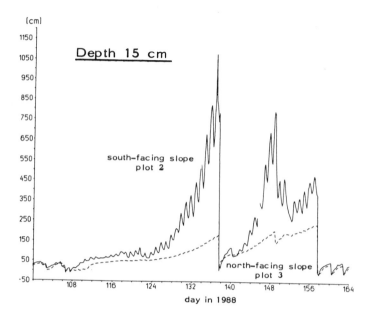

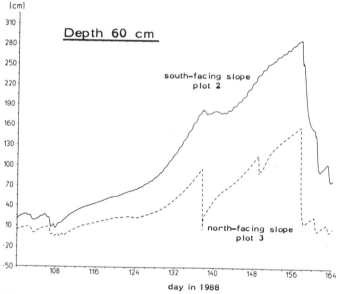

Figure 9-3. The time trends of soil water potential recorded by 15-minute intervals with data logger during 1988. The time is given as julian date and measurements are from plot 2 (south-facing) and plot 3 (north-facing) at 15 and 60 cm, respectively (note difference in scale).

physiological water stress (Hauhs, 1985). In this case the observed differences in soil water potential would be involved in the recent decline symptoms at Lange Bramke (Hauhs, 1985; Hauhs and Wright, 1986).

An evaluation of the long-term trends in areal precipitation and runoff showed an increase in water losses due to evapotranspiration from the Lange Bramke catchment until about 1976 by 200 mm yr^{-1} and a decreasing trend since then (Liebscher, 1985). This trend is absent at the Dicke Bramke catchment where no major change in forest cover occurred. Such increases in evapotranspiration following afforestation are caused by the higher net interception losses under a forest canopy (Stewart, 1977). At Lange Bramke the increase in water loss since the time of clear-cut matches present-day measurements of interception from the forest canopy (Hauhs, 1985). The hydrological explanation for the reversed trend starting in 1977 is unknown (Liebscher, 1985).

IV. Ion Cycling

A. Acid Deposition

The long-term average of wet deposition of major ions at Lange Bramke is almost identical to that at the Solling area, 100 km to the west. At Solling deposition measurements started in 1969 and comprise the longest record of this type (Ulrich, 1984). Throughfall fluxes of H^+ and SO_4, however, under the spruce stand at the valley bottom of Lange Bramke (plots 1 to 8) are only 41% H^+ and 55% SO_4, respectively of the fluxes measured at the Solling spruce stand; they correspond more closely to the results from the beech stand at Solling (Table 9-2) (Ulrich, 1984). This difference is probably due to the fact that the spruce stand at Solling is older (105 versus 35 years at Lange Bramke) and is more exposed to cloudwater interception than the sheltered valley of Lange Bramke. The exposed part of the spruce stand at the ridge of Lange Bramke (plot 9) shows for most ions similar fluxes as at the Solling spruce site. The difference among exposed and sheltered spruce of the same age at Lange Bramke is significant at the 0.001 level (water flux at the 0.05 level)* for all ions. With the exception of H^+ and SO_4, this is the same difference as measured among different species standing on adjacent sites at Solling. Like in the Solling area, most of the SO_4 and H^+ deposition occurs in the winter.

The generally high loadings of acid deposition in the Harz area are due to the contribution of dry and occult deposition (cloudwater interception) (Hauhs et al., 1988). Because these deposition pathways depend on the structure of the receiving surface, exposure effects may be enhanced and differences in height or among slope expositions may be larger than in areas where wet deposition is the major constituent of total atmospheric deposition.

---

*Tested after Dunn (1964).

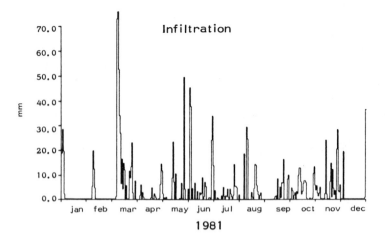

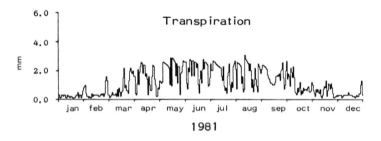

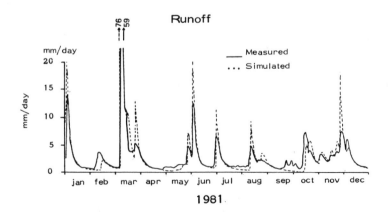

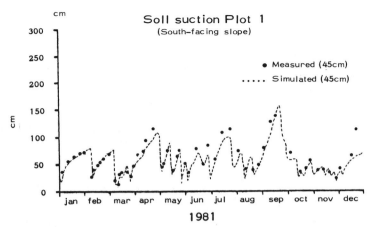

Figure 9-4. Daily sum of infiltrated water, actual transpiration, soil water potential (plot 2, depth 45 cm), and runoff for the cross section of the south-facing slope in 1981. Infiltration (upper figure, p. 284) is used as input variable whereas the latter three figures show results of the model calculations (from Hauhs, 1986).

B. Soil Chemistry

The soil profiles at plots 1, 2, 5, and 7 are brown earths by the German classification, whereas the other plots (3, 9) are podzols. The concentrations of ions in the humus layers. (O_L, O_F, and O_H) are given in Table 9-3. For comparison, the results from the sampling of needles in 1984 are included. Nitrogen contents and total pools (see Table 9-8 in section C) were significantly ($\alpha = 0.01$)* higher at plot 9 (ridge) and are given separately in Table 9-3. The content of K in the humus layers decreases relative to needle levels. This element is easily leached, as already indicated by wet deposition and the throughfall fluxes (Table 9-2). Mg is low throughout all compartments of this ecosystem. The trees suffer from severe symptoms of Mg deficiency that were first observed in 1984.

The chemical differences among soil profiles at Lange Bramke can be exemplified by plots 1, 3, and 9. The base saturation is extremely low within the root zone (0 to 60 cm; (Table 9-4). The cation exchange capacity at ambient soil pH decreases to values of about 25 μeq g^{-1} at 50-cm depth and stays constant beyond 50 cm down to the bedrock (Table 9-5). Exchangeable Mg and Ca increase beyond a depth of 130 cm in profile DD1 (south-facing slope), whereas no such increase was detected in profile DD2 (ridge). Results from the deep drilling at the north-facing slope (DD3: not shown) are intermediate to the two profiles at DD1 and DD2 (Hauhs and Dise, 1988).

The analytical method for exchangeable cations also includes water soluble salts from the sample. These can qualitatively be identified by a saturation extract

---

*Tested after Dunn (1964).

Table 9-2. Average fluxes in wet deposition.

[kg ha⁻¹ yr⁻¹]	H_2O[a]	H	Na	K	Ca	Mg	SO_4-S	Cl	NO_3-N	NH_4-N
wet deposition										
LB (81-88)[b]	1327	0.72	8.1	3.4	6.1	1.5	22.0	15.9	8.9	13.2
SO (69-85)[c]	1032	0.82	7.8	3.7	9.8	1.7	23.2	16.7	8.7	11.9
throughfall										
LB-1-8 (81-)[d]	1172	1.3	10.4	24.3	18.0	2.9	45.4	23.6	10.0	9.3
LB-1-8 (86-)[e]	1234	1.4	10.7	23.7	15.4	3.2	45.6	24.1	10.8	9.1
LB-9 (86-)[f]	1282	2.4	20.4	32.8	28.7	5.4	70.0	42.4	18.2	12.8
SO-B (69-85)[g]	736	0.8	10.4	19.1	18.6	3.1	33.4	23.9	10.0	11.6
SO-S (69-85)[h]	752	3.2	17.0	28.0	31.4	4.7	83.1	38.6	15.7	15.5

[a](mm yr⁻¹).

[b]Lange Bramke, 1981–1988.

[c]Solling, 1969–1985.

[d]Lange Bramke spruce at the valley bottom, plots 1, 2, 3, 7, and 8; 1981–March 1988.

[e]Lange Bramke spruce at the valley bottom, plots 1, 2, 3, 7, and 8; January 1986–March 1988.

[f]Lange Bramke spruce at the ridge, plot 9; January 1986–March 1988.

[g]Solling beech, 1969–1985.

[h]Solling spruce, 1969–1985.

Table 9-3. Contents of major ions in the humus layer (Sept. 1987) and the ranges in needles of three-year-old shoots (April 1984).

$(mg\ g^{-1})$	O_L	O_F	O_H	needles
C	493	444	401	n.m.
$N_{1,2,5}$	17.5	18.5	14.8	12.9–15.4
$N_{3,9}$	19.5	20.0	16.9	n.m.[a]
P	1.03	1.02	0.89	0.77–1.07
Si	22.9	61.2	144.3	4.1–7.2
Na	0.05	0.08	0.10	n.m.
K	0.78	0.85	1.18	5.1–7.5
Ca	1.9	1.4	0.8	.9–2.2
Mg	0.23	0.31	0.29	0.16–0.40
Fe	2.2	6.7	10.7	n.m.
Al	1.1	2.3	5.0	n.m.

Averages for all plots (1, 2, 3, 5, and 9) are given. Only when significant ($\alpha = 0.01$) differences between plots 1, 2, 5 (south-facing) and plots 3, 9 (north-facing and ridge, respectively) occur are these values shown separately.

[a] Not measured at plot 9.

(Tables 9-6 and 9-7). The high values of exchangeable K in Table 9-3 (deep-drill profiles) are probably derived from such a water-soluble solid phase. Although K is a minor fraction of the total cations in the water extract in the topsoil samples and in soil solution it parallels high sulfate values in all three deep-drill profiles (Dise and Hauhs, 1988; Tables 9-6 and 9-7). Nitrate levels in the saturation extracts from the profiles at the south-facing slope are always below the nitrate levels in the other profiles. This difference is exemplified by the profiles in Tables 9-6 and 9-7.

Table 9-4. C, N contents and cation exchange characteristics of the soils at Lange Bramke 1983–1984. Plot 1 is at the south-facing, plot 3 at the north-facing slope, and plot 9 at the ridge.

cm	C	N	CEC	H	Na	K	Ca	Mg	Fe	Mn	Al
plot 1	%		$(\mu eq\ g^{-1})$				%				
0–10	2.57	.14	149.	23.	.6	.8	1.3	.5	8.9	1.8	62.7
10–20	2.99	.16	101.	0.	.9	1.6	1.2	.6	.9	5.0	90.0
22–60	1.24	.11	32.7	0.	5.5	4.9	1.1	.7	.3	2.4	85.0
60–70	.03	.0	17.2	0.	6.0	5.4	1.3	.6	.6	3.4	82.8
plot 3											
0–10	1.51	.08	89.7	25.	1.1	.9	.5	.4	4.8	.9	66.0
20–30	3.05	.16	43.6	0.	2.0	2.1	1.1	.4	.7	1.2	92.5
40–50	2.05	.10	27.5	0.	3.4	3.6	1.5	.5	.8	1.1	89.1
plot 9											
0–10	0.85	.02	37.1	40.4	2.7	2.8	4.6	1.2	3.8	.1	44.4
10–20	3.20	.13	166.9	11.9	.6	.7	1.3	.5	3.5	6.8	74.6
25–35	1.80	.10	92.3	3.6	1.1	1.2	1.0	.4	.8	6.2	85.8
50–60	0.73	.05	28.6	0.	3.0	2.9	1.3	.4	.4	2.5	89.6

Table 9-5. Exchange characteristics of deep-drill profiles at Lange Bramke 1985.

cm	CEC	H	Na	K	Ca	Mg	Fe	Mn	Al
DD1[a]	(μeq g^{-1})					%			
50–70	19.8	0	3.7	23.1	6.8	7.8	0	7.0	51.8
70–90	19.2	0	4.3	27.7	4.6	4.7	0	5.3	53.5
90–110	18.6	0	4.8	15.7	4.4	2.7	0	5.6	66.9
110–130	17.7	0	4.6	23.5	6.3	6.7	0	7.5	51.4
150–170	20.7	0	4.4	34.9	5.1	13.7	0	9.6	33.3
200–230	18.7	0	4.0	22.8	16.9	43.6	0	5.6	7.1
230–260	23.0	0	3.0	19.0	16.3	57.2	0	4.4	0
310–330	22.0	0	4.0	30.0	17.0	41.6	0	7.4	0
330–340	19.0	0	3.8	25.3	18.9	44.6	0	7.4	0
DD2[b]									
15–25	96.8	12.9	3.6	1.1	1.9	.8	5.0	14.8	63.1
25–35	81.5	6.0	1.2	2.0	1.9	1.1	2.5	13.8	71.5
35–50	55.1	0	1.3	3.3	2.3	.9	.8	10.5	80.8
50–70	18.3	0	3.1	10.6	3.7	1.6	0	5.6	75.4
70–100	27.7	5.9	2.3	11.4	2.2	1.4	0	3.6	73.1
140–170	28.2	2.3	1.9	10.8	2.1	1.4	0	4.5	77.0
230–260	27.3	1.0	5.1	13.5	2.6	1.6	0	3.2	73.0
300–320	28.6	2.3	3.8	10.3	2.6	1.3	0	2.7	77.1
400–420	24.8	3.2	3.9	12.8	3.1	2.0	0	3.5	71.6
490–500	14.8	0	5.3	43.1	7.0	9.3	0	8.3	27.1

[a]South-facing slope.
[b]Ridge.
From Hauhs and Dise, 1988.

Table 9-6. Soil water extracts from the same samples shown in Table 9-4 (topsoil profiles).

cm	pH	K	SO$_4$	NO$_3$	Cl
plot 1			μeq l^{-1}		
0–10	3.68	118	132	17	101
10–20	4.16	35	152	<1	86
22–60	4.28	32	134	<1	72
60–70	4.29	47	128	<1	76
plot 3					
0–10	3.85	39	310	8	108
20–30	4.45	19	218	42	30
40–50	4.35	10	180	40	24
plot 9					
0–10	3.45	93	180	49	164
10–20	3.61	72	158	280	101
25–35	3.84	54	168	255	133
50–60	4.36	124	190	170	158

Table 9-7. Soil water extracts from the same samples that were shown in Table 9-5 (deep-drill profiles).

cm	pH	K	SO_4	NO_3	Cl
DD1			μeq l^{-1}		
50–70	4.59	558	696	14	39
70–90	4.81	803	954	15	52
90–110	4.87	470	686	13	34
110–130	4.78	706	917	10	46
150–170	4.81	923	1160	20	86
200–230	4.87	402	598	15	58
230–260	4.83	450	842	13	210
310–330	4.97	670	804	13	161
330–340	5.18	499	680	5	80
DD2					
15–20	3.95	51	150	148	63
25–35	4.06	59	115	209	39
35–50	4.18	55	150	211	44
50–70	4.80	312	344	58	84
70–100	4.56	432	510	51	58
140–170	4.61	292	383	75	40
230–260	4.72	532	600	65	78
300–320	4.44	471	383	48	346
400–420	4.62	492	589	48	80
490–500	5.04	540	568	65	127

From Hauhs and Dise, 1988.

C. Total Pools for Ca, Mg, Na, and K in the Humus Layer and as Exchangeable Cations in the Mineral Soil

The amount of organic dry matter in the O_H layer at plots 3 and 9 (91.9 t ha^{-1}) is significantly ($\alpha = 0.01$) higher than at plots 1, 2, 5 (31 t ha^{-1}; Table 9-8). No differences among plots were found in the O_L (13.6 t ha^{-1}) and O_F layers (23.2 t ha^{-1}). This difference among plots at the south-facing slope and plot 3 and 9 is of the same magnitude as the difference between organic matter accumulation at the beech and spruce stands at Solling.

Table 9-8. Total storages of Na, K, Ca, and Mg in the humus layer at plots 1, 2, 3, 5, and 9. Because of different pools of organic dry matter in the O_H layer plots 1, 2, 5 and 3, 9 are summed separately.

(t ha^{-1})	O_L	O_F	O_H (plots 1, 2, 5)	Σ 1, 2, 5	O_H (plot 3, 9)	Σ 3, 9
Na	0.68	1.86	3.16	5.7	9.2	11.7
K	10.6	19.5	37.3	67.4	108.4	138.5
Ca	24.7	32.5	25.3	82.5	73.5	130.7
Mg	3.1	7.2	9.2	19.5	26.7	37.

The spatial differences in organic matter accumulation at Lange Bramke may be due to a lower input of fine root litter to the humus layers at the south-facing slope (section V) and differences in the decomposition rates due to temperate and moisture differences. The latter effect was indicated by the delay in snowmelt at the north-facing exposition (see section II). This temperature difference was probably maximum during the years immediately following clear-cutting in 1947.

The total storages of Ca, Mg, Na, and K for the uppermost 60 cm of the mineral soil were calculated from the two inventories in 1974 and 1983 to 1984 (Table 9-9). The profiles sampled in these two surveys are not identical (Figure 9-1). Thus the differences that are indicated for exchangeable Ca and Mg storages should be interpreted with caution (Figure 9-9). The total amount of these cations in the mineral and humus layers, however, clearly limits further growth at Lange Bramke. These storages are equivalent to the estimated uptake for a period of only 5 to 10 years.

D. Soil Solution Chemistry

At Lange Bramke we have continuous measurements of soil solution chemistry over a 10-year period. The lysimeters at all monitoring plots 1, 2, and 3 revealed time trends in the soil solution concentrations. Means and standard deviations for the first and last 3 years of the total period (Table 9-10) were calculated from the means of single lysimeter nests at each plot. The years 1977 to 1979 and 1984 to 1986 were hydrologically similar.

The pH values in all lysimeter nests increased from 1977 until 1983 to 84 and then decreased since. At plot 1, Ca dropped to about 50% of its level in 1977 to 1979. The decreasing Ca levels at the other two plots (2, 3) are either not significant (plot 2) or are less pronounced due to a lower initial level (plot 3). The decrease in Ca (and K) at plot 1 is accompanied by an increase in Al concentration. It doubled also at plot 3 (north-facing slope). The long-term trends of NO_3 and Al are similar at plot 3 (depth 80 cm) (Figure 9-5). During the first two years (1977 and 1978), soil solution nitrate levels dropped at some lysimeter nests during the vegetation period (Figure 9-5). During the following years, sampling was restricted to the frost-free period due to technical problems. A more detailed analysis of seasonal variations over the year is thus precluded. At plot 9

Table 9-9. Storages of exchangeable base cations at Lange Bramke (0 to 60 cm).

	1974 Median (min–max) (n = 5)	1983 to 1984 Median (min–max) (n = 7)
(t ha^{-1})		
Na	not meas.	60 (48–92)
K	78 (66–109)	90 (62–160)
Ca	98 (52–244)	42 (20–56)
Mg	24 (14–52)	8 (6–12)

From Hauhs and Dise, 1988.

Table 9-10. Mean concentrations in soil solution (depth 80 cm) at plots 1, 2, and 3. All aluminum is assumed to be of the form Al^{3+}. The water fluxes were estimated from results of the hydrological model.

(μeq l^{-1})	plot 1		plot 2		plot 3	
	1977–79	1984–86	1977–79	1984–86	1977–79	1984–86
H$_2$O[a]	600	645	600	645	785	843
H	77 ± 15	61 ± 14	61 ± 10	36 ± 5	66 ± 11	41 ± 4
Na	81 ± 15	118 ± 24	63 ± 19	48 ± 8	54 ± 8	47 ± 5
K	31 ± 29	9 ± 3	11 ± 3	10 ± 3	12 ± 8	9 ± 3
NH$_4$	5 ± 4	<2	3 ± 1	<2	5 ± 1	<2
Ca	111 ± 17	59 ± 11	64 ± 41	31 ± 13	45 ± 7	30 ± 4
Mg	57 ± 13	56 ± 11	30 ± 13	18 ± 5	25 ± 5	20 ± 3
Mn	29 ± 15	24 ± 2	20 ± 8	14 ± 3	12 ± 6	9 ± 2
Al	206 ± 68	397 ± 51	186 ± 96	185 ± 71	133 ± 45	277 ± 51
SO$_4$	418 ± 113	560 ± 43	352 ± 178	300 ± 98	247 ± 60	255 ± 64
Cl	96 ± 9	134 ± 26	71 ± 20	43 ± 6	63 ± 15	53 ± 5
NO$_3$	<2	5 ± 5	7 ± 7	11 ± 4	29 ± 28	117 ± 49
N$_{org}$[b]	.81 ± .033	.36 ± .05	.52 ± .17	.40 ± .13	.54 ± .03	.30 ± .04
Σ cations	597	724	438	342	352	433
Σ anions	514	699	430	354	339	425

[a](mm yr^{-1})
[b](mg·l^{-1})

From Hauhs, 1985.

Plot 3

north-facing slope

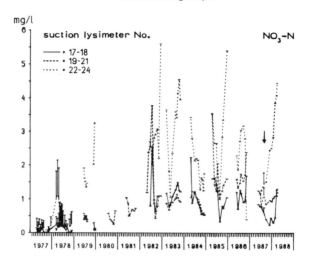

Plot 3

north-facing slope

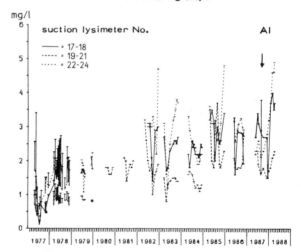

Figure 9-5. Nitrate (upper figure) and aluminum (lower figure) at plot 3 (north-facing slope) in soil solution at 80-cm depth. The plot is equipped with eight suction lysimeters (number 17–24) and all data from this period are shown. In 1977–79 all lysimeter were analyzed separately, in 1980–81 all samples were combined into one sample before analysis, in 1982–88 samples were combined into three replicates for the lysimeter nests: 17 + 18, 19–21, and 22–24. In August 1987 all samples were analyzed separately for inspection of changes in plot variability (arrow).

(established in 1986), we have continuous sampling of soil solution from 40- and 80-cm depths. The first 16 months of sampling revealed seasonal trends for NO_3 (with maximum in autumn) and strong variations of Cl, SO_4, and K (Figure 9-6).

Occult deposition of Cl shows maximum difference between the valley bottom and the ridge at Lange Bramke during the winter months of 1986 and 1987. Cl in 40- and 80-cm depths at plot 9 followed a clear seasonal trend in 1987. In the coarse soil of this site, the input Cl is rapidly transported through the profile. The SO_4 concentration shows a similar time trend at a 40-cm depth but at 80-cm depth the concentration of SO_4 seems to be buffered. Within this depth interval, the concentration of K in soil solution drops. The results from the samplings of soil profiles at plot 9 and DD2 (ridge), however, indicate for this depth interval an increase in an easily soluble salt fraction containing K and SO_4 (Tables 9-6 and 9-7). These findings support the hypothesis of a precipitation- and/or dissolution-controlled mechanism for sulfate retention in Lange Bramke soils (Dise and Hauhs, 1988). In this case the composition of the precipitate appears to be similar to alunite $(KAl_3(OH)_6(SO_4)_2)$.

E. Seepage Water Fluxes

The fluxes of major ions in seepage water from 1977 to 1979 were calculated from monthly average concentrations at each monitoring plot and water fluxes that were estimated by the model of slope hydrology (Hauhs, 1986). For subsequent years these fluxes were based on annual mean concentrations and seepage water estimates that were derived from measured runoff and model results from the first three years 1977 to 1979 (Table 9-10). For Na and Cl, the fluxes in throughfall, seepage water, and runoff are similar (Hauhs and Dise, 1988). For SO_4 these fluxes decrease from 45.4 kg ha^{-1}yr^{-1} in throughfall to 35 to 40 kg ha^{-1}yr^{-1} in seepage water and further to 25 kg ha^{-1}yr^{-1} in runoff.

F. Spring and Groundwater Chemistry

The spring water of the Lange Bramke catchment is acid, whereas stream water at the weir has a positive alkalinity and rarely falls below pH 5.5. The analysis of surface water chemistry close to plot 4 (Figure 9-1) never (1982 to 1985) exceeded pH 4.9. These chemical gradients along a short stretch of stream channel are relatively constant in time. The flow volume at the upper sampling point is much lower than the corresponding fluxes at the weir of Lange Bramke. Most of the seepage water from upstream catchment areas pass this sampling point (plot 4) as groundwater flow. In parts of this upstream area, however, seepage deeper than about 1.5 m is prevented by shallow bedrock.

The chemistry of surface water close to plot 4 is similar to the soil solution chemistry of the monitoring plots 1 to 4. The shallow groundwater from a well in the middle of the catchment showed a composition that corresponds with runoff water at the weir (Table 9-11) (Hauhs, 1986). The hydrological work at Lange Bramke indicated that during base and stormflow the flow pattern was spatially

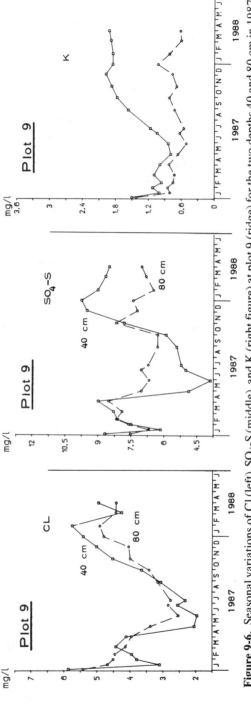

Figure 9-6. Seasonal variations of Cl (left), SO$_4$-S (middle), and K (right figure) at plot 9 (ridge) for the two depths 40 and 80 cm in 1987. Each point represents the mean of four lysimeters.

Table 9-11. Mean concentrations of major ions in headwater of the Lange
Bramke catchment near plot 4 (Figure 9-1), and groundwater from the well shown
in Figure 9-1 (from Hauhs, 1985), and runoff in 1984. All Al in the headwater is
assumed to be of the form Al^{3+}.

(μeq l^{-1})	Headwater 8/1983–9/1984	Groundwater 1984 (n = 4)	Runoff 1984
H	34	2	1
(pH)	(4.47)	(5.70)	(6.12)
Na	65	73	72
K	17	25	16
NH$_4$	<2	<2	<2
Ca	105	179	176
Mg	73	159	140
Mn	12	2	2
Al	111	<2	<2
SO$_4$	271	215	228
Cl	89	88	87
NO$_3$	65	41	46

similar and consisted of unsaturated vertical flow through the first 1 to 2 m of soil
at the slopes (section III). The depth to which water infiltrates into the subsoils is
thus likely to be the critical variable for surface water chemistry at Lange Bramke.
Such a flow model can explain the chemical gradients within the soil (Table 9-5)
and along the stream channel.

G. Runoff

The concentrations of major ions in stream water at Lange Bramke and Dicke
Bramke are independent of flow, with the exception of pH and HCO$_3$ (Hauhs,
1985). The annual fluxes in runoff were calculated from daily runoff totals and
weekly* analysis of streamwater chemistry. Runoff from Dicke Bramke has been
disturbed since 1970 by a tunnel that passes the catchment at about a 300-m depth.
Before that time the water yield from the two catchments was similar, and we used
the measured flow from Lange Bramke for the calculation of exports from Dicke
Bramke. Nitrate levels in stream water increased steadily between 1977 and 1986
at both catchments. Because of this increase, flux averages for the first and last
three years of the total period are shown separately (Table 9-12). The HCO$_3$ export
rates were estimated by difference in sums of cations and anions. Organic anions
are neglible (TOC < 2 mg L^{-1}). There has been a loss of alkalinity at both
catchments mainly due to the increased nitrate levels. We have no explanation for

---

*After 1980 only 50% of the streamwater samples were analyzed for all major ions.
Samples were selected for analysis due to flow volume changes and pH.

Table 9-12. Average fluxes in runoff from Lange Bramke (LB) and Dicke Bramke (DB) in 1977–79 and 1984–86. The values for HCO_3 were estimated by the difference between sum of cations and sum of anions.

$(kg\ ha^{-1}\ yr^{-1})$	H_2O[a]	H	Na	K	Ca	Mg	SO_4-S	Cl	NO_3-N	HCO_3-C
						runoff				
LB 77–79	586	.010	9.8	4.6	19.6	10.0	22.3	19.1	1.9	3.5
LB 84–86	632	.010	10.8	4.7	23.5	11.5	25.0	20.3	5.6	2.4
DB 77–79	586	.015	16.3	5.9	46.5	27.1	52.5	33.4	6.3	9.4
DB 84–86	632	.049	17.4	7.0	51.7	30.7	58.9	34.6	20.2	0

[a] $(mm\ yr^{-1})$.

the relatively high calculated alkalinity levels in the first 3 years at Dicke Bramke (Table 9-12).

The generally higher concentrations of most ions in stream water from Dicke Bramke are likely to be the consequence of higher acid deposition under the Norway spruce stand at Dicke Bramke that is older and more wind exposed than at Lange Bramke. The range in SO_4 export rates from the two catchments (25 to 59 kg $ha^{-1}yr^{-1}$) is below the range in deposition measurements from different expositions at Lange Bramke (Table 9-2) (45 to 70 kg $ha^{-1}yr^{-1}$).

Sulfate and nitrate concentrations in runoff from Lange Bramke show parallel seasonal trends since 1982 (Figure 9-7), whereas no seasonal trends exist at Dicke Bramke (Figure 9-8). There, the similarity between sulfate and nitrate time series seems to decrease over time. In 1981 the analytical methods for SO_4 was changed.

V. Forest Effects

Forest decline in the upper Harz is manifested in at least two distinct types of symptoms (Hartmann et al., 1985). The yellowing of 2-year-old (or older) sun-exposed needles of Norway spruce is termed type I, whereas the thinning of crowns by loss of brown or green needles is termed type II. Needle losses without yellowing (type II) from exposed trees at the ridges of Lange and Dicke Bramke started at least 10 years ago. By 1985, Norway spruce had died in about 10% of the catchment area of Dicke Bramke (Table 9-1).

In 1982 yellowing (type I) appeared first in dominant trees on the slopes of the Lange Bramke Valley, mainly in south-facing exposition. By 1983 one-third of

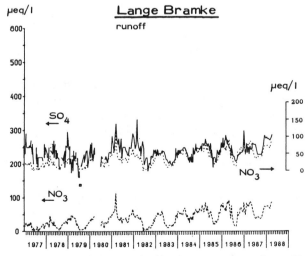

Figure 9-7. The time trends of nitrate and sulfate in stream water at Lange Bramke (weir). The nitrate scale to the right was elevated by 180 μeq L^{-1} for comparison with sulfate.

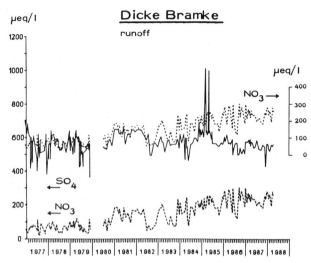

Figure 9-8. The time trends of nitrate and sulfate in stream water at Dicke Bramke (weir). Note the differences in scales to Figure 9-7. The nitrate scale to the right was elevated by 500 μeq L^{-1} for comparison with sulfate.

the population was affected. As in other Norway spruce stands in Germany, this yellowing was accompanied by Mg deficiency (Table 9-3). The 1984 reconstruction of basal area and height growth histories of 104 trees revealed growth reductions for affected trees since 1981 (basal area) and 1982 (height) (Dobertin, 1985). During the following years, the yellow needles were shed, and the symptom spread to younger needles. Dong and Kramer (1987) also reported growth reductions for trees affected by yellowing at Lange Bramke. The growth reductions were linearly related to the amount of needle loss. The classification of needles from 26 trees cut close to plots 1 and 3 respectively showed that the trees at the south-facing slope kept their needles longer but were more affected by yellowing (Dobertin, 1985).

The two types of damages can be separated in infrared aerial photography, and their spatial distribution was assessed separately for the western Harz (Hartmann et al., 1985; Stock, 1988). An additional detailed analysis of the upper Harz experimental basins by the same method in 1984 showed a concentration of type I damages (yellowing) at the slopes of the Lange Bramke Valley, whereas needle losses without yellowing prevailed along the ridges (Graeber, personal communication).

Some of the trees on or adjacent to the monitoring plots at Lange Bramke were also affected by decline symptoms. In 1983 trees became yellow at plots 1, 2, and 5. In 1985 plot 9 was newly selected within an area of type II damages. Also here trees became yellow by 1987.

The fine root survey revealed significant differences between plots 1, 2, and 5 and plots 3 and 9. The plots on the south-facing slope (1, 2, 5) had the mass of fine

Table 9-13. Fine root biomass of the Norway spruce stand at Lange Bramke (Sept. 1987).

(t ha^{-1})	Plot 1[a]	Plot 2[a]	Plot 5[a]	Plot 3[b]	Plot 9[c]
living roots					
humus layer	2.38	2.48	1.85	3.30	4.05
0–60 cm	4.27	3.82	5.58	1.83	1.28
dead roots					
humus layer	0.62	0.25	0.88	0.18	0.86
0–60 cm	1.41	0.89	1.32	1.38	0.78
NO$_3$ [%]	1.0	1.5	3.0	20	5

The table includes the median of the NO$_3$ fraction of the sum of anions in the saturation extracts in 32–37 cm depth.
[a]South-facing slope.
[b]North-facing slope.
[c]Ridge.

roots in the mineral soil down to 60 cm. At the other two plots (3, north-facing; 9, ridge), fine roots concentrate in the humus layers (Table 9-13). Profiles with relatively high fine root densities in the mineral soil (20 to 40 cm) showed low nitrate levels in saturation extracts (32 to 37 cm) (Table 9-14). The opposite—low fine root density implies high nitrate—was not true. As discussed below, this may be due to different time constants for the response of roots and microorganisms to soil chemical changes.

VI. Discussion

The observed changes of terrestrial and aquatic ecosystems at Lange Bramke are related to acid deposition and soil acidification. We first discuss the current rates of soil acidification and the corresponding changes in streamwater chemistry. Then we present a hypothesis that interprets the current results from these catchments as a response to transient forest decline in which soil acidification is the key variable.

Soil acidification is defined as a net loss of acid-neutralization capacity (van Breemen et al., 1984). It can be identified by a loss of exchangeable base cations

Table 9-14. Comparison between the mass of fine roots in cores between 20- and 40-cm depth and the concentration of NO$_3$ in a saturation extract (32–37 cm) from the same profiles. The number of cases is given separately for plots 1, 2, and 5 and plots 3 and 9.

	NO$_3$ <30 (μeq l^{-1})	NO$_3$ >30 (μeq l^{-1})
fine root > 300 mg	8/0[a]	1/0
fine root < 300 mg	14/3	6/15

[a]Sum of cases in plots 1, 2, and 5/ sum of cases in plots 3 and 9.

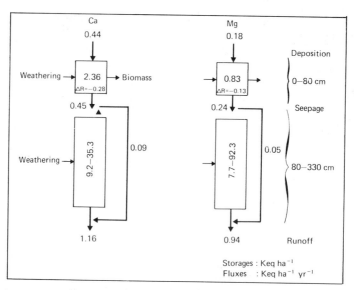

Figure 9-9. A model of ion transport through the soil of the Lange Bramke catchment (from Hauhs and Dise, 1988). Soil storages of exchangeable Ca and Mg were separately estimated for upper (0–80 cm) and deeper layers of soil (80–350 cm). The rates of changes in storages in the upper boxes are calculated from the differences between the two inventories in 1974 and 1983/84 (Table 9-9). Seepage flow that bypasses the deeper layers and emerges directly from soil depth above 150 cm into the stream is estimated to occur on about 20% of the catchment area. Fluxes are given in keq ha^{-1}yr^{-1} and storages (1983 to 1984) in keq ha^{-1}. The upper and lower bounds in the lower boxes are from the total exchangeable pools of these ions in the three deep drillings (DD1–DD3).

(Ca, Mg, K, Na).* In Figure 9-9 we have compiled the information from flux measurements and soil inventories for Ca and Mg. These two ions show the largest net export from the Lange Bramke catchment. The net Ca and Mg export can stem from silicate weathering and/or depletion of the pool of exchangeable bases on the soil. The fluxes indicate that the depletion of storages stems from subsoils below 80 cm. In the deep profile DD1 (south-facing slope), base saturation may exceed 90% (Table 9-7). Because of the relatively steep gradients in the profiles of base saturation, these profiles may be regarded as transient. Even with constant or decreasing deposition rates, the zone of base cation depletion may eventually reach the region of saturated, lateral transport (Hauhs and Wright, 1988). It is difficult to estimate the rate for this process as long as silicate weathering rates are unknown. In the rooting zone we can give upper bounds for the weathering rate of Ca and Mg (<0.5 keq ha^{-1}yr^{-1}). Because of the decreasing trend in storages, this

---

*I will term these cations *base cations* because in soils they are usually bound to weak acid anions such as clay minerals and silicates.

input has to be lower than the current uptake rates into biomass (see Figure 9-9). The current rates of acidification at Lange Bramke are driven by uptake of Ca and Mg into biomass and the input of strong acids from the atmosphere. The mobility of the accompanying SO_4 and NO_3 ions in the soils of this catchment controls how acidification is shared among soils and surface waters.

The alkalinity export from deeper soil layers (1.5 to 3.5 m) still prevents a breakthrough of toxic H^+ and Al into the lower stretches of the streams of Lange and Dicke Bramke. In the headwater areas, this breakthrough has already occurred, and here the acidity of the stream water corresponds closely to that of the seepage water from the upper soil zones.

The aquatic biota of the Dicke Bramke stream about 100 m below the weir still resembles an intermediate situation with respect to acidification-sensitive species (Lessmann, personal communication). A survey in 1985 revealed some species of brown trout above the weir of Lange Bramke, but no juvenile forms were found (Lessmann, personal communication). At Dicke Bramke, however, the alkalinity of the stream water had dropped below zero by 1987 (Table 9-11). The volume-weighted average for the pH is now 5.1. The recent losses in alkalinity at Lange and Dicke Bramke are almost entirely due to increased nitrate levels in stream water (Table 9-11). Similar nitrate levels in surface waters are reported from regional surveys of other forested catchments in the Harz (Heinrichs et al., 1986).

The observed trends at the upper Harz experimental basins can be accelerated if seasonal pulses of nitric acid are involved in the retention and release processes for sulfate. Chloride in throughfall shows about the same variation between winter and summer months as sulfate. The chloride levels in runoff, however, are not affected by seasonal variations in either of the two catchments. The mean amplitude for SO_4 and NO_3 in runoff at Lange Bramke is of similar magnitude, and it is unlikely that biologically controlled processes are directly involved. If, however, the observed sulfate retention is indeed controlled by alunite precipitation in subsoil layers beyond 50-cm depth this process will be limited by Al (and K) availability. A seasonal flush of HNO_3 from the root zone of the north-facing slope may disturb the sulfate retention at this slope and could superimpose the observed signal in streamwater sulfate. The severely damaged spruce stand at Dicke Bramke has apparently reached nitrate saturation, and a seasonality in export due to forest uptake no longer exists. Here sulfate levels also lack any seasonal variation (Figure 9-8).

The severe dieback and subsequent removal of spruce from the exposed ridges of Dicke Bramke may explain the opposite trends in sulfate and nitrate concentrations since 1982 (Figure 9-8). A reduced occult deposition will lower SO_4 inputs whereas NO_3 responses are probably dominated by the decreased uptake and increased mineralization rates. Acid deposition has thus contributed to water acidification at Lange and Dicke Bramke directly via the input of strong acids and indirectly via the changes in ion cycling within the forest ecosystem. The links between changes in the terrestrial and aquatic ecosystems make predictions about the reversibility of such changes extremely difficult.

The following hypothesis tries to integrate the complicated picture of spatial gradients and time trends for several key variables of forest decline at Lange Bramke:

> The forest ecosystem is in a transient state of decline. At the north-facing slope decomposition rates are lower and have been involved in the development of podzols. Today plot 3 has the lowest Ca/Al molar ratio in the soil solution. Here roots were probably affected earliest by the increase of soil solution acidity (Table 9-10). Damage to the fine root system caused lowered uptake of water and other nutrients such as nitrate from the mineral soil. Microorganisms may have delayed such a decrease in uptake by trees for several months or years. This is indicated by the observations at plots 1 and 2, where during the winter months (e.g. 1987 to 1988) nitrate in soil solution is below 0.1 mg L^{-1}. After a couple of years, however, the shift toward a more superficial rooting system at the north-facing slope led to increased nitrate leaching in seepage water (Figure 9-8). This hypothesis would explain the decreasing trend in areal evapotranspiration losses in the midseventies and the subsequent increase in nitrate export from the two catchments (Figures 9-7 and 9-8). The current distribution of nitrate in water extracts from soil profiles is related to the amount of fine roots (Table 9-13). Profiles with low nitrate and deep rooting (at south-facing plots 1, 2, 5) are likely to resemble the initial situation, profiles with low fine root concentration and also low nitrate would be intermediate (mainly south-facing), whereas the low fine root—high nitrate case would be the end of such a development (north-facing slope). The increased probability of water stress due to the change in rooting pattern may be involved in the loss of green and brown needles and crown thinning (type II). Plot 3 could be regarded as an initial phase of such a decline. Plot 9 at the ridge suffers from the highest atmospheric load and the poorest soil conditions. Here the rooting pattern has probably been restricted to the humus layer since the planting of these trees. Needle losses had already started 10 years ago at this plot. Rising nitrate levels in stream water would be a sensitive early warning of incipient forest damages of this type.
>
> The second type of damage occurs at sites where the root system is fully developed in the mineral soil and potentially toxic ions such as Al are still at sublethal levels. Such systems will benefit from the increased nitrogen deposition, and the enhanced growth will accentrate the ongoing depletion of Ca and Mg from the soil. The result would be an Mg deficiency that is likely to occur during phases of maximum annual growth. It would be typical for such stands that the rooting pattern and water and nitrogen uptake are not affected prior to visible symptoms of decline. Plots 1 and 2 are examples of this type of decline.

This hypothesis implies that the development of type I and II decline symptoms in Norway spruce should follow spatial gradients in acid deposition and soil sensitivity (Hauhs and Wright, 1986). This was indeed observed at the Harz Mountains (Stock, 1988). The two decline symptoms—yellowing and crown thinning without yellowing—were independently mapped from individual classification of more than 60,000 trees in infrared aerial photography (Hartmann et al., 1985; Stock, 1988). The symptom of yellowing prevails in 30- to 60-year-old stands, whereas crown thinning dominates in older (>60 years) stands. Furthermore, the maxima of the symptoms are spatially separated: yellowing of spruce (type I) is concentrated on granites and sandstones, whereas type II occurs on quarzite and silicate-poor graywackes (Stock, 1988). Damages of the latter type show a tendency to concentrate on western slope aspects, the main wind direction in this area.

These data fit the hypothesis that damages of type I (yellowing) are not linearly correlated to the external stress by acid deposition. The Mg deficiency (type I) symptom is apparently restricted to relatively moderate stress conditions: acid deposition $< 50 \, kg \, ha^{-1} yr^{-1}$ as S and moderate bedrock quality in combination with high growth rates. The second type of damage (type II) prevails under more adverse site conditions. Several investigations have shown that root damages are involved in this type of decline (Ulrich, this series).

Soil acidification as the predisposing factor seems to control the long-term changes in forest ecosystems at Lange Bramke and thereby the spatial distribution of decline symptoms. Other stress factors that act directly on the trees, such as ozone, may be involved in the final development of both types of decline. The above hypothesis provides a testible explanation for the complicated spatial pattern of forest dieback in central Europe in terms of gradients of known stress factors such as site sensitivity, acid deposition, ozone, and climate.

VII. Future Research at Lange Bramke

In 1988 the Lange Bramke catchment became part of the long-term monitoring network in Lower Saxony (FRG), and the measurements of ion fluxes will continue. In addition, several process-oriented studies are planned:

Investigation of the differences in the soil water budgets at plots 1, 3, and 9 by micrometeorological, physiological, and soil physical methods

Identification of the mineral phases that control sulfate retention and release in Lange Bramke soils

Modeling exercises to reconstruct and link the history of acid deposition, growth, and soil and water acidification at Lange Bramke

The data records from Lange and Dicke Bramke offer perhaps the clearest examples for the empirical links between acid deposition, forest decline, and water acidification in the Federal Republic of Germany. We regard soil acidification as the critical variable for these links. A large amount of data has been collected at this site from the time of planting until the spreading of decline symptoms. This data base can serve as a critical test for many of the currently discussed hypotheses of forest decline in central Europe.

Acknowledgments

Among the many persons who contributed to research at the upper Harz experimental basins and deserve acknowledgment, I want to give my special thanks to the following individuals: the hydrological studies at Lange Bramke were initiated and coordinated from 1948 to 1985 by A. Wagenhoff. In 1974 B. Ulrich, R. R. van der Ploeg, and P. Benecke started the ion budget project at Lange Bramke. H. Mühlhan has been responsible for the construction and the maintenance of field installations and sampling within these projects from 1974 to 1988.

I thank A. Hermann for the use of the groundwater well at Lange Bramke. The project is currently funded by the Umweltbundesamt (prj. no.: 10204350).

References

American Health Association. 1976. *Standard methods for the examination of water and wastewater*. American Health Association, Washington, D.C., 661 p.

Anderson, E. A. 1973. National weather service river forecast system. NOAA Technical Memorandum, NWS Hydro 17.

Delfs, J., W. Friedrich, H. Kiesekamp, and A. Wagenhoff. 1958. Der Einfluss des Waldes und des Kahlschlages auf den Abflussvorgang, den Wasserhaushalt und den Bodenabtrag. Aus dem Walde, Bd. 7, 283 p.

Dise, N., and M. Hauhs. 1988. Sulfate retention characteristics of an acid forest soil at Lange Bramke, West Germany. Geoderma (submitted).

Dobertin, M. 1985. Zuwachskundliche Untersuchungen von jungen Fichten mit und ohne neuartigen Vergilbungssymptomen. Master thesis, Forstliche Fakultät Göttingen. Büsgenweg 2, D-3400 Göttingen, FRG.

Dong, P. H., and H. Kramer. 1987. Allg Forst u Jgd-Ztg 158:122–125.

Dunn, O. J. 1964. Technometrics 6:241–252.

Eliassen, A., H. Øystein, T. Iversen, J. Saltbones, and D. Simpson. 1988. Estimates of airborne transboundary transport of sulphur and nitrogen over Europe. EMEP/MSC-W 1/88. Norwegian Meteorological Institute, Blindern, Oslo.

EMEP/CCC. 1987. Summary report from the chemical co-ordinating centre for the third phase of EMEP. EMEP/CCC-report 3/1987, Norwegian Institute for Air Research, Lillestrøm.

Geenen, S. C. 1976. Das Harz-Projekt—eine Bodenkartierung und einige erste bodenphysikalische Ergebnisse des Untersuchungsgebietes Lange Bramke. Master thesis, Landbouwhogescholl Wageningen, Abt. Bodenkunde und Geologie, The Netherlands.

Görz, H. 1962. Beitr Minerl Petrogr 8:232.

Hartmann, G., R. Uebel, and R. Stock. 1985. Forst- und Holzwirt 40:286–292.

Hauhs, M. 1985. Wasser- und Stoffhaushalt im Einzugsgebiet der Langen Bramke (Harz). Berichte des Forschungszentrums Waldökosysteme/Waldsterben Bd. 17, Büsgenweg 2, 3400 Göttingen.

Hauhs, M. 1986. Geoderma 38:97–113.

Hauhs, M., and N. Dise. 1988. Depletion of exchangeable base cations in an acid forest soil at Lange Bramke, West Germany. Geoderma (submitted).

Hauhs, M., T. Paces, B. Vigerust, K. Rost-Siebert, and G. Raben. 1988. Evaluation of sulfur and nitrogen fluxes in European forests (in preparation).

Hauhs, M., and R. F. Wright. 1986. Water, Air and Soil Pollution 31:463–475.

Hauhs, M., and R. F. Wright. 1988. Reversibility of water acidification—a literature review. Commission of the European Communities—Air Pollution Research Report 11, 41 p.

Heinrichs, H., B. Wachtendorf, K. H. Wedepohl, B. Rössner, and G. Schwedt. 1986. Neues Jahrbuch Minerl Abhdlg 156:23–62.

Herrmann, A., J. Koll, C. Leibundgut, P. Maloszewski, R. Rau, W. Rauert, and W. Stichler. 1986. Deutsche Gewässerkdl Mttlg 30:85–93.

Herrmann, A., J. Koll, M. Schöninger, and W. Stichler. 1987. A runoff formation concept to model water pathways in forested basins. *In Proceedings of the Vancouver Symposium, August 1987*. IAHS-AISH Pub. no. 167:519–529.

Liebscher, H. J. 1985. Nationalpark Bayr Wald 5/2:505–516.

McNaughton, K. G., and P. G. Jarvis. 1983. Predicting the effects of vegetation changes on transpiration and evaporation. In T. T. Kozlowski, ed. *Water deficits and plant growth*, vol. 7, 1–47. Academic Press, New York.

Meiwes, K. J., M. Hauhs, H. Gerke, N. Asche, E. Matzner, and N. Lammersdorf. 1984a. Die Erfassung des Stoffkreislaufes in Waldökosystemen. Berichte des Forschungszentrums Waldökosysteme/Waldsterben, Bd. 7 (3400-Göttingen Büsgenweg 2, West Germany), 142 p.

Meiwes, K. J., König, P. K. Khanna, J. Prenzel, and B. Ulrich. 1984b. Chemische Untersuchungsverfahren für Mineralboden, Auflagehumus und Wurzeln zur Charakterisierung und Bewertung der Versauerung von Waldböden. Berichte des Forschungszentrums Waldökosysteme/Waldsterben, Bd. 7 (3400-Göttingen Büsgenweg 2, West Germany), 142 p.

Murach, D. 1988. Judgement of the applicability of liming to restabilise forest stands. *In* P. Mathy, ed. *Proceedings of an International Symposium held in Grenoble, May 1987. Air Pollution and Ecosystems*, 445–451, D. Reidel Publ. Co. Dordrecht, The Netherlands.

Richards, L. A., ed. 1954 Diagnosis and improvement of saline alkali soils. U.S. Dept. Agric. Handbook 60.

Stewart, J. B. 1977. Water Resour Res 13:915–921.

Stock, R. 1988. Forst und Holz 43:283–286.

Ulrich, B. 1984. Atmos Environ 18:621–628.

van Breemen, N., C. T. Driscoll, and J. Mulder. 1984. Nature 307:599.

Wagenhoff, A., and von Wedel. 1959. Mittlg Schw Anst Forstl Versuchsw 35:127–138.

Index